Dynamic Behavior of Concrete and Seismic Engineering

Dynamic Behavior of Concrete and Seismic Engineering

Edited by
Jacky Mazars
Alain Millard

ISTE

WILEY

First published in France in 2004 by Hermes Science/Lavoisier entitled: *Comportement dynamique des bétons et génie parasismique* © LAVOISIER, 2004
First published in Great Britain and the United States in 2009 by ISTE Ltd and John Wiley & Sons, Inc.

ISTE Ltd
27-37 St George's Road
London SW19 4EU
UK

www.iste.co.uk

John Wiley & Sons, Inc.
111 River Street
Hoboken, NJ 07030
USA

www.wiley.com

Library of Congress Cataloging-in-Publication Data

Comportement dynamique des bétons et génie parasismique. English
Dynamic behavior of concrete and seismic engineering / edited by Jacky Mazars, Alain Millard.
p. cm.
Includes bibliographical references and index.
ISBN 978-1-84821-071-4
1. Earthquake resistant design. 2. Structural dynamics. 3. Concrete--Plastic properties. 4. Buildings, Reinforced concrete--Earthquake effects. I. Mazars, Jacky. II. Millard, Alain.
TA658.44.C653 2009
624.1'762--dc22

2009004399

British Library Cataloguing-in-Publication Data
A CIP record for this book is available from the British Library
ISBN: 978-1-84821-071-4

Printed and bound in Great Britain by CPI Antony Rowe, Chippenham and Eastbourne.

Table of Contents

Preface xi

Chapter 1. Dynamic Behavior of Concrete: Experimental Aspects . . . 1
François TOUTLEMONDE and Gérard GARY

1.1. Introduction. 1
1.1.1. Meaning of the word "dynamic" 1
1.1.2. Reminders about dynamic experimentation 3
1.1.3. Identifying the behavior of concrete under fast dynamic loadings . 8
1.2. Tests in which the transient rate has little influence 10
1.2.1. Tests involving deviatoric behavior 11
1.2.2. Tests with prevailing spherical behavior 17
1.3. Tests with transient phase conditioned interpretations 20
1.3.1. Tests involving mainly traction behavior 20
1.3.2. Tests implementing compression behavior 25
1.4. Other tests. 29
1.4.1. Tests adaptable to an energetic approach 29
1.4.2. Validation tests on structures requiring an inverse analysis 30
1.5. Synthesis of the experimental data on concrete and associated materials 33
1.5.1. Data on cement paste mortar and concrete 33
1.5.2. Data available for reinforced concrete. 40
1.5.3. Data about fiber-reinforced concretes 42
1.6. Conclusion 46
1.7. Bibliography 47

Chapter 2. Dynamic Behavior of Concrete: Constitutive Models 55
Patrice BAILLY

2.1. Dynamics of concrete structures 55
2.1.1. Macroscopic phenomena 55
2.1.2. Perforation 57

2.1.3. Ejection of fragments . . . 58
2.1.4. Loading range . . . 59
2.1.5. Loading path . . . 60
2.2. Fast dynamics applied to concrete . . . 62
2.2.1. Impacts and waves . . . 62
2.2.2. Impact and shock polar curve . . . 66
2.2.3. Shock between two solids . . . 67
2.3. Scabbing . . . 68
2.4. Effect of a shock wave on the structure of materials . . . 69
2.5. Modeling types . . . 70
2.5.1. Behavior description theoretical frames . . . 70
2.5.2. Integrating sensitivity to the strain rate . . . 71
2.5.3. Elasto-plasticity and criteria . . . 72
2.5.4. Damage . . . 74
2.5.5. Notion of a state law . . . 74
2.5.6. Location limiter and time sensitivity . . . 75
2.6. Models . . . 76
2.6.1. Elasticity-based model . . . 76
2.6.2. Models based on the theory of plasticity . . . 77
2.6.3. Models based on damage mechanics . . . 81
2.6.4. Model coupling damage and plasticity . . . 82
2.6.5. Model coupling damage and mechanics of porous media . . . 83
2.6.6. Model deriving from a hydrodynamic approach . . . 84
2.6.7. Endochronic models . . . 87
2.6.8. Discrete element method . . . 88
2.7. Conclusion . . . 90
2.7.1. Main features of the models . . . 90
2.7.2. Contribution of distinct elements . . . 91
2.8. Bibliography . . . 92

Chapter 3. Seismic Ground Motion . . . 95
Pierre-Yves BARD

3.1. Introduction . . . 95
3.2. Measuring seismic motions . . . 96
3.2.1. Differences between seismological and accelerometer networks . . . 96
3.2.2. Accelerometer networks . . . 97
3.2.3. Accelerometer data banks . . . 98
3.3. Quantitative characterization of seismic movements . . . 98
3.3.1. Time maximum values . . . 98
3.3.2. Spectral characterizations . . . 99
3.3.3. Features of hybrid characterizations . . . 106
3.3.4. Caveats regarding differential motions . . . 107
3.4. Factors affecting seismic motions . . . 108
3.4.1. Spectral signature of the seismic source . . . 109
3.4.2. Effects of propagation in the Earth's crust . . . 111

3.4.3. Site effects 113
3.5. Conclusions. 120
3.6. Bibliography 121

Chapter 4. Soil Behavior: Dynamic Soil-Structure Interactions 125
Alain PECKER

Introduction 125
4.1. Behavior of soils under seismic loading 126
4.1.1. Influence of the nature of soils on seismic movements. 126
4.1.2. Experimental description of soil behavior 128
4.2. Modeling soil behavior 131
4.2.1. An experimental description of elastic soil behavior ($\gamma \leq \gamma_s$) 134
4.2.2. Linear visco-elastic models for medium strain domains where $\gamma_s \leq \gamma \leq \gamma_v$ 134
4.2.3. High strain domain non-linear models where $\gamma > \gamma_v$. 141
4.3. Linear soil-structure interactions 143
4.3.1. Illustration of the soil-structure interaction effect 143
4.3.2. Expression of a soil-structure problem 148
4.3.3. Superposition theorem 151
4.3.4. Practical modeling of the soil-structure interaction 154
4.4. Non-linear soil-structure interactions. 158
4.4.1. Geometric non-linearities and uplift of the foundations 158
4.4.2. Non-linearities of behavior 158
4.4.3. Modeling the non-linear soil-structure interaction 159
4.5. Bibliography 161

Chapter 5. Experimental Methods in Earthquake Engineering 165
Alain MILLARD, Pierre PEGON and Jean-Claude QUEVAL

Introduction 165
5.1. The pseudo-dynamic method 167
5.1.1. Introduction 167
5.1.2. History of the PSD method 168
5.1.3. The ELSA laboratory 168
5.1.4. Comparison with shaking tables 170
5.2. The conventional pseudo-dynamic method 170
5.2.1. Algorithms 170
5.2.2. Implementation at ELSA 172
5.2.3. The sub-structuration method. 174
5.2.4. Illustration 177
5.3. Continuous pseudo-dynamic method. 178
5.3.1. Continuous method principle 178
5.3.2. Implementation at ELSA 180
5.3.3. Sub-structuration for the continuous method. 182
5.4. Final comments 183

5.5. Shaking table tests 184
5.5.1. Introduction 184
5.5.2. Characteristics and performance of shaking tables 187
5.6. Laws of similarity 193
5.7. Instrumentation 194
5.8. Loading 195
5.9. Conclusion 196
5.10. Bibliography 197

Chapter 6. Experiments on Large Structures 201
Patrick PAULTRE and Jean PROULX

Introduction 201
6.1. Instrumentation 202
6.2. Dynamic loads 205
6.3. Data processing 206
6.4. Application to buildings 208
6.4.1. The slanting tower at the Montreal Olympic Stadium 209
6.4.2. Reinforced concrete building 210
6.5. Bridge application 213
6.5.1. Pedestrian footbridge 213
6.5.2. A mixed cable-stayed/suspension bridge 215
6.6. Application to large dams 220
6.6.1. Assessment of a response spectrum on the crown 220
6.6.2. Study of foundation-ice-reservoir-dam interactions 222
6.6.3. Study of the effect of the water level inside the reservoirs 228
6.7. Conclusion 230
6.8. Acknowledgements 230
6.9. Bibliography 230

Chapter 7. Models for Simulating the Seismic Response of Concrete Structures 233
Didier COMBESCURE, Nicolas ILE, Jacky MAZARS and Jean-Marie REYNOUARD

7.1. Introduction 233
7.2. Different discretization families 234
7.2.1. Global modeling 234
7.2.2. Semi-global modeling 235
7.2.3. 2D and 3D fine models 239
7.3. Behavior laws for concrete 240
7.3.1. Semi-empirical mixed models 240
7.3.2. Damage model 241
7.3.3. Plasticity model for concrete 243
7.3.4. Cyclic models for steel 247
7.3.5. Taking construction layouts and second-order phenomena into account 248

7.4. A few examples with their validation through experiments 250
7.4.1. Application of the semi-global method to a four-storey structure 250
7.4.2. Semi-global and local models applied to concrete walls 253
7.5. Conclusions. 269
7.6. Bibliography 270

Chapter 8. Seismic Analysis of Structures: Improvements Due to Probabilistic Concepts 273
Jean-René GIBERT

8.1. Introduction. 273
8.2. The modal method. 274
8.2.1. Data about the seismic source 274
8.2.2. Calculation of structural responses using the modal method 277
8.3. Criticism of the modal method 279
8.4. A few reminders about random processes 280
8.4.1. Definition and characterization of a time random process 280
8.4.2. Second order characterization 280
8.4.3. Response of a linear system to random stress 282
8.4.4. Using stochastic equations 285
8.4.5. Extrema statistics in a stationary process 286
8.5. Improvements to the modal method 292
8.5.1. Complete quadratic combination. 294
8.5.2. Peak factor effect 295
8.6. Direct calculation of the floor spectra 297
8.6.1. Representation of non-stationary processes 297
8.6.2. Adjusting a separable process from the ORS data. 298
8.6.3. Determination of the floor spectra 299
8.7. Creation of synthetic signals and direct numerical integration 301
8.8. Seismic analysis of non-linear behavior structures. 304
8.8.1. Introduction 304
8.8.2. Main non-linearities of seismically-loaded structures 304
8.8.3. Notion of "inelastic spectra" 305
8.8.4. Conventional method of stochastic linearization 310
8.8.5. Random parameter stochastic linearization. 316
8.9. Conclusion 323
8.10. Bibliography 323

Chapter 9. Engineering Know-How: Lessons from Earthquakes and Rules for Seismic Design 327
Philippe BISCH

9.1. Introduction. 327
9.2. Lessons from earthquakes 327
9.2.1. Pathologies linked to overall behavior 328
9.2.2. Problems linked to local under-design 330

9.2.3. Problems linked to construction layout 334
9.3. The aims of anti-seismic protection standards 336
9.3.1. Standardization of anti-seismic design 336
9.3.2. Main objectives of anti-seismic protection 337
9.3.3. Verification method 338
9.3.4. Capacity-design method 341
9.4. General design 344
9.4.1. Design principles 344
9.4.2. Regularity conditions 345
9.4.3. Calculation of seismic action effects 346
9.5. Behavior coefficients 349
9.5.1. Using behavior coefficients 349
9.5.2. Structure behavior and behavior coefficients. 351
9.5.3. Local ductility and behavior coefficients 352
9.5.4. Ductility classes and behavior coefficients 352
9.6. Designing and dimensioning reinforced concrete structure elements . . 353
9.6.1. Regulations specific to reinforced concrete in seismic areas 353
9.6.2. Main types of reinforced concrete bracing 354
9.6.3. Main frames 356
9.6.4. Reinforced concrete bracing walls 359
9.6.5. Detail designing. 364
9.7. Conclusions. 366
9.8. Bibliography 366

List of Authors 369

Index 373

Preface

The static and quasi-static behaviors of concrete have been the subject of so many works that we often consider that they are quite well known and mastered as far as modeling with a view to structure calculations is concerned. However, the same is not true of concrete's dynamic behavior, because of the complexity of the tests needed to reach pertinent loading rates.

The subject matter of Chapter 1 is divided into two parts: it presents the most widely used experimental techniques to study the dynamic behavior of concrete, drawing attention to the difficulties in interpreting the results of tests designed to identify its intrinsic parameters. It also offers a synthesis of properties that have been published in the literature dealing with concrete (chiefly its traction and simple compression strengths), as well as values for reinforced or fiber-reinforced composites. An extensive bibliography enables the reader to refer to the relevant original articles.

Dynamic loadings can generate non-linearities and a range of deteriorations in concrete (failure from bending and/or shear, traction, mechanical spalling, tearing, compression, compaction and hole perforation, etc.), all of which have to be carefully modeled to enable prediction of the behavior of a specific structure under a violent action. The variety of responses has generated several unique modeling approaches. Depending on the phenomenon under consideration, we use either the *damage approach* for cracking, the *plasticity or viscoplasticity approach* for shear, or the *still volume-pressure influence approach* for compaction. The theoretical contexts are discussed in Chapter 2, before the essential elements of several "conventional" models are described, along with their strengths and weaknesses.

In Chapter 3, the subject matter turns to the particular category of dynamic oscillations associated with earthquakes. As an introduction, Chapter 3 deals with the way seismic movement measurements – which generate the data used for structure reaction calculations – are made. Besides presenting the addresses of databases of signals measured in different countries, this chapter also introduces the concept of *spectral representation*, which plays a key role in engineering practice. A geophysical interpretation of seismic movements in connection with subjacent phenomena is proposed, which integrates the contributory effects of the site and the topography of the environment around the structure.

Though typical practice involves calculating the reaction of a structure submitted to an earthquake by considering its base to be totally embedded, the nature of some soils, coupled with the exceptional character of some structures (like dams and nuclear reactors), demands that the behavior of the structure is modeled in a particular environment. This problem is called structure-soil interaction, and forms the subject of Chapter 4. To solve this problem, it is necessary to have a model of the soil's behavior under cyclic loading. Different models exist, depending on the nature and amplitude of the loading. After modeling, the interaction problem can either be treated by superposition, by considering the soil and the structure separately for linear cases, or globally for non-linear situations.

The difficulty of conducting structure tests on full-sized models led to the development of experimental methods employing scale models. Vibrating tables, which reproduce earthquakes on a small scale, were designed for this purpose. The subsequent development of fast and powerful computers gave birth to the pseudo-dynamic method, in which the purely dynamic effects of an earthquake are simulated using calculations. These complementary techniques both have their own advantages and disadvantages. The quality of the results they can produce depends mainly on the quality of model implementations, which are described in detail in Chapter 5.

Chapter 6 is concerned with experimental techniques on large structures. Experiments play an essential role in obtaining realistic data about a structure's dynamic signature; mechanically-controlled vibration tests are not easy to implement, but they are a crucial source of information. This chapter shows how an excitation with rotating masses, coupled to adapted instrumentation and measurement processing, gives access to a vast amount of key information concerning natural modes, frequency damping, damage indications and coupling effects between the structure and its environment. This forms an invaluable database that model-makers need to calibrate their models, which is a pre-requisite of any realistic analyses of the seismic response of an existing structure.

Chapter 7 examines the structure-modeling field as applied to the seismic analysis of concrete buildings. The chapter focuses on approaches that allow engineers to simulate reactions to the application of an earthquake by exploring the non-linear field and collapse modes.

In this context, three model families are considered: global, semi-local and local models. The first rely on empirical behavior descriptions, gathering phenomena at the level of a single section or structural element with occasional brief discretization. The second type of model works out global laws from phenomenological local models, with discretization made at the multi-fiber or multi-layer beam level. The third type of model is more sophisticated and takes the responses of a building's constituent materials and their bindings into account. These demand very thorough discretization, their main disadvantage being that they are very time-consuming to implement.

Validation analyses are proposed based on experiments carried out on shaking tables or reaction walls. The results show that that modeling has reached such a sophisticated stage of development that it allows complete experimental and experiment-feedback analyses, and is therefore ready for transfer to everyday engineering.

Chapter 8 introduces a quite promising analysis procedure: probability analysis. It is clear that the uncertain nature of seismic loading must be taken into account for the dimensioning of large structures. However, though sophisticated methods are used (seismic movement correlation, structure-soil interaction, behavioral non-linearity, etc.), the models remain "deterministic".

This chapter shows that the determinist approach can sometimes lead to erroneous predictions, and that better control of phenomena makes it necessary to take into account the probabilistic character of the problem. Probabilistic seismic analysis is a new field of research that should lead to significant advances in paraseismic engineering.

Chapter 9 considers the craft aspect of engineering in the field of seismic building analysis. The subject of this chapter, experiment feedback and regulations, is important, as engineers are ultimately responsible for the safety of people as well as buildings, despite the fact that building science is not a totally exact science. Numerous theories can help analyze the behavior of structures, but the limits of the problems faced by engineers in that field remain "fuzzy" (action characterization, structure complexity, local behavior facts, soil-structure interaction, etc.). Regulations can provide a framework that makes the analysis concepts reliable. Experiment feedback gives an indication of those approaches that have worked and those that have not, and considering this when developing regulations obviously assists progress in safety control. This is the subject of this chapter, which also gives an excellent account of the spirit in which the paraseismic design of various concrete structures should be approached.

Jacky MAZARS
Alain MILLARD

Chapter 1

Dynamic Behavior of Concrete: Experimental Aspects

1.1. Introduction

1.1.1. *Meaning of the word "dynamic"*

As distinct from the term "static", "dynamic" implies the influence of time. A test is said to be "quasi-static" when the effects of time are present, but can be neglected. For a structure test, and for any real test, the effects of time are typically expressed in two ways:

– by forces of inertia resulting from the not equal to zero acceleration to which the elements of structures are submitted;

– by the behavior of each elementary volume of the material depending on the evolution in time of the elementary mechanical values (stress and strain) and possibly of their time derivatives. This dependence is described by the generic name of "viscosity".

This distinction is strictly linked to the notion of elementary volume underlying the definition of behavior. Actually, the fact that viscosity effects might be the manifestation of inertial microscopic phenomena cannot be excluded. This remark is important in the case of concrete, as considerations about material homogenity

---

Chapter written by François TOUTLEMONDE and Gérard GARY.

involve decimeter elementary volumes (but this is not the case with metals for which the elementary volume is sub-millimetric).

Thus, to be quite clear, we will consider the dynamic behavior aspect as limited to the description of the effects of time using elementary mechanical values and excluding inertia effects.

From general physical and thermodynamic considerations concerning behavior laws [MAN 67], we can deduce that the generalized mechanical variables Q (t) (stress) and q(t) (strain) can be related in the following way:

$$Q(t) = G\left[H_{-\infty}^{t}(q(\tau)); q(t), \dot{q}(t), \cdots \right] \qquad [1.1]$$

where H describes the loading history. This formulation highlights the fact that these values do not play a symmetric role. The instantaneous mechanical reaction depends on the geometric history, its current value, and the values of its higher time derivatives. Thus, it is not natural to consider stress velocity as a behavior variable.

If we limit our attention to formulations likely to be easily integrated into calculation codes, the relation expressed in equation [1.1] can be re-written in the following incremental form:

$$d\sigma = f(d\varepsilon, \dot{\varepsilon}, \ddot{\varepsilon}.., \alpha_i..) \qquad [1.2]$$

The values α_i are internal parameters that take process history into account. Their evolution has to be described as a complement to the relationship in equation [1.2]. Their dependence on the history of the process explicitly results in their loading and unloading paths being different. The values playing a part in equation [1.2] are tensors. We can see the complexity of this relation. In most cases, the simplifications carried out involve discarding strain time derivatives higher than 1, and expressing the strain speed using a scalar value. Such simplifying assumptions are justified for two types of reasons. Firstly, programming laws into codes will be simplified by doing this. Secondly, an insufficient variety of dynamic tests is available to identify more parameters. For this reason, from this point onwards, we will refer to "strain velocity" without going back over the definition.

As far as strain velocity is concerned, it is standard practice to study its effects on long time scales revealed through creep. Even though creep tests can clarify the analysis of dynamic tests, we will not be considering them. The experimental aspects of creep tests have no dynamic aspects, as typical strain velocities implemented are around 10^{-10} s^{-1}, as compared to "static" test standard velocities that range from 10^{-6} to 10^{-5} s^{-1}, and the strain velocities reached during "hard" shocks on civil engineering structures, which are usually in the range of 0.1 and 10 s^{-1}.

These elementary considerations understood, it appears that a critical factor in the experimental characterization of concrete behavior is discarding the inertia terms. The problem is more delicate with concrete (a brittle material) than it is with metals. As a matter of fact, the first manifestation of inertial effects on a sample submitted to dynamic loading is the transient response observed when the time taken by elastic waves to pass through the sample (the transfer time) is significant relative to the test's time duration. When studying this problem, the pertinent time-dependent parameter is not the strain velocity (which, in any case, is not well-defined in the transient phase), but the loading time relative to the transfer time. If sufficient strain levels are reached in very short periods of time, the sample could fail before a homogenous stress and strain state, *measurable as an average,* could be reached. In fact, low amplitude traction strains (ranging from 100 to 150 x 10^{-6}) lead to material failure. Test analysis is generally difficult. For common sized samples (centimeter scale), we cannot go beyond 1/s average strain velocities when conducting a quasi-static test analysis. This feature of brittle materials can be exploited advantageously, and is used in scabbing tests (see section 1.3.1).

This limitation is far less a problem with metals, where important local strains arise, but do not cause failure. Such a situation can only occur in concrete if particular conditions that guarantee mechanical field homogenity exist to prevent cracking. This is the case when tests are conducted in strong confinement (under which circumstances, concrete behavior is described by plasticity-type models). As far as metals and most polymers are concerned, it is also important to take thermo-mechanical coupling into account, due to the adiabatic feature of dynamic tests. This effect can only be neglected when failure occurs under low strain for which the dissipated heat remains low: with concrete, it can also be neglected in confinement tests, since we can presuppose a low thermo-mechanical coupling.

1.1.2. *Reminders about dynamic experimentation*

1.1.2.1. *Specificity of dynamic tests*

As far as statics and dynamics are concerned, it is reasonable to consider sample analysis in a separate section, along with the overall measures it involves (generally carried out on the peripheral part of the material). This is the second aspect mentioned in the introduction.

The first difficulties encountered in dynamic experimentation fall under the first category mentioned in the introduction. They are linked to transient effects inside the machine and the associated sensors: the balancing time of the machine and its sensor array (elastic waves moving back and forth several times) are not negligible relative to the length of the test. Thus, carrying out quality measures often requires a *transient analysis of the response of the machine itself*. Hence, in a real situation, characteristic testing times have to be compared with the acquisition chain and the sensor pass-band. If the acquisition frequency is not far higher than the frequency of transient signals, the observed result can be completely modified by the measuring chain, and even average values can be wrong.

1.1.2.2. *Hopkinson bar test*

For average strain rates in excess of 50/s, because the transient effect inside the test machine cannot be neglected, a way round the problem involves explicitly taking wave propagation phenomena into account, using a bar system. Whilst the transient analysis of three-dimensional structures is too complex to be taken into account efficiently, using "one-dimensional" bars makes it possible, as we will now explain.

1.1.2.2.1. A description of the bar test

To carry out a dynamic compression tests with Hopkinson bars [HOP 14] (also called the SHPB (Split Hopkinson Pressure Bar) system, or Kolsky bars [KOL 49]: named after the first person to use the system in its current configuration), a small sample is placed between two identical long bars with a high elastic limit relative to the tested material (Figure 1.1). Strain gages are glued to both bars. Due to a projectile, a compression longitudinal elastic wave is induced into the input bar. Part of this gets reflected at the sample-bar interface, whilst another part is transmitted to the sample before inducing a wave in the output bar.

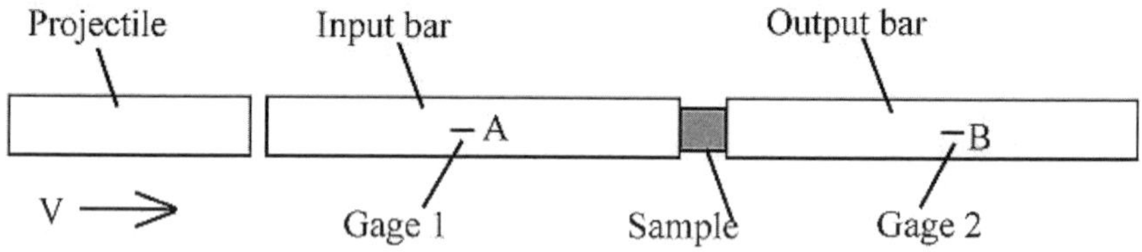

Figure 1.1. *Hopkinson bar assembly*

The waves at points A and B are determined by measuring and recording the structurally-associated longitudinal strains. The need to know A, the incident wave induced by the impact separator, and the reflected wave B, which depends on the reaction of the sample, arises because we need to find the optimal position of the measuring point at the middle of the bar. On the other hand, considering the bar as one-dimensional does not allow us to place the strain gauge too near an end. A typical recording for a concrete sample compression test is shown in Figure 1.2.

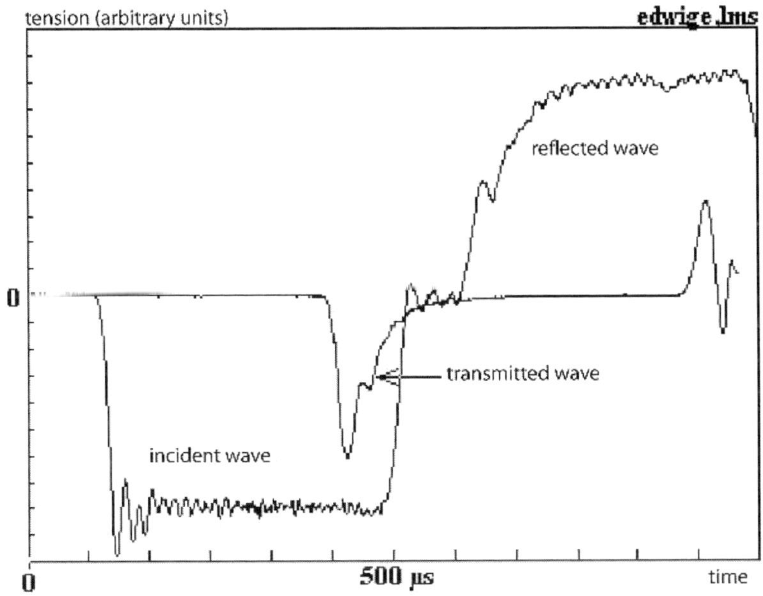

Figure 1.2. *Compression test on a concrete sample (40 mm diameter aluminum bars): basic waves*

Next, the waves have to be carried to the contacts between the sample and the bar. Then we can calculate the stresses and displacements (by integrating the velocities, which are directly accessible) on the corresponding faces.

The particulate velocities at the input and output faces can be written respectively as:

$$V_e(t) = -c\left[\varepsilon_i(t) - \varepsilon_r(t)\right]$$
$$V_s(t) = -c\varepsilon_t(t) \qquad [1.3]$$

The forces on the input and output faces are respectively:

$$F_e(t) = S_b E\left[\varepsilon_i(t) + \varepsilon_r(t)\right]$$
$$F_s(t) = S_b E \varepsilon_t(t) \qquad [1.4]$$

Measures on the two opposite faces of the sample allow estimation of strain field homogenity by comparing the forces on each face (section 1.3.2, Figure 1.12). We note that for this test, the assumption of homogenity in mechanical fields is hazardous. As a consequence, the notion of average strain velocity is also hazardous. In section 1.3.2 we will see the best way to use the available measurements. Thus, we should stress that the Hopkinson Bar leads to overall values of *loads and displacements on both sides of the sample.* All mechanical quantities are obtained by making additional assumptions completely separate from the test facilities. These have been widely reported in the literature [NIC 80].

1.1.2.2.2. Limitations of the conventional system

Accurate analysis of wave transport

To carry out a precise virtual wave transport between the measured points and the sample (forward transport for the incident wave and backward transport for the others), the three-dimensional feature of the bars need to be considered, and the *dispersal correction* must be introduced using a signal treatment technique. This parameter corresponds to signal modification during transport. An accurate time calibration (to within a micro-second) is also necessary [ZHA 96]; it is especially important for measurement of small strains, and thus for brittle materials such as concrete.

Multi-axial characteristics of the test

The uniaxial characteristic of the test is also an approximation. Let us examine this aspect in the case of compression. Whenever the material presents a Poisson effect, the longitudinal strain comes with a lateral strain (as is the case in statics if the support conditions are well controlled), which is opposed by radial inertial effects. This causes an induced confinement. The confinement explains the obvious sensitivity of concrete to strain velocity that is universally observed in dynamic compression (see section 1.3.2).

Measurement duration

The proportionality between the mechanical values associated with a wave inside a bar, on which the Hopkinson bar technique is based [1.3]-[1.4], only applies to a wave propagating in a single direction, which requires measurement of the incident wave (propagating one way) separately from the reflected wave (which propagates in the other direction). This limits the measuring duration to ΔT $(\Delta T = L / C)$, C being the propagation velocity and L the length of the input bar. ΔT is thus a function of the length of the bars. Consequently, for a behavior test, the total strain cannot exceed the product of the average strain velocity and ΔT. For instance, measuring duration will not exceed 400 μs (C≈ 5,000 m/s) for a 2 m long aluminum bar, and the total strain will be limited to 4% for a test with a 100 s^{-1} average strain velocity. Because of this limitation, even with concrete (for which high strains are unlikely to be reached), the conventional Hopkinson bar system will not allow tests at average strain velocities lower than 50 s^{-1}. On the other hand, for reasons explained in section 1.2.1, traditional machines used without specific precautions do not give reliable results at lower velocities. Besides, their superior limit is not clearly established and is determined to an extent by the material being tested (the test piece). The machine must be used in a particular way; it varies between 1 s^{-1} and about 10 s^{-1}. However, a recent experimental technique using bars [BUS 02] that covers this problem now exists.

1.1.2.2.3. Difficulties inherent to dynamic measurements

The dynamic test facilities have numerous limitations, especially for stresses other than simple compression or small strains. This limitation mostly affects low strength stressed materials (impedance adaptation and high strain problems) and brittle materials (low strain at failure).

The Hopkinson bar example illustrates the generic difficulties quite well. The very short loading times do not enable us to carry out multi-axial dynamic loadings easily, and it is not easy to synchronize loading with two (or three) orthogonal Hopkinson bars. If synchronization is tricky in dynamics, it is all the more so when piloting the test. Therefore, we cannot (for now) contemplate carrying out tests under controlled multi-axial loading (deviatoric, for example), as is required in a quasi-static mode. The need to control the loading and the difficulty in carrying out dynamic displacement measurements limits the potential tests to a very small number, which are described, along with their specific problems, in sections 1.2 and 1.3.

1.1.2.2.4. Compression tests with confinement

It is quite easy to superimpose quasi static confinement on a dynamic compression test. A cell in which a gas pressure confinement can be maintained during the compression test is described in [GAR 99]. Some authors have proposed a bi-axial loading scheme, where the secondary static stress is applied using a jack [WEE 88]. For higher confinements (necessary if we want to study compaction of concrete, for example), a metal cylinder can be used [GAR 99]. In this case, confinement pressure is not studied, but can be measured during the test by assuming the (most often elastic) response of the confinement ring is known (as in an oedometric test). Another way to carry out high confinements involves using the "plate on plate" test developed to study the high-speed spherical behavior of metals. It is a plane strain-loading test, the inverse analysis of which is based on behavior modeling. High confinement there is associated with very high strain speeds.

1.1.2.2.5. Traction tests

A conventional traction test can be carried out with a Hopkinson bar [REI 86]. If we consider only global measures, the main difficulty is due to keeping the sample in contact with the bars. To avoid having to resort to assemblies leading to impedance failures, it is reasonable to glue the sample to the bars. Some authors [TED 93] have had the idea of using the Brazilian test again in dynamics. In this case we have to check that the conditions of strain homogenity are compatible with the assumptions. Finally, the spalling test [DIA 97] allows an accurate measurement of the average stress just before failure, but its interpretation is difficult as it is between the classical traction test and the fracture test (toughness measurement).

1.1.3. *Identifying the behavior of concrete under fast dynamic loadings*

When identifying the dynamic behavior of concrete, we are confronted with a series of typical problems for each high-speed behavior identification test. Some of these problems are increased by the nature of concrete, which is the reason why we prompt the reader to be very cautious when using experimentation signals or results.

Due to its structure in aggregates, where it is mixed with sand and hardened cement paste, concrete can be a highly heterogenous material, and the size of a representative sample is not always an easy thing to state. As far as statics is concerned, a 2 slenderness cylinder, over five times as big in diameter as the aggregates, is the lowest volume necessary to obtain stable properties representing the material in these tests, particularly as far as strength is concerned, otherwise "scale effects" will be observed. Such a constraint raises several types of problems:

– for standard concretes, in which the maximum size of aggregates ranges from 20 to 25 mm, the dimensions of test samples (diameter over 10 or 12 cm, mass over 5 kg) involve resorting to important energies, particularly for high speed tests, which involves sometimes tricky technological arrangements;

– to avoid this difficulty, tests are often carried out on micro-concrete, mortar or cement paste samples. Transposing these results to structure concrete requires a critical analysis, mainly because the volume fraction of cement paste (generally considered as the viscous element of the composite) is not always constant. In the same way, the propagation of waves disrupted by the module differs between cement pastes, and aggregates can also be different depending on the composition of the studied material;

– even if we managed to identify the intrinsic properties of the material on big enough samples, for many structures, the "representative material point" size is important compared with the dimensions of the smallest pieces (building shells about 20 cm, bridge webs from 30 to 45 cm). Furthermore, significant stress variations on the scale of the structure can be discerned over short distances of the same order of magnitude as the dimensions of the test sample. What is then involved is the application of continuous medium theory, which is based on the assumption – generally not well verified – that the material point is infinitely small compared to the structure;

– a problem (which occurs in statics too) that becomes crucial as far as the dynamic interpretation of tests is concerned is that the sample is not submitted to an homogenous state of strain and stress owing to its size, and has to be considered as a structure submitted to transient loading.

Because concrete is a brittle material (like most geomaterials, concrete can only withstand very weak extension strains and its apparent "failure" takes place for a compression strength about 10 to 20 times as strong as its traction strength), most of the time, in practice, while interpreting the tests, we must consider:

– that we are dealing with an elastic homogenous material (which implies the size precautions referred to above): the assumption is necessary for relatively low velocities or low strain levels, in continuity with the quasi-static field. It is not good enough to interpret the totality of a test when the speed increases, since the maximum stress is reached when localized cracking has been reached significantly on only a part of the structure;

– that beyond the stage corresponding to localized cracking, the test sample can be modeled as a cracked structure where damage concentrates in the crack area, which corresponds to fracture models;

– that beyond a stage corresponding to a distributed deterioration (which corresponds to the bonding material crumbling away), the material can be described by combining damage and plasticity models.

Hence, at the material failure of the sample, the interpretation of the tests requires different analysis models, regardless of whether we are mainly in a deviatoric behavior with a possible extension direction allowing localized cracking, or in a mainly "spherical" behavior, and depending on the stress peak being identifiable or not.

It is important to note that because of the weak growths withstood by the material at high velocities, experimental precautions have to be taken – especially because of transient effects, in dynamic experiments where limit conditions are difficult to control. In experiments where an "energetic" approach is privileged, this aspect is also important: the inertia of the test sample cannot always be neglected with regard to that of the test machine, and the energy dissipation through damage on the support, or through contact with the impact separator can be important compared with the energy supposedly dissipated by the "normal" cracking expected in bending.

Finally, a delicate feature of concrete is its porosity: it has such a tortuous network that water exchange times with the environment are quite long (about 10 years for the representative volumes considered above). We can consider the hydration state of the sample as constant during dynamic tests, which is not the case for shrinkage or plastic flow tests. However, important relative pore moisture and mechanical state coupling, together with frequent cracking due to the stress levels reached when desiccated, begins at the sample's surface and/or their environment as soon as they are fabricated. In at least one stress and velocity field ([DAR 95] [TOU 95a]), researchers have shown that the partly water-saturated feature of the porous network explains the modification of apparent mechanical properties: these are generally called "velocity effects" in the literature.

In following sections (1.2 to 1.4), we will detail the arrangements, test type by test type, used to analyze the results and infer the indications and modifications required to calculate and understand the behavior of fast dynamic loading concrete structures. The actual and measured behaviors are summarized in a rational way in section 1.5.

1.2. Tests in which the transient rate has little influence

In this chapter, we will deal with behavior identification tests that, for reasons developed in section 1.1 can have a "quasi-static" interpretation.

Two test families can be distinguished. The first is derived from typical concrete characterization tests and emphasizes growth or cracking failures. This is called deviatoric behavior, and is the failure kind that is also, indirectly, the cause of collapse observed in compression and even in biaxial compression. The second test type corresponds to "volumic" behavior, which can seldom be observed in ordinary structures, except in relatively confined areas where specific reinforcement by the surrounding material ensures tri-axial confinement at high velocity: concrete areas directly submitted to impact and those close to an explosive charge or perforating projectile are examples.

Combining both types of information in statics enables a definition of failure or plasticity criteria closed on the hydrostatic compression axle, as opposed to the "intrinsic curves" (Coulomb criterion), the validity of which is preferentially ensured when an extension direction is possible.

1.2.1. *Tests involving deviatoric behavior*

1.2.1.1. *High-speed press machines and traction tests*

Because of the difficulties connected with carrying out dynamic tests, most authors use privileged uniaxial tests. Owing to the basic feature, traction behavior identification stands out, and has given rise to a great number of tests. In order to ensure continuity in the geometry of test samples, by controlling the loading application speed and considering its limited artifacts, a direct traction test on a cylindrical specimen has become essential. This is detailed in [HOR 87], [REI 82] and [TOU 95a]. With particular precautions, this test can actually be carried out on conventional servo-controlled machines with load build-up speeds ranging from about 0.05 MPa/s (which is the standard loading rate for standard identification tests) to about 10,000 times this load, with identification at still higher speeds of the order of 50 GPa/s possible on the same specimen type thanks to the modified Hopkinson bar (SHB).

The necessary precautions particularly involve:

– choosing to glue the specimen in place with centering and a rigid (without spherical pairs) mounting onto the press to limit looseness which is a source of interfering moments;

– choosing aluminum hard supports to limit the transversal strain divergence at both ends of the specimen;

– controlling the hydration state of the specimen [TOU 95a];

– choosing a not too important slenderness ratio (1 to 1.5) to limit potential bending;

– gauge extensometer or extensometers fixed in the middle of the specimen to avoid the deformations due to the glue joint;

– using specimens with adequately sized diameters considering the maximum size of the aggregates, and if possible core cylinders for better homogenity of the material and to avoid scaling effects [ROS 92a].

With a sufficient automatic control and oil flow unit, and potentially using a preload to carry out high velocity tests, we can consider that the load build-up speed is rather constant during the test. The propagation speed of the waves within concrete – about 4,000 m/s, the standard size of specimens (10 cm) – and the traction failure stress (4 to 10 MPa) limit the quasi-static interpretation of this kind of test, results typically showing a divergence about 10% between the specimen's input stress and output stress.

The measurements typically carried out during this test are of the applied force as a function of time, and of the average longitudinal strain at the center of the specimen (extensometer gages or sensors for which we have to check that the inertia will stay weak and the fixing will be ensured during the test). Taking into account the small size and fixedness of the assembly, we can consider that there are no differences between the measured force and the force applied to the specimen, so we can assess traction uniaxial behavior by eliminating time. In such a test, the specimen behavior corresponds quite well to brittle elastic behavior up to localized cracking. Localization brings about loss of the homogenity of the strains, and an almost instantaneous decrease of the load.

Going through these tests, which implies expressing the maximum measured stress according to the "load build-up" parameter in a logarithmic diagram, typically allows us to define a traction rate effect corresponding to the strength relative increase.

1.2.1.2. *High-speed press machines and compression tests*

The second most conventional test that can be performed at high speed is the compression test. It enables us to define a compression "rate effect" from the measurement of the maximum strain reached [BIS 91]. The size of the test sample necessary to free oneself from the size effect and to ensure the correct strain level reached lead to strict constraints on press dimensions, unit power and the jack flow rate. For this reason, a great number of the tests described in the literature were carried out on mortar, cement pastes or micro-concrete [HAR 90]. As is the case in traction, it has proved possible to look for a size compatible with the higher speed test performed with Hopkinson bars [DAR 95].

As it is difficult to stop the jack when its speed has been stabilized, few test reports have included extensometer measurements [BIS 91], measuring the load obviously remains the main data. For standard size specimens (10 cm), considering the wave propagation speed and the maximum stress reached, the load build-up rate beyond which the sample cannot be deemed to be in a stationary process is about 10 times as important as it is in traction tests, which correspond to the strength ratio. When expressed in terms of strain rate, the threshold is about 10 instead of 1 s^{-1} [MAL 98].

The incidence of superfluous interference moments is generally less important than it is in traction tests; however, the precautions to be taken to avoid restricting transversal strains are as important as in statics, especially for specimens with slenderness ratios below 2. To this end, we can mention lubricating the faces or using aluminum. The quality of the stress transmission surfaces is essential to avoid premature concentration of stresses.

We note that as in statics, and even for specimens that are simply laid, the relative displacement (including interface crush and deformations at the ends of the specimens) cannot result in a reliable indication of the strains of concrete in its standard part, the error typically ranging from 30 to 100% [BOU 99]. As in statics, failure obtained in compression tests begins with transversal extensions. The traction rate effect results in an "inertial confinement". However, the maximum stress is only reached when the cracks parallel to loading direction meet, allowing either buckling in the "small columns" formed inside the test body, or shear localization. As a consequence, interpreting the strength evolution, where load build-up speed is the only parameter, becomes complex.

We could not find any references to tests deriving from standard quasi-static identification of multi-axial behavior with prevailing deviatoric behavior (bi-traction, pure shear, bi-compression), at least not in areas where transient test characteristic can be neglected. As a matter of fact, the most frequent cases of dynamic multi-axial behavior identification use unidirectional loading with a Hopkinson bar [GAR 98], [LOU 94] and [WEE 92], while confinement or loading in the other direction is often "static". These tests will be described in section 1.3. Such a situation can indeed be explained by the difficulty in controlling and synchronizing dynamic loadings, even at the "low" speeds reached by conventional presses or jacks. Furthermore, taking the properties of concrete into consideration, the regulations rarely take multi-axial behavior into account. As a result this lowers the validation of high velocity dynamic models adapted to concrete, in situations other than simple traction, uniaxial compression or compaction.

1.2.1.3. *Tests with small plates or beams submitted to pressure loading*

Considering the difficulty in carrying out dynamic loading with mechanical application of the loads, some authors perform controlled loadings on mini-structures (small rectangular plates, beams or small plates), using a pressure loading generated by an explosion. The purpose is then to identify the bending behavior, the bend-moment law being material information directly transposable to the calculated structure, taking into account the similar nature of the tested material and the geometric and energy similarities – called Hopkinson's – on the load. Detailed experimentation of this kind will be described in [BAI 87] and [BAI 88]. The limitation on the energies that can be used in a laboratory forces the use of centimeter thick test elements, therefore generally of mortar (or possibly fiber-reinforced) rather than concrete.

The loading process links the level of the applied overpressure with its duration and load build-up speed. However, interpreting the trial remains simple in so far as the load build-up times can be considered as very short compared with the specific period of the structure. Thus, we have what is called a pulse loading: the overpressure time, which causes the structure to start vibrating, is very slightly ahead of the latter's peculiar period, which is then in a free-vibration system. As the probable area of maximum strain and even failure is known, the relevant section can be instrumented in a preferential way. Therefore, we can measure the traction by bending the final strain and the final bend. Note that shortly before failure, the strain of the compressed side is slightly inferior to that of the opposite side (the start of non-linearity which could be representative of micro-cracking). Since the structure is undergoing free vibrations, the deformations should be linked to the stresses generated from the by-pulse loading, which implies that a dynamic analysis can be used to calculate the moments to link to the bends within the scope of behavior law identification. Nevertheless as long as we stay at moderate loading levels and deal with the behavior just before a brittle material fails, an elastic analysis is satisfactory. The divergence from elastic behavior can be identified "at a quasi-static speed".

1.2.1.4. *Shock tube tests on plates*

The principle of a gas pressure by-pulse loading can also be applied by resorting to a uniform loading the value of which is controlled thanks to a tube used as a wave guide and called a *shock tube*. Using such a device is quite conventional for testing industrial equipment in the defense field. Using the device for structure elements was developed more recently [TOU 93]. Using explosives is limited and the loading profile as well as its spatial repartition is better controlled than open-air explosions. In so far as the conditions at limits can also be well controlled, we can directly access to the behavior of a bending plate, which represents "basic" data for the structure designer [KRA 93] or a simple basic situation to validate a behavior model [PON 95, SER 98a].

The innovation of this trial was that it generated loading by means of a well-controlled air shock wave (Figure 1.3). By using the closed tube, for the same plate with the same support conditions, it is possible to carry out quasi-static loadings by slowly inflating the whole tube. As an example, a 35 m long tube, 66.6 cm in diameter, was used to compile an important experimental database about concrete and reinforced concrete plate bending parameters [TOU 95a].

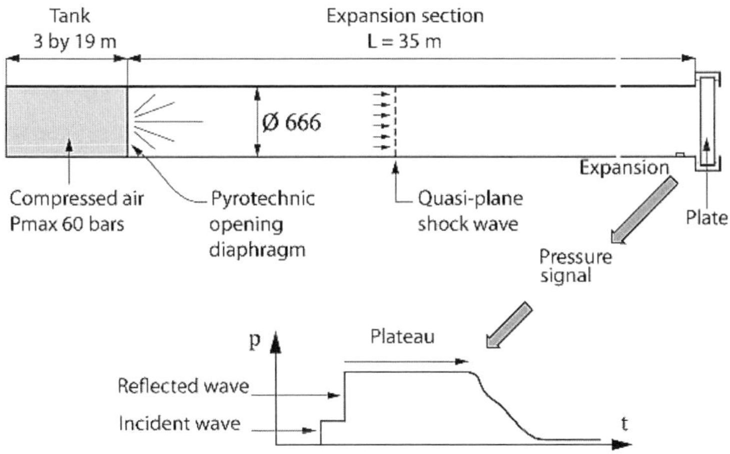

Figure 1.3. *Skeleton diagram of the shock tube trial (from [TOU 95a])*

Considering the inner diameters of the tube and the support area (82 cm), to preserve the cylindrical symmetry of the test, the test sample is a "thin plate" (thickness/span < 1/10) 900 mm in diameter and 8 cm high. It has dimensions compatible with the performances of the tube (allowing it to actually reach failure requires using a concrete with aggregates that are not too small, or realistically standard reinforcement (welded wire mesh), or fiber reinforcements). We can note the particular care taken to achieve limit conditions close to those for an ideal simple support, the circular slab being "pinched" between the humps of two massive guides, a thin rubber-steel sandwich (a 3 cm wide ring) allowing absorption of geometric defects and distribution of the clamping load. Its stiffness has been measured, and control of the displacement and acceleration on the supports during blasts enables analysis of the bending of the support slab under uniform loading on a driven reference line.

Figure 1.4. *Shock tube failure trials for a plain or reinforced concrete slab (from [TOU 95a])*

In addition to excellent loading control and a size adapted to a well controlled trial on "realistic" concrete, the advantages of this test are the realistic representativity (bending is obtained with maximum deformation speed typically ranging from 0.01 s^{-1} and 1 s^{-1}, which corresponds quite well to the "hard" shock range) and geometric simplicity (radial symmetry is preserved up to cracking) which make it possible to validate a calculation model as well as for comparing various materials. The relative ease of interpretation stems from the fast loading building up (about 10 μs for a maximum deformation reached in about 1 ms) and from the absence of pressure gradients on the loaded face. We can consider that the plate is loaded instantly (vibration setting with a first deformation peak which is particularly intense compared to static loading), but with a bearing constant loading, which allows a stationary vibration rate to be set up before unloading. A "conventional" modal analysis enables access to local stresses and strains, at least until cracking starts.

In [TOU 95a], the details about the instrumentation implemented to characterize strains in test samples in these types of trials are presented. We have seen that in a series of plain or reinforced concrete plates with strength in the range of 35 to 120 MPa (Figure 1.4), we are able to show the progressive deterioration of the modal response (frequency drop, increasing damping), the appearance of deflection, plastification of the reinforcement, crack progression (which is sometimes delayed with regard to the maximum strain rate) and the collapse mode type (shear force/bending competition) the respective appearances of which can be justified by limit analysis-inspired calculations [TOU 95a].

1.2.2. *Tests with prevailing spherical behavior*

When loading has a strong tri-axial component, concrete undergoes a global reaction resembling that of a coherent material, even when it has failed on a microscopic scale, which is the case for confinements over 10% [GAR 99]. The models used to describe this are generally plastic models (not necessarily standard and usually coupled to deviatoric and spherical behaviors). In these cases, even high strain gradients do not bring about failure or localization, and the concrete sample can be analyzed as if composed of a homogenous material.

1.2.2.1. *Slab-plate tests*

In the standard case [ZUK 82], loading is caused by the impact of two identical plates. The impact speed is known. On the unused slab, a rear face rate measurement (usually made using laser interferometers) is conducted. Another version of the trial involves applying the same type of loading (in plane strain), using an explosive. The shock induces a plane shock wave propagating at a velocity D. Discontinuities of material rate u, pressure P, volumic mass or mass volume V and inner energy E are associated with this wave. Assuming the material speed ($u_0 = 0$) and pressure ($P_0 = 0$) initial conditions are zero, we can infer from the Rankine-Hugoniot conservation equations that:

$$
\begin{aligned}
V &= V_0(D-u)/D \\
P &= Du/V_0 \\
E &= E_0 + P/2(V_0 - V)
\end{aligned}
\qquad [1.5]
$$

As the time of the shock is known (by contact measurement for example), measuring the free rear face speed allows us to locate the moment when the wave arrives and to measure *D*. It also allows us to calculate *u*. Thus, one test establishes a relationship between *P* and *V*, and also between D and u: these are called "shock polar curves". To deduce a strain-stress uniaxial relationship from them, we will have to make a hypothesis about the behavior model of the material.

For metals and high strength shocks, the elastic response is neglected, and we assume that the plastic behavior is purely deviatoric (without any volume variation). Strictly speaking, concrete behavior analysis should be different. Each test gives a point on a curve. The "Hugoniot curve" links pressure to material physical speed and the " shock polar curve" links shock speed with material speed (objective measurements). There is therefore no direct way of converting this to mechanical values that geomechanical engineers are familiar with.

The plate-plate test is a relatively pure trial. However, it has to be interpreted, is difficult to implement, and only can only inform us about concrete compaction behavior at very high strain rates (above 10^5 s^{-1}).

1.2.2.2. *Hopkinson bar tests with strong confinement*

This test was developed at the LMS in co-operation with LMT Cachan [GAR 99]. A cylindrical specimen is confined within a metal cylinder (Figure 1.5). It is loaded using a large diameter (80 mm) steel Hopkinson bar, which allows the use of test samples large enough in comparison with aggregate size to be adequately representative of the material.

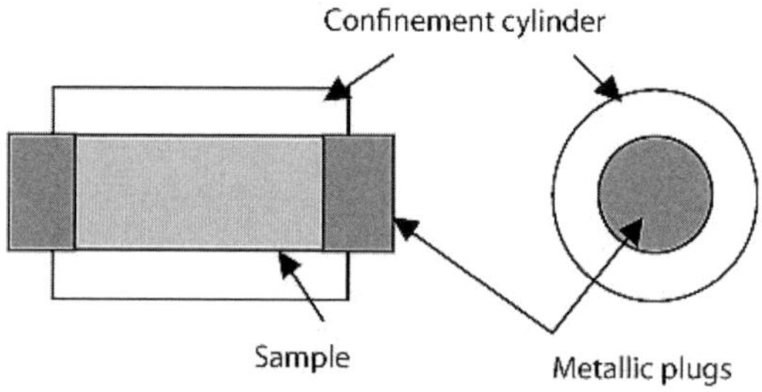

Figure 1.5. *Confined sample for Hopkinson bar test (from [GAR 99])*

The complete collection and analysis of the signals recorded on the bars (described in section 1.1.2.2) allows the measurement of the forces and displacements applied on both faces of the sample.

When the input and output forces are equal (which is the case shown in Figure 1.6) and we can assume an homogenous state of stress and strain, the stresses, strains and axial strain speeds can be deduced.

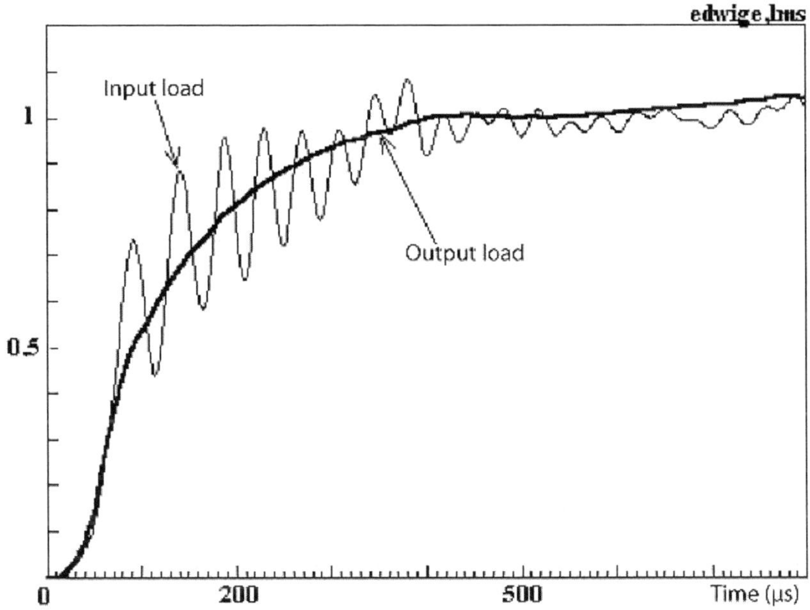

Figure 1.6. *Static and dynamic volume-pressure relationship*

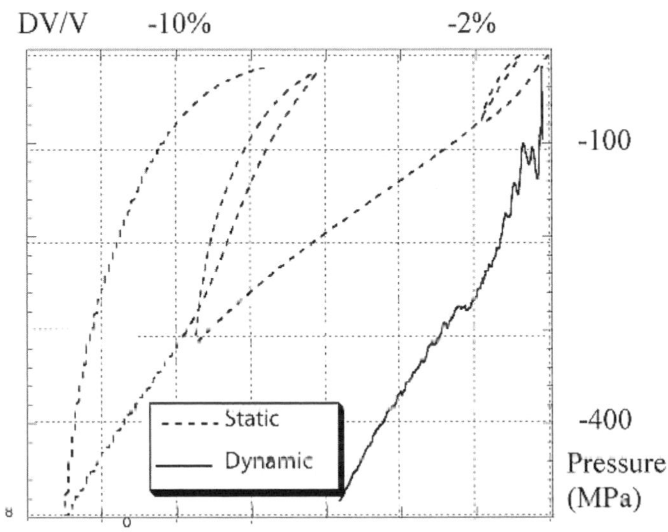

Figure 1.7. *Static and dynamic volume-pressure relationship*

The behavior law of the metallic ring is known. A thick enough ring to remain in the elastic field allows the application of strong confinements. Using a ring made of material that enters the plastic field (brass for example) will enable controlled confinement to be applied. Thus, measuring the transversal strain of the ring allows the confinement to be calculated, after which we can calculate the values that are usually dealt with in geomechanics. As an example, Figure 1.7 shows evolution of the volume-pressure relationship compared to the same relationship obtained using a static trial.

1.3. Tests with transient phase conditioned interpretations

1.3.1. *Tests involving mainly traction behavior*

1.3.1.1. *Modified Hopkinson bar*

As explained in section 1.2.1.1, traction behavior is essential for characterizing the failure of brittle geomaterials like concrete, which is why adapted tests have been designed to obtain this data for high speeds, and has been widely studied.

The design has been achieved, mainly thanks to modified Hopkinson bar configurations in which the specimen is glued between the input and output bars, where it is submitted to traction produced by a shock to a retaining shoulder at the end of the input bar. The main results with this technique were obtained on the device of the University of Technology in Delft [REI 86] and [ZIE 82] between 1980 and 1995. The tested specimens are typically core sampling specimens 74 mm in diameter (the same diameter as the bars), with a 1 to 1.5 slenderness. The duration and energy of the shock which generates the traction wave depends on the mass used, hydrostatic pressure and the number of dampers inserted between the masses whose fall is triggered and the lower input bar shoulder.

In practice, as we want the shock to be intense enough to cause specimen failure, and the loading build-up rate to be constant during the trial, the device allows loading rates ranging from 4 to 200 MN/s, about 100 to 1,000 above the rates reached with conventional press machines with similar specimen geometries.

The analysis of specimen loading uses the transient analysis described in section 1.1. The quality of glueing interfaces and the nature of the aluminum bars contributes to impedance compatibility between concrete and the loaded material, so an important part of the wave is transmitted to the specimen and the obstacles to transversal strains are limited. We have verified that the transmitted-wave signal gives a precise measurement of the average stress developed inside the sample – after conversion into stress and calibration in time.

Figure 1.8. *Direct traction tests on Delft University's Hopkinson bar, plain concrete and very high performance concrete*

The simultaneous measurement of the strains on the specimen (Figure 1.8) is made possible either by extensometers gages glued to the sample [TOU 95a] or by pre-slotted fiber concrete (where the measurements concerns crack opening), by gages fixed directly on the sample [TOU 99b]. For the speeds considered, the time delay between stress and strain signals is about 220 µs, whilst the space difference is about 1 meter. The "suitable" loading time (from 0 to maximum load) ranges from 100 to 500 µs, and sampling is carried out at 250 kHz. The excellent stress-strain linearity obtained confirms the validity of the hypotheses. Nevertheless, considering the time to go through the specimen (about 25 µs, i.e. a difference about 1 MPa), the rates reached limit the interpretation as far as sample homogenity is concerned.

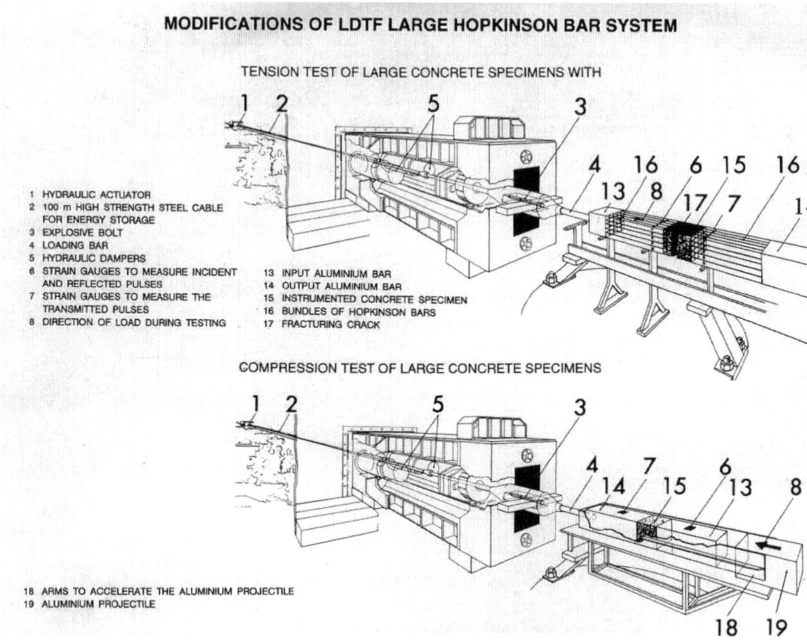

Figure 1.9. *Ispra Centre device for big sample testing*

To improve understanding of the mechanisms of traction failure and crack dynamic propagation, a specific device has been developed for effort transmission and measurement and is included in the large-scale dynamic test equipment (LDTF) at the European Research Centre Ispra [CAD 01]. To increase the capacity of the shock transmitted to the specimen at that installation (20 cm-edge cube), the shock is generated by the violent release of a tight cable. The device (a Hopkinson Bar Bundle (HBB)) consists of a prismatic Hopkinson bar beam, each bar being instrumented, which transmits the traction wave to the specimen. Potential helical reinforcements at both ends of the specimen are eliminated. It is possible to follow both the opening of a crack across the specimen and the loading transmission remaining in the not yet broken ligament, by applying a simplifying hypothesis of wave propagation and load transmission inside the breaking specimen.

Most of the significant results concerning high-speed traction concrete behavior detailed in section 1.5 were discovered using this installation (Figure 1.9) on quite large scales.

1.3.1.2. *Hopkinson bar Brazilian test*

The test is an expansion of the Brazilian test, whose traditional analysis is based on the assumption of brittle elastic behavior. We consider an elastic cylinder compressed perpendicularly to its generators: compression is applied along two diametric generators. A plane deformation elastic calculation shows that loading causes practically constant traction maximum stress along the cylinder axle, at right angles to the compression axle. We assume cylinder failure takes place when the strain reaches the ultimate value. Carrying out this test in quasi-statics is not obvious, as it requires strict respect for limit conditions and the ideal elastic model (stiff supports among others). Nonetheless, this trial is easy to carry out and gives a consistent order of magnitude for simple traction failure stress.

Extension to the dynamic situation is easy. Compression is applied using a Hopkinson bar. If we want to analyze the results in the standard way, we suppose that the situation is not too far from the quasi-static case. To do this, we have to assume that inertial effects can be neglected. They can be neglected before failure but, as is the case for simple compression, they cause an apparent increase in the maximum load after failure, so consequently it is important to detect failure by direct observation (using high-speed imaging), as it is for dynamic compression tests where localization of strains with block development does not necessarily lead to load drop immediately. We should also check that the mechanical fields are not too far away from the fields we would have in statics at the same applied force value. Thus, we have to verify that failure occurs at a time when input and output forces are quasi-equal. Such a situation will only happen when loading is slower than the homogenizing time (typically the time for the elastic waves to cover the diameter of the sample several times).

Achieving all these conditions simultaneously is difficult, but as we saw in section 1.1.2.2, the Hopkinson bar provides us with information about the loads and displacements applied to the sample all the time. Assuming this data is accurate, we can then carry out a numeric simulation of the test (assuming brittle elastic behavior), which gives a more precise assessment of failure strength [TED 93]. However, this hybrid approach (calculation-test association) is that of a structure trial, and is better suited to model validation than directly determining a behavior parameter.

1.3.1.3. *Scabbing test*

The scabbing test is a test with a fundamentally transient analysis. Actually it is based on analyzing wave propagation inside a bar made of the material itself. Concrete, a brittle material has a uniaxial compression strength that is clearly superior to its traction strength.

By using an assembly like the one in Figure 1.10, we induce a compression wave (propagating to the right in the figure), which is reflected at the free end as a traction wave [BRA 99], [DIA 97].

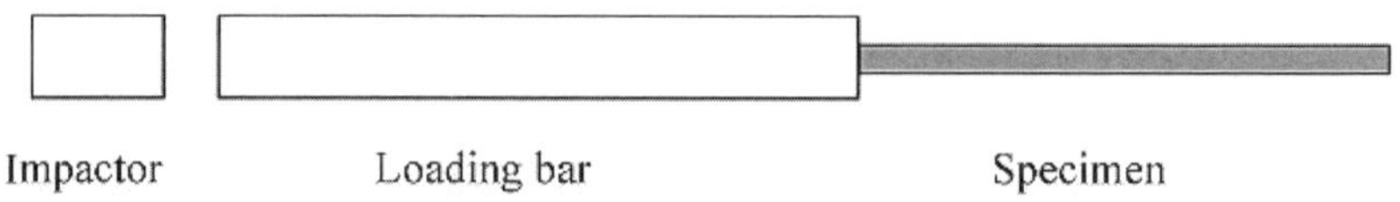

Figure 1.10. *Scabbing test diagram*

The compression pulse produced by the impactor is measured via a strain gauge glued to the bar. The elastic properties of the bar and the sample being known, we can infer the shape of the pulse induced inside the sample. We can also glue a gauge on the specimen to measure it directly. The compression wave thus produced has a lower amplitude stress than the compression concrete failure stress. The opposite amplitude reflected traction wave is sufficient to cause failure in the sample at a specific position. By applying the principle of elastic wave superposition, we can infer the stress value at the failure point. The analysis is easy because the pulse is short compared to the propagation time inside the sample. This is why we use short impactors and long specimens. Making specimens respecting homogenity conditions is therefore delicate.

The accuracy of the test analysis can be improved by additional information such as the failure instant, which can be obtained by high-speed imaging. In some cases, we can observe successive failures in the sample, analysis of which gives redundant measurements of failure stress.

This trial also gives accurate and reliable measurements of limit conditions, and the loading parameters are well-mastered. Fine interpretation still remains difficult as it is one-dimensional (as far as wave propagation is concerned), whereas failure has to propagate in the transverse direction. Moreover, the characteristic phenomenon is quite local. High strain gradients do not allow easy measurement of the strain rate characteristic of the test. This speed is usually taken as the strain time derivative near the failure point; for a one-dimensional wave, this derivative is proportional to the deformation spatial derivative.

For concrete, a very marked increase of failure stress with strain rate has been observed [BRA 99]. Between 1 and 100 s^{-1}, failure stress can be multiplied by as much as a factor of 10. The physical interpretation of this result still has to be more closely examined.

1.3.2. *Tests implementing compression behavior*

1.3.2.1. *Hopkinson bar trial*

As explained in section 1.1.2.2, the Hopkinson bar allows an accurate measurement of the forces and displacements applied on a both faces of a sample, especially in compression. Particular precautions pointed out give access to the weak strain area in the case of concrete. Figure 1.11 shows an example of the forces measured on each face, as well as the rates applied to each face of the sample (Figure 1.12) and the associated displacements (Figure 1.13).

For this test, the specimen is initially 40 mm in both length and diameter. Its relative density is 2.25 kg/m^3, with a largest aggregate diameter of 8 mm. It is loaded via an aluminum Hopkinson bar, 40 mm in diameter. The 1.3 m long impactor is projected with a speed of 14.5 m/s. When observing the speeds to be measured, we notice that the specimen absorbs little of the available energy, since the loading bar speed is roughly equal to the initial speed of the impactor at the end of the test, i.e. when the sample has failed. The induced displacements are very low, as the displacement associated with the force peaks is below 1 mm. The post-peak phase observed on the loads says a lot about the existence of inertial confinement.

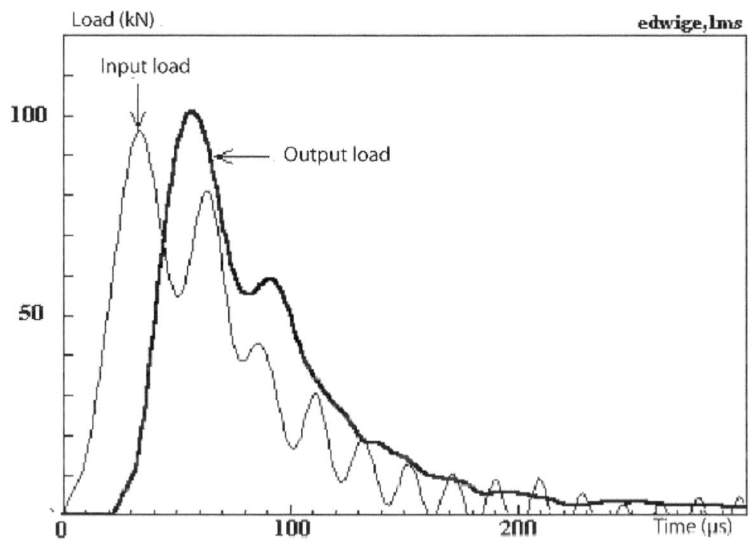

Figure 1.11. *Hopkinson bar compression test input and output loads*

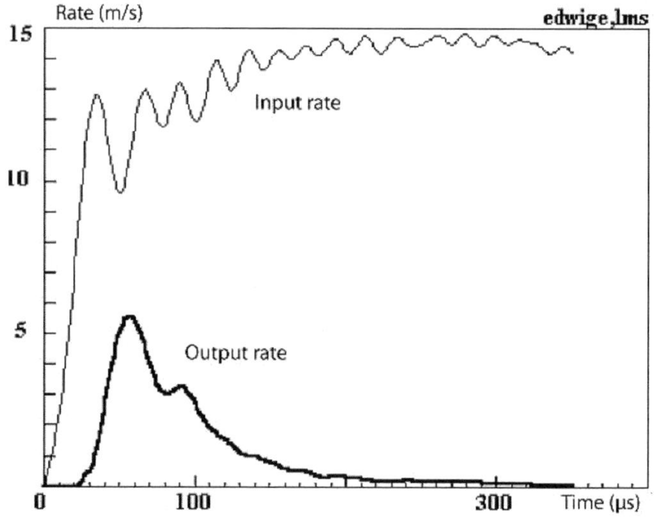

Figure 1.12. *Hopkinson bar compression test input and output rates*

A one-dimensional transient elastic calculation simulates the test quite well. The incident wave being known, we calculate the reflected and transmitted waves during the first 45 μs, after which the result of the calculation suddenly deviates from the measurements. From this, we can infer that failure takes place after 45 μs at the latest. This instant is more or less synchronous with the output load peak. For this calculation, the apparent elastic modulus is 7.8 GPa. This "modulus" is quite weak and can probably be explained by concentrated strains at the interfaces between the specimen and bar surfaces: these cannot be neglected in statics either when considering the specific strains on a specimen. If we only consider this phase of the trial, we notice a time shift between the maximum of the load, equal to about half the 45 μs.

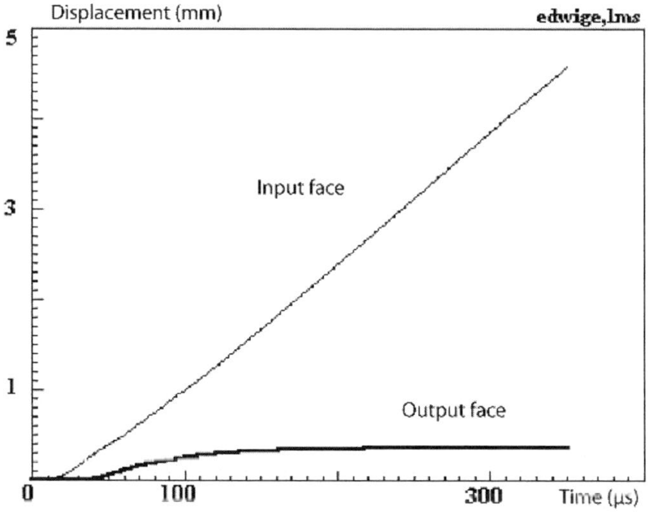

Figure 1.13. *Hopkinson bar compression test. Displacement at input and output faces*

These values taken into account, it is certainly not reasonable to suppose the mechanic fields are homogenous and deduce a stress–strain relationship from them. Nevertheless, if this simplified analysis is done to obtain an order of magnitude of the strain associated to the stress peak and corresponding strain rate, we obtain the results shown below (Figure 1.14). Depending on the way the stress is calculated (without homogenity, there is no reason to consider the average effort more than the output load), the stress peak is reached for an overall "strain" (average relative displacement between the input and output faces) ranging from 0.75% to 1%. In fact, as indicated, failure probably occurred 20 μs before, i.e. maybe for a half as low "strain", the average strain rate is about 200 m/s without being really accurate.

A transient analysis can be carried out within a one-dimensional frame ([GAR 96], [GAR 98], [ZHA 96]), by using a "simple" negative strain-hardening elasto-visco-plastic model. This approach gives results in accordance with the measurements. However, it is insufficient, because it considers the sample as the material and it does not take into account structure effects or inertial confinement. It makes more sense to use a three-dimensional model and to simulate the test using finite element dynamic calculation. This approach was developed in the GEO network [BAI 99].

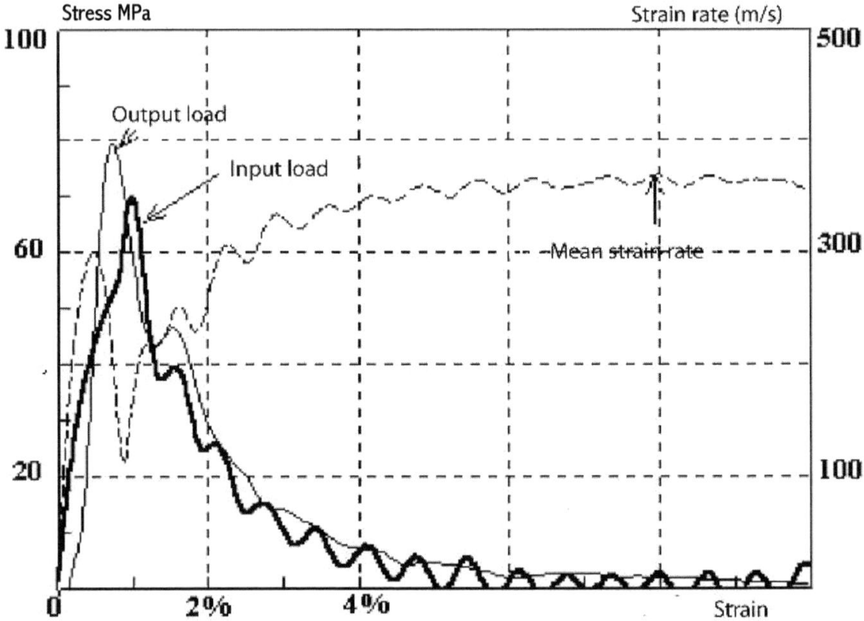

Figure 1.14. *Approximate average behavior*

At last, the same type of trial can be carried out in confinement, using confinement cells coupled to a compression Hopkinson bar system [GAR 98]. It is important to use pressure confinement with gas and a large enough chamber inside. Using an incompressible fluid actually leads to interference confinement, because it acts as a strain limiter. Typical results [GAR 98] show that the confinement effect is the same kind as the strain speed effect to which it is added. Therefore, we can infer that the main effect of strain rate in a concrete dynamic compression test is a structure effect linked to the inertial confinement. This result was confirmed by digital simulations, which are developed in [BAI 99].

1.3.2.2. *Direct impact tests (shock cannon)*

The Hopkinson bar test gives information on high-speed concrete compression behavior, however the practical limitation of transversal dimensions (less than 10 cm in general) works against good representativity of the concrete material. That is why "block bar" devices that carry out direct impacts on a concrete cylinder have been developed ([BIS 95], [DAR 95]). The impactor is guided and propelled either by direct falling or a compressed-air canon. The quality of the contact between the impactor and the sample face is essential for a good repartition of loads. In practice, the surface of the impactor and that at the rear of the reaction system can lead to a limitation of the transverse strains at both ends of the specimen, upon which it is necessary to use a specimen with sufficient slenderness (2 or more). If an efficient anti-helical reinforcement system is used, we observe prismatic failure corresponding to the cutting of angular sectors [MUR 86].

The rates reached are important, ranging from 1 to 100 s^{-1}. The strains can only be measured by extensometer gages glued to the sample. An artifact is possible due to confinement, which is different at the heart and at the periphery of the specimen. Measuring the loads also requires particular precautions, considering the inertia of the impactor. It is generally inaccurate to consider the acceleration measurement on the impactor. Bischoff and Perry developed an ultra-flat pressure cell to minimize the reflections of waves therein, the cell being inserted between the specimen and the assumed motionless reaction body at the back of the specimen.

The results obtained with these devices (mainly the maximum average stress reached) are along the same lines as those obtained on Hopkinson bars. Guidance defects (centering loads) and surface evenness (contact hard spots) could be the cause of dislocations when the results are in a continuity with those obtained with presses. Besides which, the immobility of the reaction device has to be verified, otherwise a correction for inertia becomes necessary. Anyway, at the speeds reached, the direct transformation of the maximum effort recorded into "failure stress" can only be considered as conventional, as the analysis reveals inertial effects and the "inhomogenous" divergent feature of the axial and radial stresses inside the specimen.

1.4. Other tests

1.4.1. *Tests adaptable to an energetic approach*

In all that has been said so far, we have noted the difficulties linked to finely identifying the high-speed dynamic properties of concrete. Considering these experimental difficulties, and the necessity to identify calculation parameters simple enough for engineers, some means have been developed to enhance – in a comparative way – the energy absorption properties of some concretes (especially fiber reinforced concrete). Basically the approach involves adapting the resilience test, which is standard for metals, to concrete, and which corresponds to a dynamic bending loading, the load being applied via a pendulum ram impact testing machine (Charpy test).

Interpreting the test relies on the fact that the drop energy (potential energy of the pendulum, turned into kinetic energy) is partially transmitted to the specimen as strain energy, and can even be completely consumed in the event of the specimen failing without resilience of the ram. Depending on the type of impacted sample (prism slotted or not), the failure energy determined in this way is incorporated in the hypothesis of a brittle-elastic behavior, by determining either a bending traction strength, the "global" energy of the failure per unit area, or an energy restitution critical rate. Employing the result of the "pre-dimensioning" test involves determining (by calibrating in comparison to thickness of well-known materials) a material thickness as a function of the shock energy to be absorbed, in applications where the concrete wall has to withstand bending impacts. Whatever the case, if this type of test allows a comparative analysis, it can be subject to artifacts, due to the energy dissipation sources which are not taken into account (movement of supports, frame and sample, local dissipation at the impact spot, local deterioration of the concrete and heating). The respective masses of the specimen, pendulum and frame are such that the inertia of the test body can rarely be neglected in shock.

However, the use of a pendulum system is based on a good control of the initially transmitted energy, which is why a traction device adapted from a pendulum has been developed ([BAN 91], [BAN 96]) so as to carry out traction or pull-out tests on fiber-reinforced concrete. The strain rate reached with this type of device is about 0.1 s^{-1}. The load measurements carried out are difficult to interpret, as the presence of the sample acts as a divergence from a no-load measurement. A significant part of the energy seems to be dissipated by frame vibrations.

1.4.2. *Validation tests on structures requiring an inverse analysis*

1.4.2.1. *Falling mass tests*

Generating shocks using falling mass devices, for example with pendulum systems, is a relatively simple and economical solution to produce high-speed loadings on concrete test-bodies, which is also useful in typical loadings, where concrete distortions on plates essentially operate in bending. These are no longer behavior identification tests of a material, but tests of representative structure components. These devices have been widely used ([KRA 93], [KRA 96], [MIN 87]). One advantage – theoretically at least – is that it enables a "static" test with similar limit conditions and loading application geometry to be carried out.

Nevertheless, correctly interpreting this kind of test requires a careful inverse analysis, if we want to identify details of the dynamic behavior of the concrete material (possibility of parametric studies with varying properties of the constituent materials) or of the reinforcement [DAN 01]. Actually the generally sought operation is a bending mechanism, though with some inherent complexities [KRA 93]. We need to consider:

– operation as a slab which cannot be reduced to a beam operation;

– plate operation (diaphragm effect);

– possible local mechanisms depending on the shock range and energy (local crush, punching or transverse load failure on the supports);

– sensitivity to the support conditions (problem linked to the dynamic unilateral support conditions);

– problems linked to the resilience of the impactor;

– sensitivity of the failure mode to singularities (corner effects).

As an example of the extra information that can emerge if the above points are considered, an important sensitivity of such tests' overall results to the "nose-shaped" impactor [BIN 01] has been shown, with a part of the impact energy being consumed when starting a penetration mechanism [WAT 02]. When a complete inverse analysis can be conducted, notably with well controlled loading and limit conditions, the competition between the different damage and collapse mechanisms can be highlighted according to the impact speed, which was especially the case for reinforced concrete walls submitted to the impact of a pendulum in [MIY 91a] and [MIY 91b].

As regards validating complex systems [SAT 95a] comprising a concrete structure (reinforced or pre-stressed) and/or supports interacting with the ground, tests where impact is achieved by means of a falling mass are often used [PER 01], the ability to control and modulate the incident energy (blocks ranging from a few tens to several hundreds of kilos, falling heights up to 30 m – limitations due to the sizes of cranes) must be taken into account. Re-calculating the test is often difficult because of the frequent presence of dissipative materials (ground, granular materials, energy dampers). Research into systematic empirical interpretations worthy of note (especially with a view to dimensioning rock fall devices to protect transport infrastructures) includes [LAB 96], [MIK 95], [MON 98] and [SAT 95b].

1.4.2.2. *Block fall tests*

If we want to collect exploitable information about the participation of the different materials and components in the overall strength, tests on structure elements require an inverse analysis, which is often complex in dynamics. However, in so far as the load and the test body are close to the "real" situation, some tests are used directly, to confirm and/or compare various technical solutions. In a manner complementary to block-falling tests, tests where the concrete element itself is submitted to a shock are used. This is especially the case when validating radioactive waste containers ([LAE 94], [VEC 88]). In order to anticipate the result of this kind of test, simplified analysis methods have been developed ([MAR 87], [SER 98a]), and these allow parametric study when designing and developing prototypes. This experimental configuration allows good control of energy during the shock, and confirmation of the hypothesis when the impacted area is motionless and dimensionally stable. On the other hand, fitting structures with the required instruments for this is generally expensive and difficult (on-board accelerometers); consequently, identifying the analysis should be based on *a posteriori* observable cracking state.

1.4.2.3. *Explosion resistance test*

For a number of protection structures, overall dynamic loading is more significant than a local impact. Adapted large-scale experimentation is carried out via pressure loading caused by an explosion. Control of such loading and its similarity rules is rather good, subject to limit conditions and simple structure geometry, which limits uncertainties linked to reflections. In this kind of experiment, instrumentation can be quite complex (numerous pressure sensors, gauges, displacement sensors and accelerometers). The most important factors are the data acquisition speed, and the qualities of activation and filtering. Many experiments have been carried out on typical structures, including structural walls and slabs ([GRO 90], [KR 96]), vaults and tunnels [KRA 89] and hot caves. Controlling the mechanical limit conditions is the main difficulty for obtaining a precise inverse analysis in such cases; therefore, this kind of test is often used for validating simplified regulations and abacuses for the dimensioning of protection structures.

Another category of tests carried out using blasting charges is aimed at characterizing the "compaction" behavior of a material, the latter being closest to the blasting charge in a highly confined stress state due to the pressure resulting from the explosion nearby on the one hand, and to the rest of the surrounding structure not yet hit by the loading wave. Such tests are transient, and the analysis requires re-calculation (made easier by the semi-spherical symmetry of the problem). The data is generally adapted to a volume-pressure limit curve interpretation, characteristic of the areas submitted to strong tri-axial compressions. A recent description and interpretation of such tests can be found in the works of the GEO network [BAI 99].

1.5. Synthesis of the experimental data on concrete and associated materials

1.5.1. *Data on cement paste mortar and concrete*

1.5.1.1. *Looking for a consistent interpretation*

When analyzing experimental results, the multiplicity of experimental data, obtained under different stress conditions, and the difficulty in interpreting them with frequent necessary re-calculation and transient analysis, coupled with the variety of observation scales used makes a clear presentation of the characteristics of concrete's dynamic behavior rather difficult.

We propose an interpretation based on Rossi's initial ideas [ROS 91], highlighted by many experimental programmes, the validity range of which is being discussed. The underlying idea is essentially linked to the fact that cement materials that have had evaporable water removed do not show any evolution of strength when stressed at rates ranging from 10^{-6} to 10 s^{-1} [TOU 95a]. This fact has been verified on mortar and concrete, in both traction and compression ([DAR 95], [HAR 90], [ROS 92b], [TOU 95a]), within a domain where the quasi-static interpretation of the test results is valid and thus allows a "conventional" interpretation of the behavior of the material.

We can infer from this that the "sensitivity" of the concrete material to stress rate is (within the considered domain) linked to the presence of free water inside the porous material [TOU 99a]. This fact makes the control of its hydration state crucial when its properties are being identified at high speeds, which is rather difficult to implement. Part of the variability of the results observed in literature can be explained by partial drying of specimens ([COW 66], [KAP 80]), in addition to self-stresses linked to drying, which are superimposed to the initial mechanical state of test bodies, the influence of which is all the more important due to their small size.

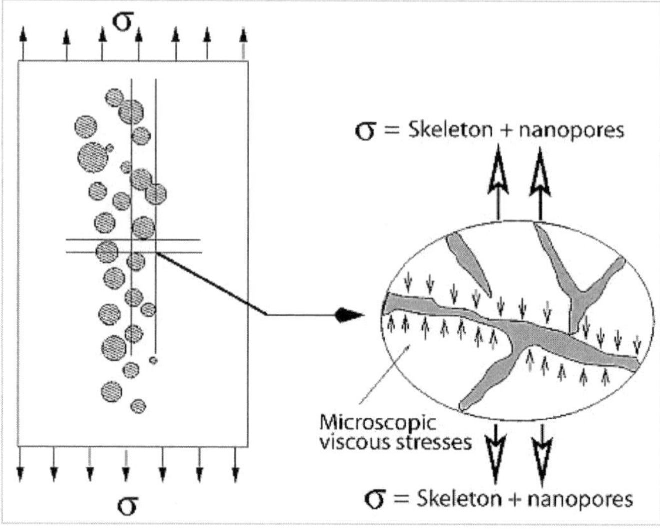

Figure 1.15. *Apparent strength of concrete in traction dynamic loading (from [TOU 95a])*

The free water present in the porous volume is stressed like a viscous fluid by the (fast) motion imposed on the sides of the skeleton, which in the ideal case of a film between two walls is known as the Stefan effect. The consequences of this can be observed in high-speed traction or compression tests as an increase (low relative value) of stiffness, and a more significant increase in the strength, called the rate effect. The macroscopic stress increase can then be interpreted by partition between the stresses borne by the skeleton and viscous stresses borne by the fluid (Figure 1.15). Things progress as if these viscous stresses cause pre-stress in the skeleton and delay either its traction failure or the failure in the extension direction induced by loading when the latter is not purely tri-axial. The partition and its effect on material failure are at the root of the elasto-plastic viscous strain-hardening model developed by Sercombe [SER 98b].

For higher-rate tests (over 1 to 10 s^{-1}), even when the hydration state is well controlled, the transient character of the test and the failure phase of the specimen take precedence over the rate effect linked to the nature of the material, qualitatively at least [WEE 98]. The relative increase in "strength", compared to the static reference value, can exceed a value of 2, even for specimens in which free water has been eliminated [ROS 96]. In fact, we can notice that a dynamic failure mechanical analysis (which takes critical crack propagation inside a material with non-zero inertia into account) is consistent with the experimental observation, which is that the relative strength increase (dynamic increase factor (DIF)) evolves with the strain rate in a logarithmic diagram, along a 1/3 slope straight line ([CHA 98], [KIP 80]), beyond a certain threshold (typically 1 to 30 s^{-1}, depending on geometry and loading) which corresponds to the limit beyond which the test has to be analyzed as a transient state ([REI 91], [WEE 89]). An analogous model taking local inertia into account [BAI 94] also justifies the "double state" obtained experimentally if the transient character of the failure is interpreted as a local property.

The latest results obtained on quite large-size concrete samples [CAD 01] are consistent with these two basic mechanisms causing the strength increases observed during high-speed dynamic tests, with the participation and viscosity of water, beyond a specific threshold, and the participation of inertia on both sides of the failure origin.

1.5.1.2. *Effect of the structure of the cement paste*

The crucial part played by free water in determining the sensitivity of concrete behavior to loading speed in a transient state has led to speculations about the relevance of conventional parameters used to describe the dependence. Actually, it appeared that the conventional definitions ([COL 88], [BIS 91], [MAL 98]) of compression or traction DIFs, as well as those of ultimate strain or Young's modulus, have led to values varying according to the static properties of concrete, including compression strength ([COL 88], [JAW 87], [ROS 95]). This is apparently responsible for the wide discrepancies in the diagrams used to describe strength evolution (Figures 1.16 and 1.17), and interferes with them being taken easily and reliably into account in a regulation context.

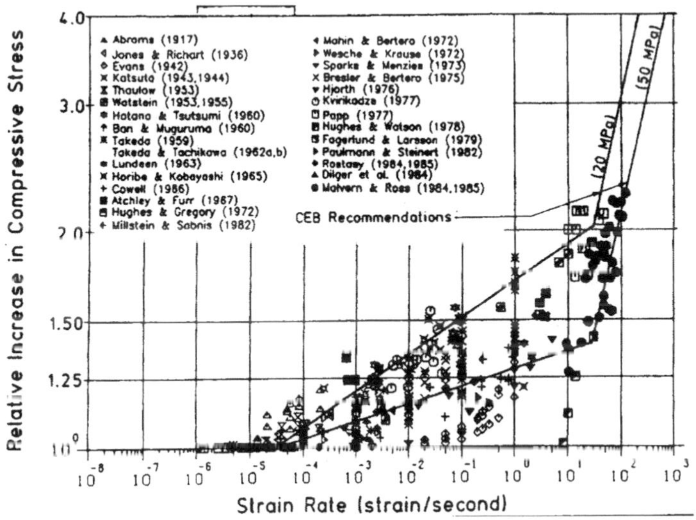

Figure 1.16. *Concrete compression strength. "Rate effects" (from [BIS 91])*

If, as a first approximation, we assert that in the rate domain (where inertial effects can be neglected), absolute traction strength increases observed in concretes of various strengths (35 to 120 MPa [TOU 95a] and 230 MPa for tests on high efficiency concretes [TOU 99b]) are not a function of the strength (Figure 1.18), with a careful identification of the specimens to avoid hydration gradients, we have proposed that this characteristic should be used [TOU 99a], rather than the relative increase [ROS 95]. A reasonable order of magnitude is obtained with a rounded down value of + 0.7 MPa/$\log_{10}$ (loading rate in MPa/s) to represent the increase in strength, for a common or high-efficiency concrete. In the same way, a + 0.9 GPa/$\log_{10}$ increase of the Young's modulus (loading rate in MPa/s) can be adopted as a first approximation. From these values, using classical hypotheses about concrete behavior in other stress states, we obtain a reasonable order of magnitude for the compression strength increase (about + 6MPa/$\log_{10}$ with loading rate in MPa/s).

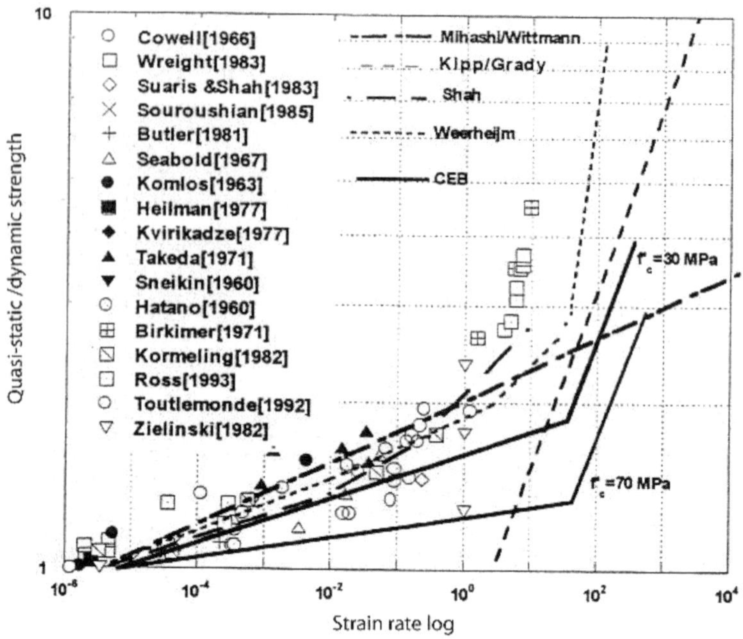

Figure 1.17. *Concrete traction strength. "Rate effects" (from [BRA 99])*

More accurate identification of the really influential parameters has been obtained from data for which the amount of hardened cement paste, the size of the biggest aggregates and the water/cement ratio (control of micrometric porosity) were used as the main variables of the experiment surface. The fact that the whole amount of hydrated calcium silicate (CSH) and the relative compactness of the aggregated skeleton [LAR 00] (a characteristic value of the part of the cement paste and its defects with regard to the part of the defects due to the paste-aggregate interfaces and the compactness defects of the skeleton) could constitute two significant factors to improve the accuracy of the description of rate effects as a function of the composition of concrete has been brought about [TOU 95a]. Within the limit where the strength is not first controlled by various defects, the sensitivity of the strength of the material at loading speed is mostly controlled by the amount of free water present in the nanopores of the material, which are intrinsic to the porosity of hydrates and become saturated when the outside relative humidity exceeds 50%.

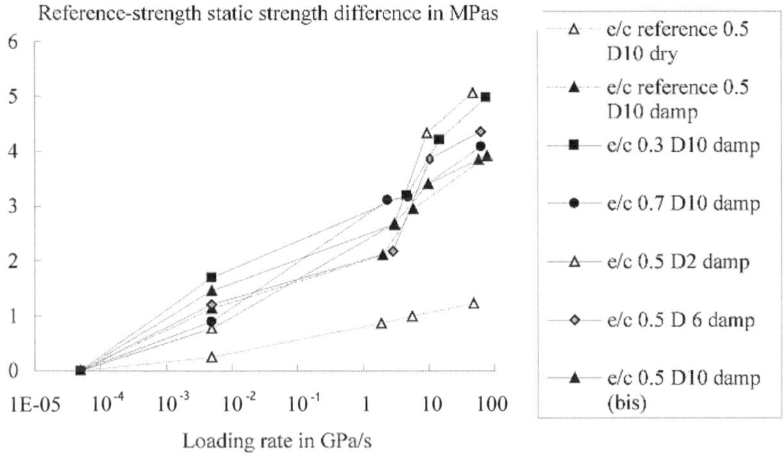

Figure 1.18. *Traction strength variation according to loading rate (from [TOU 95a])*

1.5.1.3. *Description of a plasticity criterion in terms of evolution*

The previous description from monotonous testing is based on the evolution of the maximum load applied to the specimen. For the purposes of dimensioning by extrapolation from static calculations, on verifying the section strength or the ultimate time, concrete dynamic strength (especially in compression for reinforced concrete pieces) is the main problem, and the ultimate time can be used to calibrate the oscillator that corresponds to the structure element under consideration, according to several codes or recommendations ([COL 88], [COL 86]). With such an approach, the DIF concept becomes interesting, despite the fact that the value obtained is linked, amongst other things, to the considered concrete and the stress rate.

For more comprehensive calculations, strength evolution is insufficient, as the strain speed value is not constant and cannot define the behavior characteristics alone. In addition, strain data during characterization experiments (modulus, limit strains) are not abundant and are sometimes contradictory because of difficulties with measurements and interpretations (like localization, the effects of which have been described when mobilizing inner inertial loads), as shown by [COL 88], [BIS 91] and [SER 98a].

A "unified" use of results published in the literature can lead to a description of the dynamic behavior of the material through the "evolution" of its static behavior. The advantage of this approach is that it covers *a priori* (using a conventional three-dimensional interpretation) all possible stress states, whilst also taking advantage of the (rather rare) validations of criteria in strongly tri-axial stress domains. However, from this perspective, using viscoplasticity or damage models with "standard" gradients has limitations, because experimental data coincide rather badly with the calibration of viscosity aimed at mastering numerical regularity problems ([GEO 98], [TOU 95b]).

This is why it has been necessary to explore more complex modeling by extension of a damageable elasto-plastic model, thanks to a strain-hardening variable with the same nature as a viscous strain [SER 98B]. This inner variable corresponds to an extension strain (Figure 1.19), in so far as the rate effects are principally linked to the deviatoric component of the stress state, the intervention of confinement delaying localization of failure in the potential growth direction, the same as in statics [KON 01]. The methodology for identifying the parameters of the model from a relatively low number of well controlled empirical data points (direct traction tests at various rates) have been detailed and validated by traction simulations, compression and shear tests on specimens, and bending tests on slabs [SER 98b]. These simulations have allowed it to be validated within the studied domain, for testing hypotheses about the kinematic nature of viscous strain hardening (Figure 1.20). The validation of such an approach should be continued using high-confinement tests ([GON 90], [GRA 89], [MAL 91]). As for the problem of falling containers, which was used as a basis for this development, the importance of the various sophistications of the model (damaging, taking viscous strain-hardening into account) was verified by a sensitivity study, which essentially showed the behavior of the studied structure was governed by both local compression of concrete at the impact point and propagation of induced tractions within part of the structure where confinement was weak [SER 98a].

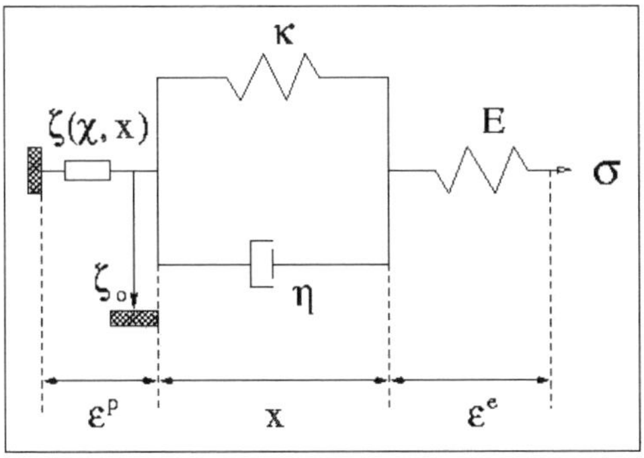

Figure 1.19. *Visco-elasto-plastic model with viscous strain hardening [SER 98a]. In this 1D diagram, the total strain ε is the sum of the 3 scalar terms ε^p, x and ε^e*

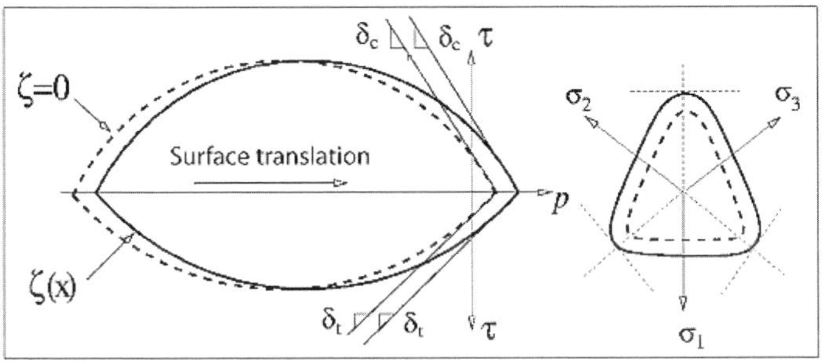

Figure 1.20. *Visco-elasto-plastic model with viscous strain hardening [SER 98a]. The plasticity criterion is translated into viscous strain hardening*

Note that such modeling is below the localization of strains integrated into the local description of the behavior. For the purpose of the study behind the development of the model [TOU 99b], this limit seemed acceptable because calculating the structure was not supposed to reveal any fracturing, and the calculated strain rates appeared to be limited to about 1 s^{-1}.

For justifications based on "ultimate limit state" calculations including possible localization of strains in part of the structure, the question of integrating inertial effects into the local behavior description can be raised [MAL 98]. We have to be aware of the theoretical difficulties caused by taking non-linearities into account, and difficulties in ensuring validation on the scale of the structure, as it has already been enhanced during test interpretation.

1.5.2. *Data available for reinforced concrete*

1.5.2.1. *Dynamic behavior of reinforcements*

To design structures, the dynamic characteristics of the steel framework are as crucial as those of concrete; indeed, the ductile character of a failure due to accidental dynamic strain can be investigated, owing to failure due to R-bars. Technical data concerning high-speed behavior of reinforced concrete framework proved to be rare, and information about pre-stressed frameworks was non-existent. A recent synthesis is available in [MAL 97]. The results are exploited to determine a relative increase in traction strength (DIF) or elastic limit. Owing to the idealization of framework behavior that is general in calculations and to the fact that no elasticity modulus variation seems to be revealed at the strain rates considered, rate effects on the elastic limit and strength are sufficient to ensure that a consistent behavior description is obtained using standard calculation methods for reinforced concrete sections.

The expression proposed by Malvar for describing the relative increase in steel strength as a power of the strain rate (value imposed in the monotonic identification tests) is consistent with the usual descriptions given for concrete ([MAL 98] for example). For about 1/s, relative increases ranging from 10 to 50% of the elastic limit can be expected, depending on the nature of the steel considered. Taking into account the mainly one-dimensional feature of strain in frameworks, it must be possible to directly calibrate a strain-hardening elasto-plastic model for frameworks from this data – with the physical meaning of the variable driving the increase of strength still to be determined.

1.5.2.2. *Identification of steel-concrete adherence in dynamics*

The R-bar-concrete adherence enables us to consider the strains of the surrounding framework and concrete as identical, and is at the root of the operation and calculation of reinforced concrete structures. The permanence of this property in dynamics and the evolution of limit shear stress are basic questions; however, experimental identification of adherence properties is relatively complex, because measuring techniques used at the interface disrupt the phenomenon itself. In practice, most of the sliding strength of reinforcement is provided by setting locks on concrete, which disconnect when traction strength is reached in the transverse direction. Adherence can then be considered to be directly linked to concrete traction strength, as stipulated in most calculation regulations. Nevertheless, transverse confinement, which delays disconnection of the rods and activates their gearing with each other, also has a significant influence [MAL 92].

In a reinforcement dynamic pull-out strain, we can expect some inertial confinement to be brought about because the concrete around the reinforcement is in traction, and traction strength increases activation. A description of the evolution of adherence limit shear with regard to speed, which uses a DIF with the same form as the equivalent coefficient for traction strength, has therefore been proposed [VOS 82]. It is based on pull-out tests carried out for an assembly using the Delft modified Hopkinson bar [REI 82]. We note that for smooth steel, the adherence increase is not significant, which confirms the proposed mechanisms.

1.5.2.3. *Repeated stresses*

The behavior of specimens submitted to repeated dynamic stresses is difficult to access experimentally, especially when it is important to achieve significant damage from the first impact. In practice, this information seems to be important if the dynamic situation is not only considered as a fortuitous action, but as likely to be subject to "replicas" (successive shocks between structure elements during earthquakes, for example) or even as a servicing action (which could be the case with rock fall protection works). As for behavior in traction repeated dynamic stresses, it appears from [REI 82] that the "rate effect" displayed is reduced in comparison to the DIF obtained for a single shock. However, the accumulation of traction cycles would cause an apparent reduction of strength that could be of the same order, even at low speeds. As regards numerous indeterminate repeated shocks, avoiding crack propagation once they have started seems out of the question. This leads us to rely on the cracking limit obtained in statics where we have to guarantee the structure must not be cracked; otherwise calculations have to take probable concrete cracking into account.

The main problem with repeated dynamic stresses concerns the operation of the R-bar framework, especially if the action causes a variation of R-bar stress sign. Without confinement, the effect of limit shear during a shock causes concrete damage around the reinforcement according to orientation, and the effect of alternating vibrations or additional shocks can keep on damaging the surrounding concrete fast and irremediably, which reduces adherence and damages the anchorage of the reinforcement. Therefore, it is advisable to cautiously take into account the "adherence increase" due to rate effects, and to associate them with constructive arrangements that will enable them to stay effective – transverse confinement, for instance.

Implementing such arrangements has particular relevance for earthquake resistance fields. Because of identification difficulties at a local level, it is often necessary to resort to experiments on structure components (see [PAU 02] and [TOR 88]). In fact, at this level it is possible to underline the effectiveness of transverse reinforcement, thereby ensuring the confinement of concrete beyond cracking, allowing loads to travel after redistribution via the creation of "joints". The amount of experimental work in this field and its complexity will not be dealt with here.

It should be noted that the difficulties arise not because of the dynamic character of stresses (some Hz frequencies with regard to the specific frequencies of the elements – in the 100 Hz order of magnitude) or the dynamic reaction of the material. The difficulties are linked to the intense and repeated feature of stresses (incursions into the plastic field, stiffness damage, crack spreading), and even to the interaction between the loading frequency and the specific frequency of the whole structure (whether sound or progressively damaged).

1.5.3. *Data about fiber-reinforced concretes*

1.5.3.1. *Post-cracking mechanisms and "rate effects"*

Thanks to metallic fibers, the favorable influence of diffused reinforcement in dissipating energy during a shock has been empirically proven [ROS 98], but is difficult to quantify. The sensitivity to pealing caused by shocks during transport of some prefabricated fiber reinforced concrete pieces, compared to corresponding concrete or reinforced concrete pieces is admitted by professionals. Using fiber reinforced concrete for particular applications where absorption of energy is important has been shown to be interesting with regard to conventional reinforced solutions [HAN 92]. Most of the time it was proved globally through experiments allowing interpretations in terms of energy. However, complete and documented experimental data on the dynamic behavior of such materials in traction for characterizing after-peak behavior and revealing the contribution of fibers [ROS 98] are quite rare.

The existing data [KÖR 88, TOU 99b] highlight major aspects of the rates of behaviors of these materials: the increase in the linearity limit corresponding to matrix traction strength, a phenomenon which can be directly compared to rate effect of all cement materials in direct traction; also, a stress increase in the after-peak phase (with regard to the load obtained in statics during this phase), the increase being all the less important as a widespread range of crack openings is considered. In other words, the relative increase of the absorbed energy and its peak value is lower than the relative strength increase of the matrix, and it is even weaker if we take crack openings into account.

Such observations are consistent with the rate effect being a function of the cement phase only. Before cracking occurs, fibers have no influence, and the linearity-limit increase (matrix cracking) is similar to that observed in a non-fiber reinforced concrete. After cracking however, the anchoring provided by fibers on both sides of the crack assists in the maintenance of strength. As for reinforcement anchoring, the strength of concrete mini-connecting rods confined by the presence of the other fibers shows a rate effect, as the anchorage is limited by the strength of the concrete around the fibers. Moreover, the wider the crack opening, the more micro-cracked the concrete where anchoring has to take place will be, which limits the effect of the viscous mechanisms underlying the strength increase.

A beneficial "synergy" could be observed by comparing the shock strength of reinforced and fiber reinforced concrete pieces to those of fibered or reinforced pieces. Considering the limited crack openings permitted by fibers for a given load, once cracking has started, some confinement seems possible around the reinforcements, which is not the case for reinforced concrete where adherence has been damaged because of alternating dynamic stresses.

1.5.3.2. *Anisotropy and its consequences*

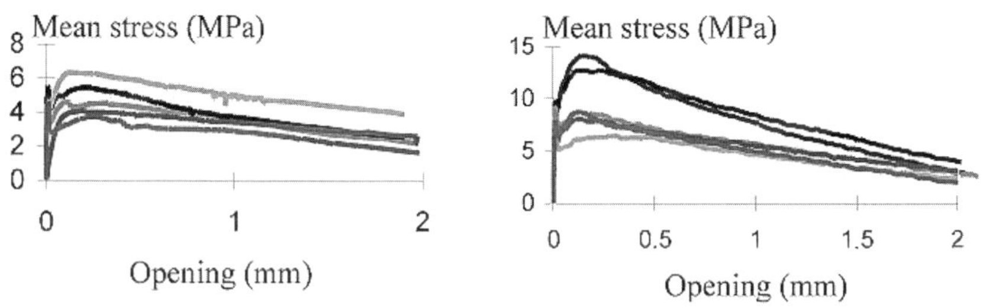

a) Opening rate 0.15 µm/s *b) Mean opening rate 3.2 m/s*

Figure 1.21a and b. *Direct traction characterization tests on very high performance concrete, at low and high rates, on pre-slotted specimens. Direction A specimens. Low efficiency of the fibers (from [TOU 99b])*

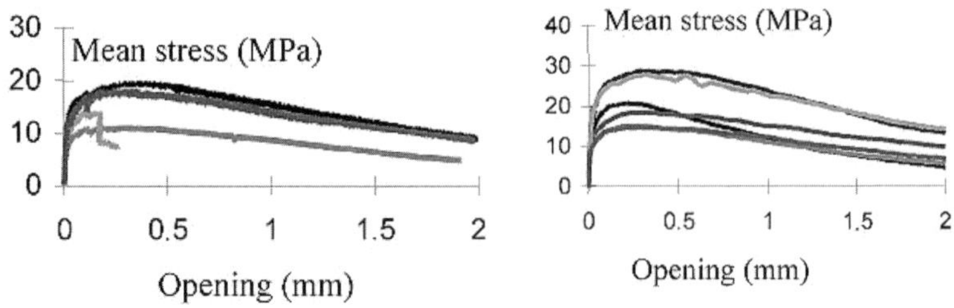

a) Opening rate 0.15 μm/s *b) Mean opening rate 2.8 m/s*

Figure 1.22a and b. *Direct traction characterization tests on very high performance concrete, at low and high rate, on pre-slotted specimens. Direction C specimens. High efficiency of the fibers (from [TOU 99b])*

	Quasi-static reference value (0.05 MPa/s) in MPa	Variation with speed in MPa/u. log.
Traction strength	8	+ 0.8
eq. 1 mm threshold stress	7	+ 0.5
Young's modulus	52,000	env. + 450

Table 1.1. *Calculation characteristics for very high performance concrete containers, derived from the traction characterization ("rounded off" values used for calculation) (from [TOU 99b])*

Value	Direction	Quasi-static value (interpolation for 0.05 MPa/s) and standard deviation	Trend of evolution with rate
Young's modulus			
	A	50.9 GPa (2 GPa)	+ 0.32 GPa/u.log.
	C	53.9 GPa (2.5 GPa)	+ 0.38 GPa/u.log.
	A + B + C	52.2 GPa	+ 0.31 GPa/u.log.
Maximum stress (unslotted specimens)			
	A	5.96 MPa (2 MPa)	+ 0.82 MPa/u. log.
	C	11.17 MPa (3 MPa)	+ 0.77 MPa/u. log.
	A + B + C	8.52 MPa	+ 0.65 MPa/u. log.
Maximum stress (slotted specimens)			
	A	4.29 MPa (2 MPa)	+ 0.70 MPa/u. log.
	C	16.35 MPa (5 MPa)	+ 0.73 MPa/u. log.
	A + B + C	9.65 MPa	+ 0.40 MPa/u. log.
Equivalent threshold stress (1 mm opening)			
	A	3.75 MPa (1.6 MPa)	+ 0.53 MPa/u. log.
	C	13.99 MPa (3.8 MPa)	+ 0.62 MPa/u. log.
	A + B + C	8.06 MPa	+ 0.45 MPa/u. log.

Table 1.2. *Traction characterization of very high performance concrete and rate effects. Results (means): A, B and C are the three perpendicular directions of the sampling within an L piece deemed representative for the project (from [TOU 99b])*

Let us recall that observing the mechanisms referred to above and obtaining characteristic properties suited for the calculation of fiber reinforced concrete structures demands respect for the strict regulations concerning possible anisotropies in the behavior and constitution of the material, which may arise due to the manufacturing mode of the structure [ROS 98]. It was thus possible to enhance (Figures 1.21 and 1.22) rate effects (strength absolute increase), according to the traction maximum stress or equivalent plastic stress absolute increase, which are comparable as they are linked to the cement matrix for the different directions considered in a fiber concrete piece.

Nevertheless, the absolute values of these strengths and stresses proved to be widely different according to the direction (Tables 1.1 and 1.2), due to preferential orientating of fibers during manufacturing [TOU 99b]. Therefore, the problem of a potential taking into account of the anisotropy as far as fiber-reinforced structure modeling is concerned remains in dynamics as well as in statics. The questions related to the dispersion of properties are also the same.

1.6. Conclusion

The accumulated knowledge available for understanding and describing the high-speed behavior of concrete material remains at a complex overall stage and leaves both the structural engineer and the mechanic dissatisfied. This can be explained by several factors: experimental difficulties in accessing the intrinsic behavior of materials in dynamics tests, difficulties linked to the heterogenity range of the concrete "material", its sensitivity to the water environment, its brittleness as a geomaterial which involves crack propagation effects within the specimens, and the wide range of materials actually corresponding to the generic term "concrete". Besides this, we also have to note that part of the difficulties reflecting on mechanical modeling problems are also present in the usual quasi-static field, even if a standardized corpus valid for engineering common needs often avoids having to ask too many questions.

After recalling the different experimental techniques that allow us to explore concrete dynamic behavior, and after taking a few precautions, we described the main established facts, i.e. the noticeable increase in strength and slight apparent increase in the Young's modulus, which can be explained by the viscosity of the interstitial water present inside the nanopores (the finest pores within the cement hydrates). This viscous inner phenomenon is inherent to porous solids, and can be observed separately in direct traction tests over a standard range from 10^{-6} to 1 s^{-1}. It also explains the rate effects induced in other stresses (compression, adherence, fiber concrete behavior) reasonably well. At higher rates, interpreting the tests involves a transient analysis of the loading and failure phases of the specimen, as inertial phenomena (in terms of measured loads) that oppose critical crack propagation become predominant.

Different empirical description levels of the mechanisms have been developed, together with the underlying theoretical support and its potential limits: the DIF, which can vary widely depending on the rate and strength of concrete, absolute traction strength increase, which enables us to calibrate a visco-elasto-plastic model with viscous strain-hardening, and application of dynamic failure mechanics when high-rate failure propagation is at stake. The choice of the behavior description level seems to be consistent with a clear choice of other modeling hypotheses: speed range of the considered strains, unique or repeated characteristic of the dynamic stress, presence or absence of uniaxial behavior, presence or absence of reinforcements, possibility or impossibility of limiting the justification validity in a domain where concrete can still be considered as continuous, etc. Depending on the circumstances, we will naturally turn to a different description level with a more or less important integration of chance in the properties of the material.

After taking these considerations into account, as well as considerations dealing with dimensioning an important number of structures and increasingly considering fortuitous situations ([PER 01], [TRO 01]), we underlined a few shortcomings. First we have seen how difficult it is to access the material under high tri-axial stress. However, it is a crucial problem for direct impacts, areas close to an explosive charge or in cases of potential penetration. As for validating "viscous strain-hardening" hypotheses, and in order to better model the combined effects of inertial confinement [UNO 02], more complete information in the field would be quite helpful as well.

While understanding the mechanisms in the case of a single dynamic stress can be considered as correct and reliable for calculations, the problem of repeated impacts remains difficult, both for validating the potential progressive damage predictions models are likely to supply, and accurately taking into account the evolution of adherence for reinforced concrete structures. This field seems to represent very important stakes for engineers, in the frequent cases of fortuitous dynamic stresses.

Finally, the increasing range of materials coming under the definition of "concrete", including fiber reinforced concretes, high and very high-performance concretes and ultra-efficient fiber-cement composites will demand more diversified validation of the indications and mechanisms highlighted in this synthesis.

1.7. Bibliography

[ALE 94] ALEXANDRE M., "Etablissement du cahier des charges d'un conteneur de haute intégrité pour l'entreposage de longue durée des déchets B en vrac", *Rapport de la commission pour les questions scientifiques et techniques relatives à la gestion des déchets radioactifs auprès du conseil scientifique du CEA*, March 1994.

[BAI 87] BAILLY P., BROSSARD J., DESROSIER C., RENARD J., "Approche expérimentale du comportement et des caractéristiques dynamiques des mortiers armés de fibres", *Cinquièmes journées universitaires de Génie Civil*, Bordeaux, 1987.

[BAI 88] BAILLY P., Effets des explosions sur les constructions. Chargement et réponse de la structure, Thesis, University of Orleans, June 1988.

[BAI 94] BAILLY P., "Une modélisation d'un matériau fragile avec prise en compte d'effets dynamiques ", *C.R. de l'Académie des Sciences de Paris*, Series II, vol. 318, pp. 1–6, 1994.

[BAI 99] BAILLY P., "Comportement des ouvrages en dynamique rapide", *Recueil de communications du groupe de travail Géomatériaux Environnement Ouvrages (GEO)*, Aussois, France, 1–3 February 1999 (*Ouvrages et Actions accidentelles* project).

[BAN 91] BANTHIA N., TROTTIER J.-F., "Deformed steel fiber – cementitious matrix bond under impact", *Cement and Concrete Research*, Vol. 21, pp. 158–168, 1991.

[BAN 96] BANTHIA N., MINDESS S., TROTTIER. J-F., "Impact resistance of steel fiber reinforced concrete", *ACI Materials Journal*, Vol. 93, N° 5, pp. 472–479, 1996.

[BIN 01] BINDIGANAVILE V., BANTHIA N., "Machine effect in the drop-weight impact testing of plain concrete beams ", *3rd International Conference on Concrete under Severe Conditions: Environment and Loading, CONSEC'01*, Vancouver, BC, Canada, Proceedings Banthia, Sakai and Gjørv (Ed.), E&FN SPON, Vol. I, pp. 589–596, 18–20 June 2001.

[BIS 91] BISCHOFF P.H., PERRY S.H., "Compressive behavior of concrete at high strain rates", *Materials and Structures*, 24, pp. 425–450, 1991.

[BIS 95] BISCHOFF P.H., PERRY S.H., "Impact behavior of plain concrete loaded in uniaxial compression", *J. of Eng. Mechanics*, Vol. 121, N° 6, pp. 685–693, 1995.

[BOU 99] BOULAY C., LE MAOU F., RENWEZ S., "Quelques pièges à éviter lors de la détermination de la résistance et du module en compression sur cylindres de béton", *Bulletin des Laboratoires des Ponts et Chaussées*, N° 220, pp. 63–74, March–April, 1999.

[BRA 99] BRARA A., Etude expérimentale de la traction dynamique du béton par écaillage, PhD Thesis, University of Metz, 1999.

[BUS 02] BUSSAC M.N., COLLET P., GARY G., OTHMAN R., "An optimization method for separating and rebuilding one-dimensional dispersive waves from multi-point measurements. Application to elastic or viscoelastic bars", *J. Mech. Phys. Solids,* Vol. 50, N° 2, pp. 321–350, 2002.

[CAD 01] CADONI E., ALBERTINI C., LABIBES K., SOLOMOS G., "Behavior of plain concrete subjected to tensile loading at high strain-rate", in *Fracture Mechanics of Concrete Structures*, de Borst *et al.* (Ed.), Swets & Zeitlinger, Lisse, The Netherlands, pp. 341–347, 2001.

[CHA 98] CHANDRA D., KRAUTHAMMER T., "Rate-sensitive micro mechanical model for concrete", in *ACI SP-175 Concrete and Blast effects*, ACI International, W. Bounds (Ed.), paper SP 175-14, pp. 281–305, 1998.

[COL 86] COLLECTIF, Tm5-1300, *Structures to resist the effects of accidental explosions*, Department of the Army, the Navy and the Air Force, technical manual, NAVFAC P-397, AFM 88-22, 1986.

[COL 88] COLLECTIF, CEB, "Concrete Structures under impact and impulsive loading", *Bulletin d'information*, N° 187, Lausanne, Switzerland, 1988.

[COW 66] COWELL W. L., "Dynamic properties of plain Portland cement concrete", *Techn. report R447*, DASA 130181, US Naval Civ. Eng. Lab., Port Hueneme, California, 1966.

[DAN 01] DANCYGIER A.N., YANKELEVSKY D.Z., "The role of concrete in resisting projectile penetration into concrete barriers", *3rd Int. Conf. on Concrete under Severe Conditions: Environment and Loading, CONSEC'01*, Vancouver, BC, Canada, Proc. by Banthia, Sakai and Gjørv (Ed.), E &FN SPON, Vol. I, pp. 612–619, 18–20 June, 2001.

[DAR 95] DARVE F., HICHER P.Y., REYNOUARD J.M., *Les géomatériaux* (3 volumes), Hermès Science Publications, Paris, 1995 (particularly Vol. 1 Ch. 1, 5, 3; Vol. 2 Chs. 1 and 4, Vol. 3 Paragraphs 2.2.2, by Bailly, Gary, Klepaczko, Hamelin, Rossi, Toutlemonde).

[DIA 97] DIAMURAYA M., KOBAYASHI H., NONAKA T., "Impact tensile strength and fracture of concrete", *Journal de Physique* IV France 7, Colloque C3 (Supplement of *Journal of Physics* III), pp. 253–258, 1997.

[GAR 96] GARY G., ZHAO H., "Measurement of the dynamic behavior of concrete under impact loading", *Proceedings of 2nd ISIE'96,* Beijing, China, *Chinese Journal of Mechanics Press*, Jin, Shi and Ding Editions, pp. 208–213, 1996.

[GAR 98] GARY G., BAILLY P., "Behavior of a quasi-brittle material at high strain rate. Experiment and modeling", *European Journal of Mechanics*, A/Solids, 17, N° 3, pp. 403–420, 1998.

[GAR 99] GARY G., BAILLY P., GATUINGT F., "Testing concrete at high strains and high rates of strain", *Proc. 3rd Int. Symp. on Impact Eng*, Singapore (December 1998), published in *Impact Response of Materials & Structures*, OUP, pp. 241–246, V.P.W. Shim, S. Tanimura, C.T. Lim (Ed.), National University of Singapore, Singapore, 1999.

[GEO 98] GEORGIN J.-F., Contribution à la modélisation du béton sous sollicitation de dynamique rapide. La prise en compte de l'effet de vitesse par la viscoplasticité, Thesis, INSA de Lyon, 203 pages, 1998.

[GON 90] GONG J.C., MALVERN L.E., "Passively confined tests of axial dynamic compressive strength of concrete", *Experimental Mechanics*, pp. 55–59, 1990.

[GRA 89] GRAN J.K., FLORENCE A.L., COLTON J.D., "Dynamic tri-axial tests of high-strength concrete", *J. of Engineering Mechanics*, Vol. 115, N° 5, pp. 891–903, 1989.

[GRO 90] GROS J.-F., Etude du comportement dynamique de dalles de béton sollicitées par des charges explosives, ENPC, 1990.

[HAN 92] HANUS F., FAURE J.-C., PEIFFER G., "Béton léger hautes performances – utilisations dans les structures devant résister aux explosions ", in *Les bétons à hautes performances. Caractérisation, durabilité, applications*, Y. Malier (Ed.), Presses de l'ENPC, pp. 651–665, 1992.

[HAR 90] HARSH S., SHEN Z., DARWIN D., "Strain-rate sensitive behavior of cement paste and mortar in compression", *ACI Mat. Journal*, Vol. 87 N° 5, pp. 508–516, 1990.

[HOP 14] HOPKINSON B., "A method of measuring the pressure in the deformation of high explosives by the impact of bullets", *Phil. Trans. Roy. Soc.*, A213, pp. 437–452, 1914.

[HOR 87] HORDIJK D.A., REINHARDT H.W., CORNELISSEN H.A.W., "Fracture mechanics parameters of concrete from uniaxial tests as influenced by specimen length", *Proc. of Int. Conf. on the Fracture of Concrete and Rocks*, Houston, Texas, USA, Shah and Swartz (Ed.), pp. 138–149, 1987.

[JAW 87] JAWED I. *et al.*, "High strain rate behavior of hydrated cement pastes", *Cement and Concrete Research*, Vol. 17, pp. 433–440, 1987.

[KAP 80] KAPLAN S.A., "Factors affecting the relationship between rate of loading and measured compressive strength of concrete", *Magazine of Concrete Research*, Vol. 32 N° 111, pp. 79–88, 1980.

[KIP 80] KIPP M.E., GRADY D.E., CHEN E.P., "Strain rate dependent fracture initiation", *International Journal of Fracture*, Vol. 16 N° 5, pp. 471–478, 1980.

[KOL 49] KOLSKY H., "An investigation of the mechanical properties of materials at very high rates of loading ", *Proc. Phys. Soc.*, B62, pp. 676–700, 1949.

[KON 01] KONO S., WATANABE F., KAJITANI A., "Stress-strain relation of confined concrete under dynamic loading", in *Fracture Mechanics of Concrete Structures*, de Borst *et al.* (Ed.), Swets & Zeitlinger, Lisse, The Netherlands, pp. 585–592, 2001.

[KÖR 88] KÖRMELING H.A., "Experimental results of plain and steel fiber reinforced concrete under uniaxial impact tensile loading", Delft University of Technology, The Netherlands, Report 5-84-8, July 1988.

[KRA 89] KRAUTHAMMER T., FLATHAU W.J., SMITH J.L., BETZ J.F., "Lessons from explosive tests on RC buried arches", *Journal of Structural Engineering*, Vol. 115 N° 4, pp. 809–825, 1989.

[KRA 93] KRAUTHAMMER T. (Ed.), Structural concrete slabs under impulsive loads, Proceedings of the Workshop held in Trondheim, Norway, Forsvarets Bygningstjeneste Forsøkskontoret, Fortifikatorisk Notat Nr 211/93, 14–16 June 1993.

[KRA 96] KRAUTHAMMER T., JENSSEN A., LANGSETH M. (Ed.), Precision testing in support of computer code validation and verification, Workshop report, Norwegian Defence Construction Service, Fortifikatorisk Notat Nr 234/96, 1996.

[LAB 96] LABIOUSE V., DESCOEUDRES F., MONTANI S., "Experimental study of rock sheds impacted by rock blocks", *Structural Engineering International*, IABSE (3), pp. 171–176, 1996.

[LAR 00] LARRARD F. de, "Structures granulaires et formulation des bétons", *Etudes et recherches des Laboratoires des Ponts et Chaussées*, OA34, 2000.

[LOU 94] LOUW J.-M., KOCH B.C., KOEN F.J., "Direct shear strength of concrete under impact", in *Structures under Shock and Impact III*, Madrid, Spain, 1–3 June, Bulson (Ed.), pp. 147–154, 1994.

[MAL 91] MALVERN L.E., JENKINDS D.A., TANG T., MAC LURE S., "Dynamic testing of laterally confined concrete", in *Micromechanics of Failure of Quasi-brittle Materials*, Elsevier Applied Science, pp. 343–352, 1991.

[MAL 92] MALVAR L.J., "Bond of reinforcement under controlled confinement", *ACI Materials Journal*, Vol. 89 N° 6, pp. 593–601, November-December 1992.

[MAL 97] MALVAR L.J., "Review of static and dynamic properties of steel reinforcing bars", *ACI Materials Journal*, Vol. 95 N° 5, pp. 609–616, 1997.

[MAL 98] MALVAR L.J., ROSS C.A., "Review of strain rate effects for concrete in tension", *ACI Materials Journal*, Vol. 95 N° 6, pp. 735–739, November-December 1998.

[MAN 67] MANDEL J., "Application de la thermodynamique aux milieux viscoélastiques à élasticité instantanée nulle ou restreinte", *C.R.A.S*, 264, A, pp. 133–134, 1967.

[MAR 87] MARTI J., GOICOLEA J., KUMAR R., ALDERSON M., "Impact analysis of transport flasks", *Int. Conf. on Nuclear Containment*, Cambridge, UK, 6–8 April 1987.

[MIK 95] MIKAMI H., TAMURA T., SATO M., KISHI N., "Shock absorbing performance of a three-layered cushion system using RC core slab reinforced with AFRP rods", Proceedings of the *Concrete under Severe conditions CONSEC '95* Conference, Sapporo, Japan, Sakai, Banthia, Gjørv (Ed.), E & FN Spon, pp. 1633–1643, 2–4 August 1995.

[MIN 87] MINDESS S., BANTHIA N., YAN C. "The fracture toughness of concrete under impact loading", *Cement and Concrete Research*, 17, pp. 231–241, 1987.

[MIY 91a] MIYAMOTO A. KING M. W., FUJII M. "Non-linear dynamic analysis of reinforced concrete slabs under impulsive loads ", *ACI Structural Journal*, Vol. 88 N° 4, pp. 411–419, 1991.

[MIY 91b] MIYAMOTO A. KING M. W., FUJII M., "Analysis of failure modes for reinforced concrete slabs under impulsive loads", *ACI Structural Journal*, Vol. 88 N° 5, pp. 538–545, 1991.

[MON 98] MONTANI STOFFEL S., Sollicitation dynamique de la couverture des galeries de protection lors de chutes de blocs, Thesis N° 1899, Ecole Polytechnique Fédérale de Lausanne, Switzerland, 1988.

[MUR 86] MURIA VILA D., Contribution à l'étude du comportement au choc des bétons, Thesis, INSA de Lyon, 1986.

[NIC 80] NICHOLAS T., "Material behavior at high strain rates", *Mechanical Properties at High Rates of Strain*, pp. 277–328, Institute of Physics, London, 1980.

[PAU 02] PAULTRE P, MITCHELL D., "Seismic response of high-strength concrete structures", *Proc of 6th Int. Symp. on High Strength/High Performance Concrete*, Leipzig, Germany, König *et al.* (eds.), Leipzig University, pp. 457-471, June 2002.

[PER 01] PERROTIN P., MOMMESSIN M., MOUGIN J.-P., TONELLO J., "Etude expérimentale du comportement d'une dalle pare-blocs", XIXes rencontres universitaires de Génie Civil, *COSS'2001*, Lille, pp. 69–75, 3–4 May 2001.

[PON 95] PONTIROLI C., Comportement au souffle des structures en béton armé, PhD Thesis, Ecole Normale Supérieure de Cachan, 1995.

[REI 82] REINHARDT H.W., "Concrete under impact loading. Tensile strength and bond", *Heron*, Vol. 27, N° 3. Delft, The Netherlands, 1982.

[REI 86] REINHARDT H.W., KÖRMELING H.A., ZIELINSKI A.J. "The Split Hopkinson Bar, a versatile tool for impact testing of concrete", *Materials and Structures*, 19, pp. 55–63, 1986.

[REI 91] REINHARDT H.W., WEERHEIJM J., "Tensile fracture of concrete at high loading rates taking account of inertia and crack velocity effects", *International Journal of Fracture*, Vol. 51, pp. 31–42, 1991.

[ROS 91] ROSSI P., "Influence of cracking in the presence of free water on the mechanical behavior of concrete", *Mag. of Conc. Res.*, Vol. 43, N° 154, p. 53–57, 1991.

[ROS 92a] ROSSI P., WU X., LE MAOU F., BELLOC A., "Effet d'échelle sur le comportement du béton en traction", *Bull.Liais. Labo. P. et Ch.*, 182, pp. 11–20, 1992.

[ROS 92b] ROSSI P., VAN MIER J. G. M., BOULAY C., LE MAOU F., "The dynamic behavior of concrete: influence of free water", *Materials and Structures*, Vol. 25, pp. 509–514, 1992.

[ROS 95] ROSS C.A., TEDESCO J.W., KUENNEN S.T., "Effects of strain rate on concrete strength", *ACI Materials Journal*,.Vol. 92, N° 1, pp. 37–47, 1995.

[ROS 96] ROSS C.A., JEROME D.M., TEDESCO J.W., HUGHES M.L., "Moisture and strain rate effects on concrete strength", *ACI Mat. J.*, Vol. 93 N° 3, pp. 293–300, 1996.

[ROS 98] ROSSI P., *Les bétons de fibres métalliques*, Presses de l'ENPC, Paris, 1998.

[SAT 95a] SATO M., NISHI H., SUGATA N., KISHI N., "Full-scale impact test of PC rock-shed with shock absorbing system", Proceedings of the *Concrete under Severe conditions CONSEC '95* Conference, Sapporo, Japan, Sakai, Banthia and Gjørv (Ed.), E and FN Spon, pp. 1623–1632, 2–4 August 1995.

[SAT 95b] SATO M., NISHI H., NAKANO O., KISHI N., MATSUOKA K.G., "Full-scale test on impact resistance of PC girder", Proceedings of the *Concrete under Severe Conditions CONSEC '95* Conference, Sapporo (Japan), Sakai, Banthia and Gjørv (Eds.), E & FN Spon, pp. 1644–1652, 2–4 August 1995.

[SER 98a] SERCOMBE J., "Modélisation du comportement du béton en dynamique rapide. Application au calcul des conteneurs à haute intégrité", *Etudes et recherches des LPC*, OA 30, 1998.

[SER 98b] SERCOMBE J., ULM F.-J., TOUTLEMONDE F., "Viscous hardening plasticity for concrete under high rate dynamic loading", *Journal of Engineering Mechanics*, ASCE, Vol. 124 N° 9, pp. 1050–1057, September 1998.

[TED 93] TEDESCO J.W., ROSS C.A., KUENNEN S.T., "Experimental and numerical analysis of high strain rate splitting tensile tests", *ACI Materials Journal*, Vol. 90 N° 2, pp. 162–169, 1993.

[TOR 88] TORO R. del, Comportement des nœuds d'ossature en béton armé sous sollicitations alternées, Thesis, Ecole Nationale des Ponts et Chaussées, Paris, 1988.

[TOU 93] TOUTLEMONDE F., BOULAY C., GOURRAUD C., "Shock-tube tests of concrete slabs", *Materials and Structures*, Vol. 26, pp. 38–42, 1993.

[TOU 95a] TOUTLEMONDE F., Résistance au choc des structures en béton. Du comportement du matériau au calcul des ouvrages, Research report of the LCPC, July 1995.

[TOU 95b] TOUTLEMONDE F., PIERMATTEI E., ULM F. J., ROSSI P., "Modélisation du béton en dynamique rapide. Résultats d'une première expérience à l'aide d'un modèle d'endommagement avec gradient", *Bulletin de liaison des Laboratoires des Ponts et Chaussées*, 198, p. 39–52, July–August, 1995.

[TOU 99a] TOUTLEMONDE F., ROSSI P., Discussion of the 95-M73 article published in the November-December 1998 volume of *ACI Materials Journal*, p. 735 "Review of strain rate effects for concrete in tension", L.J. Malvar and C.A. Ross, published discussion in *ACI Materials Journal*, Vol. 96 N° 5, p. 614–615, September-October, 1999.

[TOU 99b] TOUTLEMONDE F., SERCOMBE J., TORRENTI J.-M., ADELINE R., "Développement d'un conteneur pour l'entreposage de déchets nucléaires : résistance au choc", *Revue Française de Génie Civil*, vol. 3, n° 7–8, p. 729–756, December 1999.

[TRO 01] TROUILLET P., "Truck impacts on French toll-motorways bridge's piers", IABSE conference *Safety, Risk, Reliability – Trends in Engineering*, Malta, 2001.

[UNO 02] UNOSSON M., Constitutive equations for concrete materials subjected to high rate of loading, Linköping studies in science and technology, Thesis N° 936, Linköpings

[VEC 88] VECCHIO F.J., SATO J.A., "Drop, fire and thermal testing of a concrete nuclear fuel container", *ACI Structural Journal*, pp. 374–383, July–August 1988.

[VOS 82] VOS E., REINHARDT H.W., "Influence of loading rate on bond behavior of reinforcing steel and prestressing strands", *Materials and Structures*, Vol. 15 N° 85, pp. 3–10, 1982.

[WAT 02] WATSON A.J., "Loading from explosions and impact", in *Dynamic Loading and Design of Structures*, A.J. Kappos (Ed.), Spon Press, 2002.

[WEE 88] WEERHEIJM J., REINHARDT H.W., "Biaxial loading for dynamic concrete response". *J. de Physique*, Coll. C3 (Suppl with No.9, Vol.49), pp. 733–738, 1988.

[WEE 89] WEERHEIJM J., REINHARDT H.W., "Structural response in impact tests on concrete", 4th *Int. Symp. on the Interaction of Non Nuclear Munitions with Structures*, Panama City Beach, Florida, USA, pp. 17–21, 1989.

[WEE 92] WEERHEIJM J., Concrete under impact tensile loading and lateral compression, PhD Thesis, Delft University of Technology, November, 1992.

[WEE 98] WEERHEIJM J., "Prediction of dynamic tensile strength", in *ACI SP-175 Concrete and Blast Effects*, W. Bounds (Ed.), paper SP 175-14, pp. 215–239, 1998.

[ZHA 96] ZHAO H., GARY G., "On the use of SHPB technique to determine the dynamic behavior of the materials in the range of small strains ". *Int. J. Solid. & Structure*, vol. 33, pp. 3363–3375, 1996.

[ZIE 82] ZIELINSKI A.J., REINHARDT H.W., "Stress-strain behaviour of concrete and mortar at high rates of tensile loading", *Cem. Conc. Res.*, vol. 12, pp. 309–319, 1982.

[ZUK 82] ZUKAS J.A., *Impact Dynamics*, John Wiley & Sons, 1982.

Chapter 2

Dynamic Behavior of Concrete: Constitutive Models

2.1. Dynamics of concrete structures

2.1.1. *Macroscopic phenomena*

We can distinguish two structural response levels to a dynamic load: the local level and the overall level.

2.1.1.1. *Local effect of an impact*

This is the stress and strain state created inside the impact area. This state develops simultaneously with loading and for very high strain rates. Studying the local effect requires taking the inertia effects, wave strain propagation and tri-axial stress states into account. Whenever failure criteria are reached during this phase, scabbing and shear phenomena with chip projections will take place.

2.1.1.2. *Overall effect on a structure*

If loading is globally distributed on the structure, or if, in the case of an impact, local failure conditions have not been fulfilled, then the whole structure responds dynamically. The inertia effects of the structure are such that the reaction immediately follows loading. If loading results from an explosion or a soft shock,

---

Chapter written by Patrice BAILLY.

the response is initially of transient type, then a quasi-stationary vibration regime develops if the structure stress is exerted for a period of time that is long relative to its natural frequencies, as is the case during an earthquake. The strain rates reached are lower than those observed at the location of an impact. The behavior of the structure can be described by the strength of materials theory, i.e. the structure can be modeled by means of plates and beams subjected to traction, compression or bending. However, the strain modes can be completely different to those observed under static loading.

2.1.1.3. *Analyses of the failure elementary phenomena*

We consider a reinforced concrete wall receiving a dynamic loading on one side (Figure 2.1). The observed phenomena are the following:

– concrete compaction near the loaded side;

– scabbing on the non-loaded side;

– failure of the structure under shear or bending.

Here we describe the mechanisms separately; actually, they are linked and simultaneous.

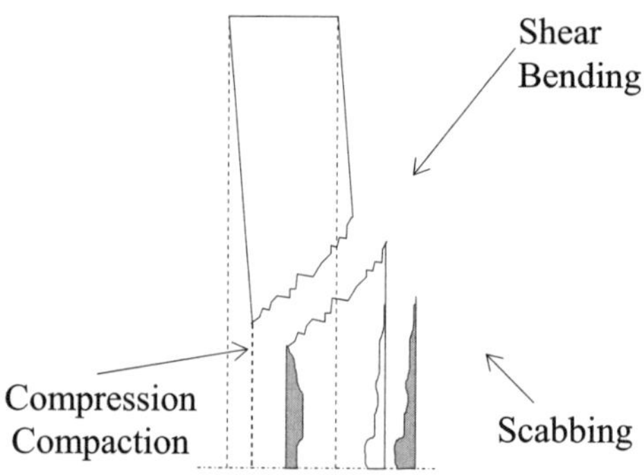

Figure 2.1. *Fracture mechanism*

2.1.1.4. *Compaction of the loaded side*

In the first microseconds and in the case of high-intensity loading (impact from a solid or strong pressure), the loaded side is subjected to high pressure and the superficial concrete is reduced to powder. A volume reduction caused by the porous structure of the material collapsing occurs. The stress state is close to that of a uni-axial strain and to approximately model that behavior, we resort to the so-called "plate-plate" test [LEV 96], which gives the pressure level necessary for dynamic loading compaction.

2.1.1.5. *Scabbing caused by wave reflection on the non-loaded side*

Scabbing of reinforced concrete elements results from fracture due to traction perpendicular to their free surface. This phenomenon is linked to wave propagation inside the material. Typically, for a 20 cm thick concrete wall, the wave arriving time onto the "free open surface" is about 0.6 μs. From experience, we can notice that the surface ejected by direct scabbing corresponds to the cover between the last layer of the reinforcement and the free surface.

2.1.1.6. *Bending of the structure*

The wall's bending phenomenon is slower, as it depends on the structure more than on the local behavior of the material. It important to note that shear failure of slabs due to shear loads are possible in dynamics, whereas it is extremely rare under static loading.

2.1.1.7. *Influence of the reinforcements*

Reinforced concrete reinforcements have an effect on the development of macro-cracks. The cracks, which separate blocks during failure, preferably meet with the reinforcements (Figure 2.2). Within structures specially designed to withstand impacts, so-called "lacing" reinforcements limit chip projections.

Figure 2.2. *Shape of fragments on a plate submitted to a shock tube (from [PON 95])*

2.1.2. *Perforation*

Perforation of a concrete wall generally reveals three areas (Figure 2.3):

– crater formation; this is the extremely damaged zone with fragment ejection on the impacted side;

– "tunneling"; the projectile progresses through the wall while causing big shear strains and compaction inside concrete;

– scabbing; at the output of the projectile, traction stresses lead to an ejection of fragments.

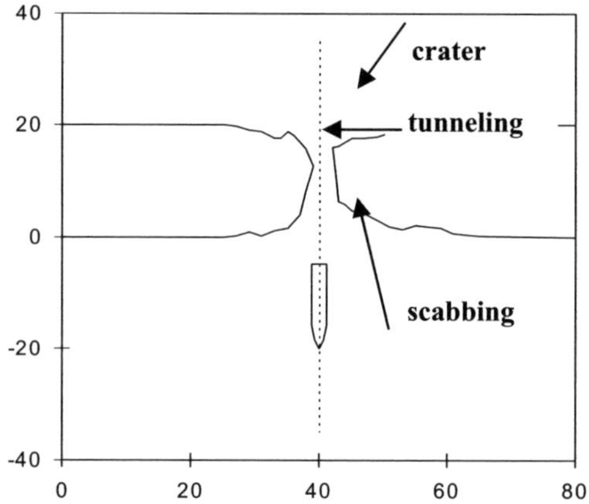

Figure 2.3. *Concrete slab perforated by a projectile (from [BUZ 98])*

2.1.3. *Ejection of fragments*

When the dynamic loading causes projections of fragments of the concrete structure, we can classify them into three types.

The scabs are formed due to concrete traction failure according to a process which will be detailed in section 2.3. If we assume a perfect brittle failure, we can estimate the ejection speed from equation [2.1], which also gives usual values for concrete:

$$\sigma_r = \rho c_e V_e \quad \left.\begin{array}{l} \sigma_r = 10\,MPa \\ \rho = 2{,}200\,kg/m^3 \\ c_e = 3{,}000\,m/s \end{array}\right\} \Rightarrow V_e \approx 1.5\,m/s \qquad [2.1]$$

The ejection of a block follows bending or shear failure. The order of magnitude of the ejection speed is closely linked to the speed of the structure element at the moment of failure, therefore to its transient dynamic response. The formation of small rubble and dust type fragments is linked to the characteristics of the dynamic failure of the material. If the failure of the structure is accompanied by a blast wave, the latter will carry the fragments along. The effect of that blast dragging along is inversely proportional to the fragment size.

2.1.4. *Loading range*

2.1.4.1. *Pressure level*

The main feature of dynamic loading is that it involves very high-pressure levels, completely out of the typical range of structures under quasi-static loading. The shock of a motor vehicle against a concrete structure, for example, can generate a several hundredths of a MPa pressure inside the latter. The consequence on a material such as concrete is compaction, i.e. an irreversible volume strain. Figure 2.4 shows the behavior of a compacted concrete (the concrete called MB 50 has been the subject of various studies within the frame of the "GEO" laboratories network [BAI 99]).

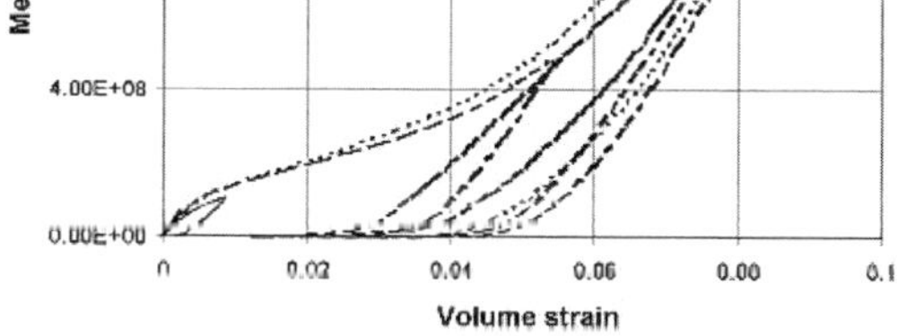

Figure 2.4. *Hydrostatic compression tests of MB50 concrete. Mean stress-volume strain diagrams (from [BUZ 99])*

2.1.5. *Loading path*

Dynamic loading induces strains and stresses which are described mainly locally as waves. The propagation and reflection of the waves lead to stress tensor components that vary within a wide range. For impacts from a solid, the normal stress components can range from a tensile value to a compressive value causing compaction. During perforation and mostly during the "tunneling" phase, shears reach large absolute values and large rotations of the main stress tensor axes may take place locally. Thus, the concrete shear behavior with a wide range of pressures or mean stresses has to be known.

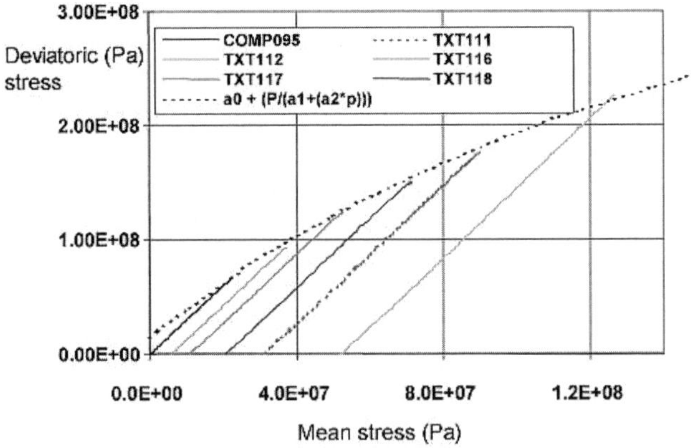

Figure 2.5. *Triaxial compression tests on MB50 concrete. Loading paths in the deviatoric- mean stress plane (from [BUZ 99])*

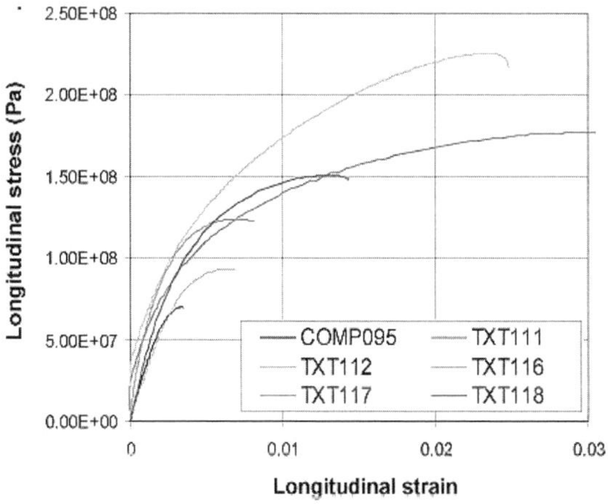

Figure 2.6. *Triaxial compression tests on MB50 concrete. Strain-longitudinal stress diagrams*

Figure 2.5 shows loading paths for concrete tri-axial tests. Figure 2.6 presents the relationship between axial stress and axial strain. We can note the brittle-ductile transition for a confinement pressure of about 50 MPa.

2.1.5.1. *Strain rate ranges*

Is it possible to establish a relationship between a dynamic loading and the strain rate it would impose on the structure material? Generally not. The strain rate is meaningful for a model describing the structure's movement. It is necessary to make a difference between the scale of the structure and the local scale as they have been defined above.

At the scale of the structure, the model describes a beam or a slab element under bending, for instance. The model implies a specific time scale that of the period of the bending mode taken into account. Thus, we implicitly admit that the propagation time for a distance corresponding to the size of the element is much shorter. This modeling scale is consistent for seismic calculations and for the overall effect of an impact on a structure. Within that frame, the strain rates can reach the 1/s order of magnitude.

On the local scale, we can observe the wave propagation in a continuum. The strain rate can be quite high (10 to 1,000/s). The strain rate is a function of the strain level (which depends on the loading) and the wave velocity (which depends on the material). The strain rate notion is interesting if it can be associated with a volume element where it is almost uniform. The higher the rate, the smaller the corresponding volume. Therefore, we have to note that in continuum mechanics, a very high strain rate can be incompatible with a given representative elementary volume, especially for a material like concrete.

The notion of strain rate becomes meaningless if we consider a shock wave (finite variation of the strain amplitude for an infinitesimal period of time).

2.2. Fast dynamics applied to concrete

This chapter presents the fundamental concepts that are useful to study a structure subject to dynamic loading. Many studies and applications have been made for metals, but here, the emphasis is placed on the specificities of applications to concrete.

2.2.1. *Impacts and waves*

2.2.1.1. *Uniaxial strain state*

In a uniaxial strain state, which can correspond to certain impact conditions, the stress and strain tensors have the following forms:

$$\underline{\underline{\varepsilon}} = \begin{pmatrix} \varepsilon_1 & 0 & 0 \\ 0 & 0 & 0 \\ 0 & 0 & 0 \end{pmatrix} \quad \underline{\underline{\sigma}} = \begin{pmatrix} \sigma_1 & 0 & 0 \\ 0 & \sigma_2 & 0 \\ 0 & 0 & \sigma_2 \end{pmatrix} \qquad [2.2]$$

To study such a strain state, the most commonly used hypothesis is that the material can be described by means of perfect elasto-plastic behavior and that the plastic flow is incompressible. Thus, the partition between elastic and plastic strains is made:

$$\underline{\underline{\varepsilon}} = \underline{\underline{\varepsilon}}^e + \underline{\underline{\varepsilon}}^p \Leftrightarrow \varepsilon_i^e + \varepsilon_i^p = 0 \left(i = 2,3 \right) \qquad [2.3]$$

and:

$$trace \underline{\underline{\varepsilon}}^p = 0 \Leftrightarrow \varepsilon_1^p + 2\varepsilon_2^p = 0 \qquad [2.4]$$

If the plasticity criterion is of the Tresca or von Mises type:

$$\sigma_1 - \sigma_2 = Y_0 \qquad [2.5]$$

In the elastic phase the relationships between stresses and strains are as follows:

$$\sigma_1 = (\lambda + 2\mu)\varepsilon_1 = \frac{E(1-\nu)}{(1-2\nu)(1+\nu)}\varepsilon_1 = \left(K + \tfrac{4}{3}G\right)\varepsilon_1$$

and [2.6]

$$\sigma_2 = \frac{\nu}{1-\nu}\sigma_1$$

At the plasticity threshold:

$$\sigma_1 - \sigma_2 = \left(1 - \frac{\nu}{1-\nu}\right)\sigma_1 = Y_0 \text{ and } \sigma_1 = \frac{1-\nu}{1-2\nu}Y_0 = \left(\frac{K}{2G} + \frac{2}{3}\right)Y_0 \qquad [2.7]$$

This elastic limit in a uniaxial strain state is also called Hugoniot's elastic limit (HEL). Beyond the flow threshold:

$$\sigma_1 = K\varepsilon_1 + \frac{2}{3}Y_0 \qquad [2.8]$$

The observable behavior is shown in Figure 2.7, together with the hydrostatic strain behavior and the uniaxial strain state behavior (simple compression).

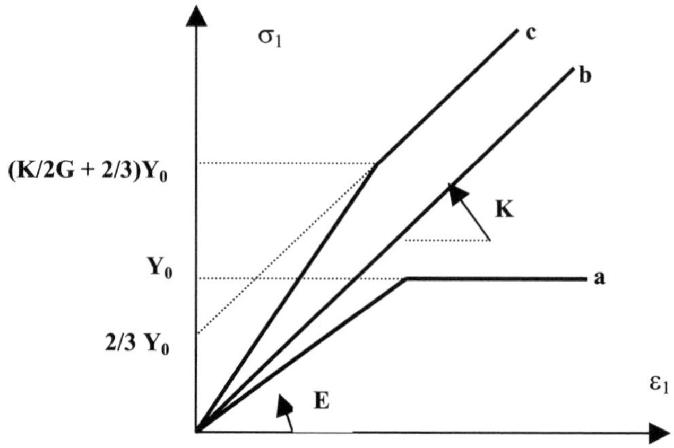

Figure 2.7. *Axial stress-axial strain relationship: a) uniaxial stress state (simple compression; b) isotropic or hydrostatic state; c) uniaxial strain state (oedometric)*

Beyond the elasticity threshold, we can see that the uniaxial strain behavior curve is parallel to the hydrostatic strain state behavior. By extrapolation, it is often admitted that the behavior obtained in a uniaxial strain state, apart from the elasticity threshold, is identical to the relationship between the pressure and the volume variation in a hydrostatic case. However, we must not forget that this result relies on such hypotheses as the von Mises criterion and plastic strain incompressibility.

Moreover, we can note the velocity of an "elastic" wave:

$$c_e = \sqrt{\frac{K + \frac{4}{3}G}{\rho}} \qquad [2.9]$$

This conventional chain of reasoning can be extended to a Drucker Prager (or Mohr Coulomb) criterion, which is more realistic with a material such as concrete. The conclusions will then be different. The plasticity criterion is written the following way:

$$\sigma_1 - \sigma_2 = A + B(\sigma_1 + \sigma_2) \qquad [2.10]$$

Under oedometric loading, the plastification threshold (HEL) appears for the stress level as:

$$\sigma_1 = \frac{A(1-v)}{1-2v-B} \qquad [2.11]$$

Note that this threshold only exists if $A > 0$ *and* $B < 1 - 2v$. Beyond that threshold, the lateral stress equals:

$$\sigma_2 = \frac{1-B}{1+B}\sigma_1 - \frac{A}{1-B} \qquad [2.12]$$

If we retain the hypothesis of an incompressible plastic flow, which in that case corresponds to a non-associated flow:

$$\sigma_1 = \frac{3K(1+B)}{3-B}\varepsilon_1 + \frac{2A(1+B)}{(1-B)(3-B)} \qquad [2.13]$$

In this case, we can observe that the conclusion considering the von Mises criterion established above is no longer valid. The volume-pressure relationship cannot be derived through a simple translation from the oedometric path strain-stress relationship. Nevertheless, if the deviatoric behavior is well known, the deduction is still possible with the hypothesis of an incompressible plastic flow.

Here we have been considering perfect plasticity models. Yet, whenever there is strain hardening, the plastic flow is either dilating or contracting, and although it is not impossible, expressing the volume-pressure relationship from the strain-stress relationship can be quite difficult.

2.2.1.2. *Creation of a shock wave*

A shock wave will be created during an impact if the impact speed is sufficient for the stress and strain levels to be located within the "positive curvature" part of the axial stress axial strain curve. The remark means that a shock wave is possible in uniaxial strain, but is impossible in uniaxial stress. The stress level is located in the upper part of Hugoniot's curve, as shown in Figure 2.8. In such conditions, a shock wave is created with a velocity of:

$$v = \sqrt{\frac{R}{\rho}} \qquad [2.14]$$

R is the slope of the straight line linking the material's representative points before and after the shock. We notice that there may be two different situations; if the shock is not too violent (Figure 2.8(i)), the shock wave will be preceded by an elastic precursor, as the shock front velocity is lower than that of the elastic waves. If the shock is very violent (Figure 2.8(ii)), its velocity can be higher than that of the elastic waves, and there is no longer a precursor.

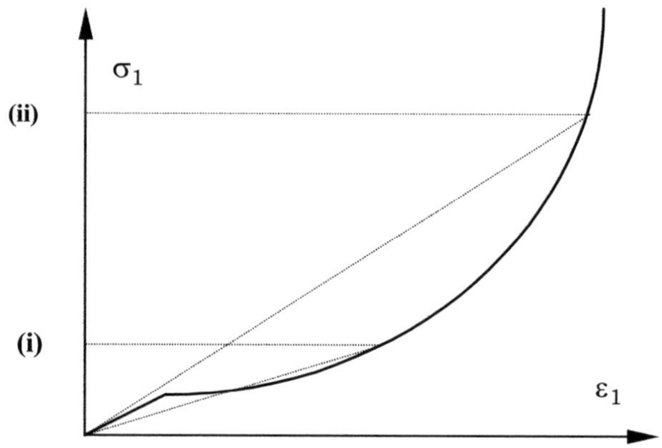

Figure 2.8. *Strain-stress relationship and stress levels involving a shock wave, with or without precursor (i and ii respectively)*

2.2.2. *Impact and shock polar curve*

2.2.2.1. *Conservation equations*

A shock wave is a non-stationary speed discontinuity that is necessarily associated with a stress discontinuity. $\underline{D}$ is the propagation speed of the discontinuity surface. On a specific point of the discontinuity surface (of normal line $\underline{n}$), the mass and momentum conservation equations take on a particular shape (the brackets represent the jump of the considered value when passing the discontinuity surface):

$$\text{Mass conservation } [\![\rho(\underline{v}-\underline{D})]\!]\underline{n}=0 \qquad [2.15]$$

$$\text{Momentum conservation } [\![\underline{\underline{\sigma}}]\!]\ \underline{n}=\rho[\![\underline{V}]\!](\underline{V}-\underline{D})\ \underline{n} \qquad [2.16]$$

It is interesting to write these equations in the case of a one-dimensional medium that is at rest after the shock (Figure 2.9). The equations are:

Mass conservation $\rho(D-V)=\rho_0 D$ [2.17]

Momentum conservation $\sigma=\rho_0 DV$ [2.18]

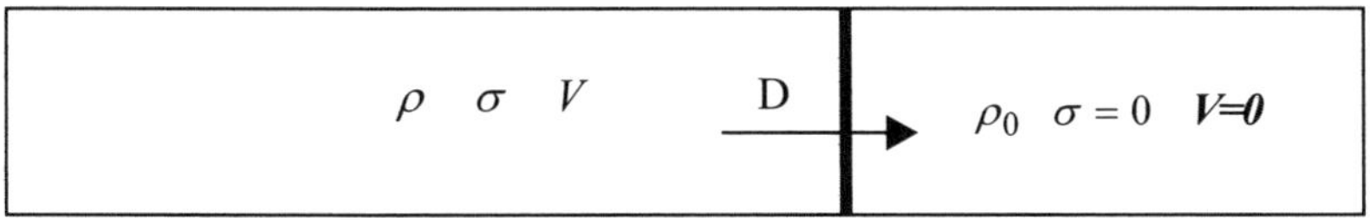

Figure 2.9. *Shock wave propagating with a velocity D within an initially at rest medium*

The strain after the shock can then be expressed (the compressions are positive):

$$\varepsilon=1-\frac{\rho}{\rho_0} \quad \varepsilon=\frac{V}{D} \qquad [2.19]$$

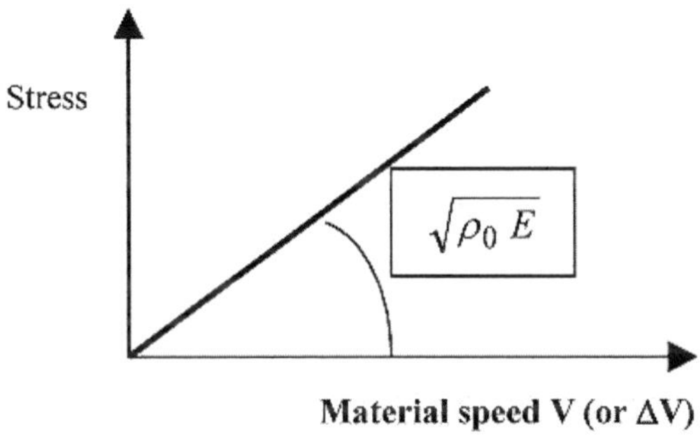

Figure 2.10. *Shock polar curve for an elastic material*

It is interesting to express the stress level reached as a function of the material speed of the solid (the corresponding diagram is often called shock polar curve). For an elastic material, the result is a straight line (Figure 2.10).

2.2.3. *Shock between two solids*

During a shock between two solids, waves are generated within each of them. The shock polar curve is a convenient way to determine the stress level associated with those waves. For instance, let us consider the shock of a metal projectile on a concrete element (Figure 2.11). Experience shows that it is reasonable to do a representation in a uniaxial strain state. The shock polar curves (Figure 2.12) show that the interface stress necessary equilibrium at the time of the shock can only be reached for values V_3 and σ_3.

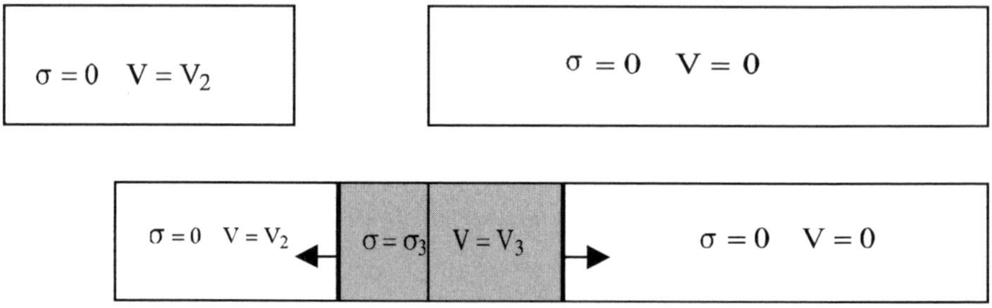

Figure 2.11. *Metal projectile hurled at speed V2 against a concrete element; image before and after the impact*

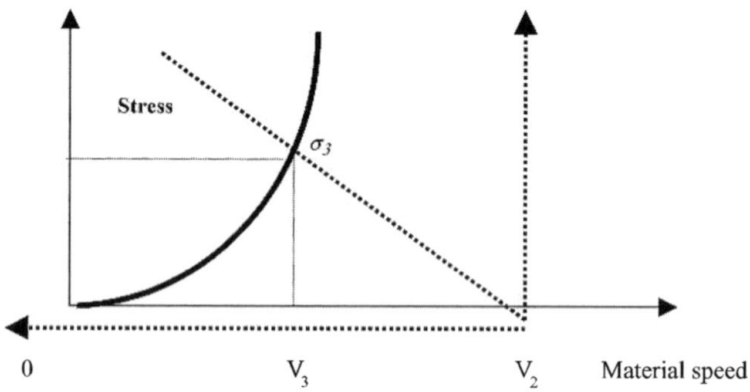

Figure 2.12. *Shock polar curve diagram (full line for concrete and dotted line for metal)*

2.3. Scabbing

The scabbing failure phenomenon occurs with materials whose tensile strength is definitely lower than their compression strength, like concrete and rocks. Dynamic loading, such as on impact, generates a compression wave like that seen previously. The compression duration corresponds to a propagation and reflection movement of the waves inside the projectile. Then the stress level is back to zero and an unloading stress wave propagates throughout the material. Besides, when the compression wave reaches a stress-free side, an unloading stress wave propagating from the opposite direction is also emitted to respect the zero-stress condition. These two unloading waves meeting cause traction of the same intensity as the compression of the initial wave. It is easy to understand that such an intensity, which can be borne in compression leads to failure when it is exerted in traction (Figure 2.13). A strain wave is then reflected on a free end by changing signs.

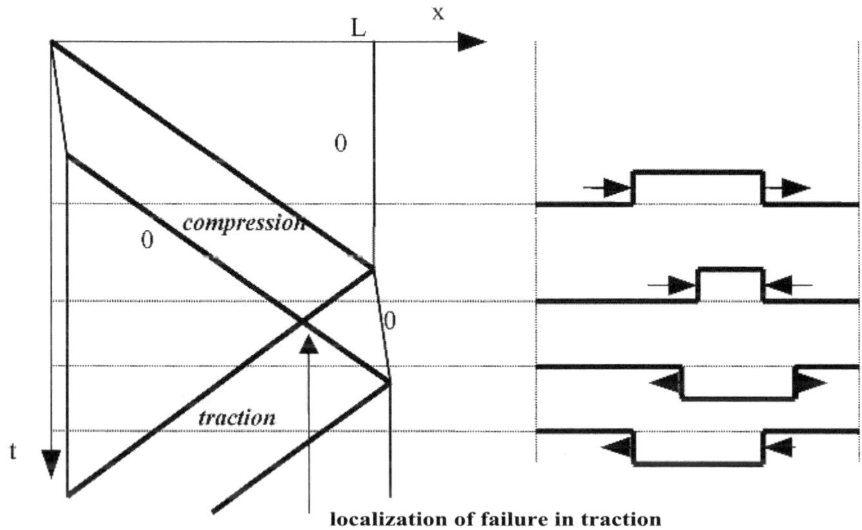

Figure 2.13. *Lagrange's diagram representing the reflection of a strain increment on a free side*

2.4. Effect of a shock wave on the structure of materials

A shock wave passing through a material can induce irreversible modifications to its characteristics. The most obvious variation involves the relative density – this is called compaction. For a dusty material, such as a highly damaged concrete, a model example of the relationship between pressure and the relative density can be the following:

$$\rho = \rho_0 \left(a + \frac{1-a}{1+b} P \right)^{-1} \qquad [2.20]$$

where a defines the initial density and b is a flexibility coefficient.

Figure 2.14 represents the pressure-volume relationship. In high intensity compaction, the powder becomes a solid. When a shock wave passes through, there is a pressure jump up to the value P. Due to the dynamic and viscous effects, the evolution follows the so-called Rayleigh line. Unloading occurs in parallel to the solid behavior curve (there is no decompaction).

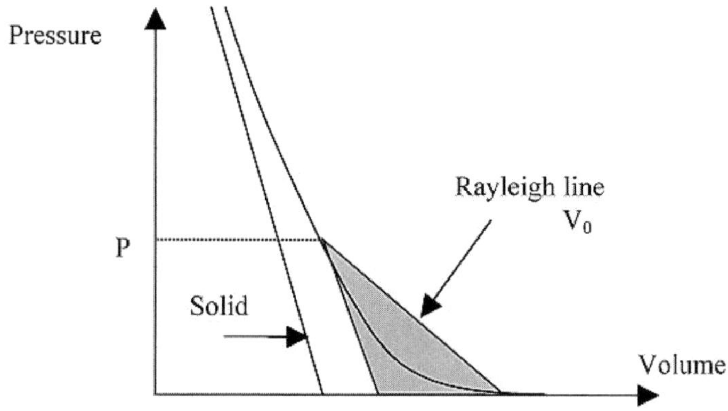

Figure 2.14. *Compacting caused by a shock wave passing through a dusty material*

The expression of the dissipated energy is the following:

$$W = \frac{1}{2} P V_0 \left(1 - \frac{\rho_0}{\rho} \right) \qquad [2.21]$$

2.5. Modeling types

2.5.1. *Behavior description theoretical frames*

To describe the dynamic behavior of concrete, engineers still mostly use continuum mechanics. Then the resolution of structure problems is achieved by means of the finite element method. Therefore, behavior models or constitutive models are developed within that frame.

Taking the previous remarks about loading paths into account, we are led to consider three model types (Figure 2.15):

– damage mechanics, which aims to describe the stiffness loss due to cracking;

– plasticity or viscoplasticity theory, which describes sliding and shear;

– pressure-volume relationships or hydrodynamic state laws for compaction.

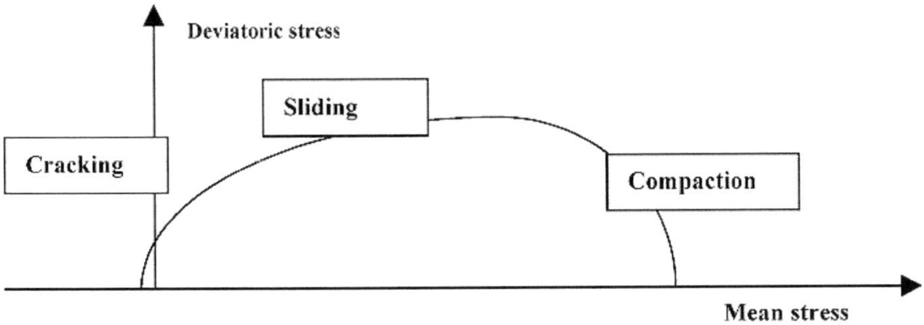

Figure 2.15. *Diagram of the elastic domain for concrete and of the different irreversible phenomena in the plane of the first two invariants of the stress tensor*

2.5.2. *Integrating sensitivity to the strain rate*

2.5.2.1. *Models with time dependence*

If we want to integrate the observed sensitivity to strain rate into the behavior model, the latter can intervene within the framework of conventional models.

Linear or non-linear visco-elasticity will reveal an apparent stiffness increase with the strain rate. If failure modeling is of the perfect brittle type and if the failure criterion is expressed as maximum strain, we will find that strength increases with the strain rate.

Visco-plasticity will show up as an apparent strength increase, in terms of the maximum stress reached rising with the strain rate.

Rate effects can also be introduced within the evolution of damage.

It is necessary to justify the introduction of the dependence on rate. Two reasons justify the use such a process:

– a physical argument, because we wish to use the model to describe a real event that has been observed experimentally. We must refer to Chapter 1 and see how cautiously these apparent dynamic strength increases must be interpreted;

– a theoretical argument, because dependence on rate within a viscoplastic or "viscous-damage" framework allows us to preserve a well-posed structure calculation problem, even in the presence of softening.

2.5.2.2. *Phenomenological law*

The apparent strength of numerous materials depends on the loading rate used to test it. The experimental results related to strength in a uniaxial stress state are generally represented in a diagram of strength versus strain rate on a logarithmic scale. For some rate ranges, strength varies more or less linearly with the strain rate logarithm. The observation has led to empirical modeling:

$$\sigma = K\left(\dot{\varepsilon}\right)^{\beta_\varepsilon} \qquad [2.22]$$

By analogy, models have been designed for application to concrete, with the intention of deriving the strain rate effect from tests carried out at different rates. The main problem with this approach, apart from the precautions to be taken when interpreting the results of the tests, involves converting to a tensorial formulation. A rather simplistic approach in this field could lead to models that have dubious thermodynamic acceptability (dissipation not necessarily demonstrated outside the radial loading paths).

2.5.3. *Elasto-plasticity and criteria*

Most of the models have been built on an elastoplastic theoretical framework. If the plasticity is perfect, the plasticity criterion corresponds to the failure criterion, but many models differentiate the elastic limit criterion from the failure criterion. The loading surface evolves from the first to the second with the evolution of a strain-hardening variable.

These criteria are expressed in terms of the stress state and the strain-hardening variables:

$$f\left(\underline{\underline{\sigma}}, k_i\right) = 0 \qquad [2.23]$$

The material independence leads us to express the criterion as a function of the three stress invariants:

– the first invariant (or pressure);

$$I_1 = \text{trace}\ \underline{\underline{\sigma}} = -3p \qquad [2.24]$$

– the second invariant, the von Mises equivalent stress;

$$\sigma_{eq} = J_2\left(\underline{\underline{\sigma}}\right) = \sqrt{\frac{3}{2}\underline{\underline{s}} : \underline{\underline{s}}} \quad [2.25]$$

– the third invariant, i.e. Lode's angle;

$$J_3 = \frac{1}{3}\left(\text{trace }\left(\underline{\underline{\sigma}}^3\right)\right)^{\frac{1}{3}} = \frac{1}{3}\left(\sigma_{ij}\sigma_{jk}\sigma_{ki}\right)^{\frac{1}{3}} \quad \cos\left(3\theta\right) = \frac{3\sqrt{3J_3}}{2J_2^{3/2}} \quad [2.26]$$

The criteria are generally isotropic, but they express asymmetry between traction and compression. Among the criteria used for concrete are Coulomb's and Drucker Prager's criteria [DRU 52].

$$f\left(\underline{\underline{\sigma}}\right) = \sqrt{\frac{1}{2}\underline{\underline{s}} : \underline{\underline{s}}} + \alpha\left(\text{trace }\underline{\underline{\sigma}} - Y_h\right) \quad \left(0 < \alpha < \frac{\sqrt{3}}{2}\right) \quad [2.27]$$

This criterion is now used as a reference for the shear failure of geomaterials. More elaborate criteria, taking the third invariant into account, for example, are used for other models. We cite those defined by Ottosen, Argyris, and Willam Warnke as examples.

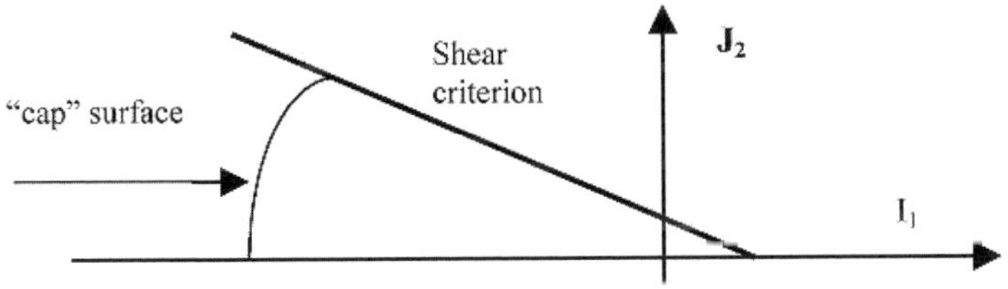

Figure 2.16. *Representation of a "cap" surface*

The criteria linked to shear failure such as those below, do not express high-pressure irreversible phenomena. In the plane of the first two invariants, the representation of the criterion gives an "open" curve on the pressure axis. The eponymous "cap" surface closes the elastic domain (Figure 2.16); it is associated with compaction.

2.5.4. *Damage*

The subject of damage mechanics seeks to represent the irreversible micro cracking and stiffness-degradation processes which occur in some materials. The threshold above which this phenomenon appears can be represented by one criterion, as in the case of plasticity. The criterion can be expressed in terms of strains instead of stresses, which is not the same in dynamics in the presence of viscosity. It can also be asymmetric, i.e. a function of extensions and not compressions. For example, Mazars' criterion [MAZ 84], which has been used for concrete, is a function of the principal strains if they are positive. The most widely used model is the scalar isotropic damage model [LEM 85]. The behavior of damaged real material is expressed by the law used for the undamaged material, except the usual stress is replaced by the effective stress.

$$\varepsilon = \frac{\sigma}{E\ (1-D)} \qquad [2.28]$$

There are also damage models that depend on time.

$$\dot{D}\left(\underline{\underline{\varepsilon}}, D, \cdots\right) \qquad [2.29]$$

This type of formulation derives its justification from the fact that the creation and propagation of cracks are phenomena that cannot be instantaneous [SUA 84].

2.5.5. *Notion of a state law*

A common hypothesis on which to build up a material behavior model involves splitting the stress tensor and the strain tensor into their spherical and deviatoric parts.

$$\underline{\underline{\sigma}} = -p\ \underline{\underline{I}} + \underline{\underline{s}}\circ \text{ and } \underline{\underline{\varepsilon}} = \theta\ \underline{\underline{I}} + \underline{\underline{e}}' \qquad [2.30]$$

Then we can define separately the relationship between the spherical parts (pressure and volume variation) called isotropic behavior or state law by abusing the language, and the relationship between the deviatoric parts (which represents the deviatoric behavior). This approach has proved quite interesting in the case of metals, as both relationships are then independent, and two independent experiments allow us to identify a behavior model. For a cohesive material like concrete, the irreversible volume and deviatoric phenomena are coupled. Nevertheless, taking the techniques developed for metal materials into account, the same partition has been carried out in certain models for use with concrete, and the technique is then enriched using coupling.

In the case of very high intensity dynamic loadings such as those generated by solid explosives, it is accepted that interior loads can be represented by a pressure scalar field, as in a perfect fluid. This means that the order of magnitude of the shear stress is smaller than those of normal stresses: in this case, we resort to so-called "hydrodynamic" codes.

Within this framework, concrete is treated as a porous material, and the state law involves the relationship between pressure and porosity, with potentially other thermodynamic variables such as temperature and phase states being used.

$$P(\mu, T, \cdots) \quad \mu = \frac{\rho}{\rho_0} - 1 \qquad [2.31]$$

The relationship between pressure and density is often expressed as a polynomial. In calculation codes, an example of such a law is "P-α model" [HER 69], α being the porosity index.

This model type is generally derived from plate impact tests. During these tests a shock wave passes through a sample of the material and the shock velocity D and the material speed u are measured [MCQ 70]. It seeks the coefficients of a law of the type:

$$D = A + Bu^n \qquad [2.32]$$

The relationship between pressure and volume variation can be deduced according to the hypotheses in section 2.2.1.

2.5.6. *Location limiter and time sensitivity*

Softening, which is also known as "cohesion loss" or "negative strain-hardening", is a characteristic shown by concrete in its post-failure behavior. In a transient dynamic calculation, having a good description of that phase is important, for, even if the concerned area is small, the global reaction of a structure is quite strongly dependent. Location limiters (either explicitly or implicitly) induce the dependence of behavior on the strain rate.

For instance, in visco-plasticity, owing to a study of wave dispersion, we can establish a characteristic length (η is the coefficient of viscosity and c_e the velocity of elastic waves):

$$\ell_c = \frac{2\eta c_e}{E} \quad [2.33]$$

We can then wonder if this time-dependence can be mixed up with the rheological aspect of the behavior. As a matter of fact, both aspects are separated by the values of the viscosities in play. In fact, the value of a characteristic length is centimetric [PIJ 87] and therefore must be related to a specific coefficient of viscosity.

$$\ell_c \approx 10^{-2} m \leftrightarrow \eta \approx 10^4 Pa.s \quad [2.34]$$

Such a coefficient of viscosity value (taken as a visco-plasticity coefficient in Perzyna's model) is without any noteworthy effect on the macroscopic behavior (for example, the apparent strength as a function of stress/speed). If we are to apply a coefficient of viscosity to express a rheological aspect, like the dependence of compaction on the strain rate [GAR 98b], We will have to introduce a coefficient about 10^8 Pa.s.

2.6. Models

An illustrative selection of models that are used in fast dynamics codes to represent concrete behavior are presented below. This collection is obviously restricted and does not pretend to be exhaustive. It is not a critical or comparative study of the models. For each model, a reference is provided, affording the reader the opportunity to find details related to each, as explained by its author. The "basic" models, such as Prager's model, are not included.

2.6.1. *Elasticity-based model*

2.6.1.1. *Cedolin*

– Reference: [CED 77].

– Principle: non-linear elasticity, spherical and deviatoric parts separated.

– Application field: low pressure level.

– Can model: sliding.

– Cannot model: negative strain-hardening, compaction, damage.

– Triaxial stress state:

$$\sigma_1 < 60 \text{ MPa}; \ \sigma_1 > \sigma_2 > \sigma_3 \quad [2.35]$$

– Strain-stress relationship:

$$\sigma_{oct} = 3K_s \varepsilon_{oct} \text{ and } \tau_{oct} = G_s \gamma_{oct} \quad [2.36]$$

$$\varepsilon_{oct} = \frac{1}{3}\left(\varepsilon_1 + \varepsilon_2 + \varepsilon_3\right) \quad [2.37]$$

and

$$\gamma_{oct} = \frac{2}{3}\left[\left(\varepsilon_1 - \varepsilon_2\right)^2 + \left(\varepsilon_2 - \varepsilon_3\right)^2 + \left(\varepsilon_3 - \varepsilon_1\right)^2\right]^{1/2} \quad [2.38]$$

$$K_s = K_s\left(\varepsilon_{oct}\right) \cdot G_s = G_s\left(\gamma_{oct}\right) \quad [2.39]$$

2.6.2. *Models based on the theory of plasticity*

2.6.2.1. *Elwi, Murray*

Reference: [ELW 79].

– Principle: elasto-plasticity.

– Application field: low pressure.

– Can model: shear.

– Cannot model: compaction and strain-hardening.

– Model data: Argyris failure surface (5-parameter surface, not closed on the hydrostatic axis).

2.6.2.2. *"CONC"*

– Reference: model implemented into "Radioss" code by "Mecalog".

– Principle: elasto-plasticity with non-associated flow.

– Application field: monotonous-dynamic loading.

– Can model: shear and dilatancy.

– Cannot model: negative strain hardening.

– Model data: failure surface (Ottosen type).

$$f\left(\sigma_m, r, \theta\right) = 0,$$

with:

$$\sigma_m = \frac{I_1}{3},\ r = \sqrt{2J_2},\ \cos(3\theta) = \frac{3\sqrt{3}J_3}{2J_2^{3/2}} \qquad [2.40]$$

The loading surface contained between the initial elastic surface and the failure surface.

$$r_{elasticity} = r\left(\sigma_m, \theta\right) k\left(k_y, \sigma_m\right) \qquad [2.41]$$

According to the value of k we can represent different behaviors of concrete, including the traction region, transition region and compression region.

2.6.2.3. *Bicanic, Zienkiewic*

– Reference: [BIC 83].

– Principle: elasto-viscoplasticity.

– Application field: cyclic loading.

– Can model: damage.

– Cannot model: compaction.

– Model data: uses two surfaces (Mohr/Coulomb):

- the threshold surface, F_D, limit between the elastic zone and the plastic zone;

- the stress limit surface F_F.

– F_D and F_F vary with accumulated visco-plastic work;

$$F_D\left(\sigma, W_p, k\right)=0 \quad F_F\left(\sigma, W_p\right)=0 \quad k=W_p-W_p^* \qquad [2.42]$$

where W_p = viscoplastic work, W_p* = energy dissipated at time t*', t* = time when F_F is reached, and where k controls energy dissipation and represents the deterioration of threshold surface;

2.6.2.4. *Han and Chen*

– Reference: [HAN 85].

– Principle: non-associated plasticity;.

– Application field: high pressure.

– Can model: elasto-plastic phase, strain hardening.

– Cannot model: damage.

– Model data: 4-parameter failure surface (Ottosen, Hsieh-Ting-Chen), and 5-parameter elastic threshold surface (Willam-Warnke).

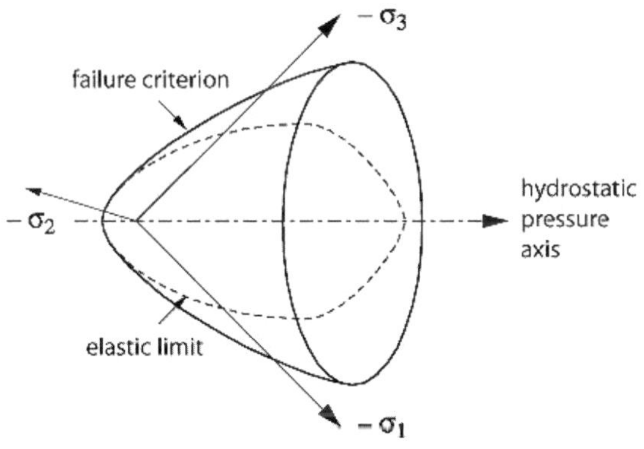

Figure 2.17. *Elastic threshold surface and failure surface of the Han and Chen model (from [HAN 85])*

The shape of the failure surface is different from the shape of the elasticity initial surface. As is the case with Ottosen, the equation of the loading surface is:

$$f=r-kr_f=0 \qquad [2.43]$$

where k is a factor the form of which is $k = k\left(\sigma_m, k_0\right)$.

2.6.2.5. *Pietruszczak*

– Reference: [PIE 88].

– Principle: non-associated plasticity.

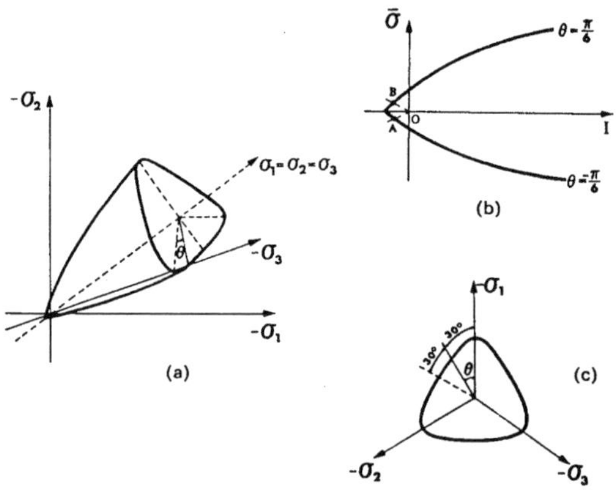

Figure 2.18. *Pietruszczak failure surface, (a) space of the main stresses; (b) meridian plane; (c) deviatoric plane (from [PIE 88])*

– Application field: low or medium confinement pressure, monotonous loading.

– Can model: the sensitivity of the material to confinement pressure, contraction-dilatancy transition, brittle-ductile transition and strain hardening.

– Cannot model: compaction, unloading.

– Model data: failure surface.

2.6.2.6. *Sandler et al. (CAP)*

– Reference [SAN 76].

– Principle: associated plasticity.

– Application field: behavior under impact, high pressures.

– Can model: compaction.

– Cannot model: damage.

– Model data: shear failure criterion and “cap” surface.

2.6.3. *Models based on damage mechanics*

2.6.3.1. *Mazars*

– Reference: [MAZ 84].

– Principle: isotropic damage (explicit formulation).

– Application field: low pressures (structures, beams), traction.

– Can model: damage, negative strain hardening.

– Cannot model: compaction, sliding.

– Model data: coefficients of elasticity, threshold strain and parameters of calculation of D:

- behavior law: $\underline{\underline{\sigma}} = (1-D)\left[\lambda\left(\text{trace } \underline{\underline{\varepsilon}}\right)\underline{\underline{\Im}} + 2\mu\underline{\underline{\varepsilon}}\right]$; [2.44]

- equivalent strain: $\tilde{\varepsilon} = \sqrt{\langle\varepsilon_1\rangle_+^2 + \langle\varepsilon_2\rangle_+^2 + \langle\varepsilon_3\rangle_+^2}$. [2.45]

If $\varepsilon_{D0} < \tilde{\varepsilon}$ there is damage, and the threshold surface is written:

$$f(\varepsilon, D) = \tilde{\varepsilon} - \varepsilon_{D0} = 0 \qquad [2.46]$$

The variable D integrates traction damage as well as compression damage:

$$D = \alpha_t D_t + \alpha_c D_c \qquad [2.47]$$

2.6.3.2. *La Borderie*

– Reference: [LAB 91]

– Principle: scalar damaging (implicit formulation).

– Application field: cyclic loading, structures.

– Can model: stiffness loss (deterioration under traction stresses, remnant strains, stiffness recovery).

– Cannot model: compaction.

– Model data: decomposition of the stress tensor, within the principal axis, in a positive part and in a negative part, to take into account the independence of stiffness in traction and compression. Two damage variables are defined by:

$$\underline{\underline{\varepsilon}} = \frac{\underline{\underline{\sigma}}^{+}}{E_0\left(1-D_1\right)} + \frac{\underline{\underline{\sigma}}^{-}}{E_0\left(1-D_2\right)} + \frac{\nu}{E_0}\left(\underline{\underline{\sigma}} - \left(trace\underline{\underline{\sigma}}\right)\underline{\underline{\Im}}\right) + \frac{\beta_1 D_1}{E_0\left(1-D_1\right)} f'\left(\underline{\underline{\sigma}}\right) + \frac{\beta_2 D_2}{E_0\left(1-D_2\right)} \quad [2.48]$$

where D_1 and D_2 are the damage variables, the other parameters being characteristics of the material.

2.6.3.3. *Dubé*

– Reference: [DUB 94].

– Principle: visco-damage (implicit formulation).

– Application field: cyclic loading, dynamic problem.

– Can model: stiffness loss (= deterioration under traction stress), residual strains, stiffness recovery and rate phenomena.

– Cannot model: compaction.

– Model data: the formulation is close to La Borderie's model, but it introduces, in the same way as Perzyna, time dependence in the evolution of the damage variables (f being the damage threshold function)

$$\dot{D} = \left(\frac{\langle f \rangle}{m}\right)^{n} \quad [2.49]$$

2.6.4. *Model coupling damage and plasticity*

2.6.4.1. *Ulm*

– Reference: [ULM 93].

– Principle: associated elastoplastic model + damage (implicit formulation).

– Application field: cyclic loadings, low-rate dynamic and monotonous loadings.

– Can model: monotonous loadings, contracting plastic behavior for high hydrostatic pressures, dilating plastic behavior for low hydrostatic pressures.

– Model data: criterion of plasticity (4-parameter modified Willam-Warnke's criterion). Coupling takes place in the following way:

- the coefficients of elasticity (K and G) evolve with the inelastic strain;

- a strain-hardening parameter depends on cracking.

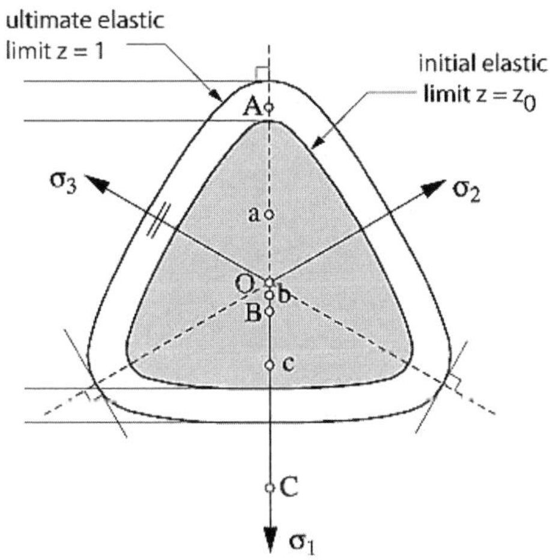

Figure 2.19. *Trace of Willam Warnke's criterion in the deviatoric plane (from [ULM 93])*

2.6.5. *Model coupling damage and mechanics of porous media*

2.6.5.1. *Burlion*

– Reference: [BUR 97].

– Principle: porous media [COU 91], visco-plasticity and 1-parameter scalar damage.

– Application field: high-pressure dynamic loadings.

– Can model: compaction, damage.

– Cannot model: sliding (explicitly).

– Model data: the threshold surface of elasticity used is adapted from Gurson's criterion [GUR 77] initially used for some metals.

$$F\left(\sigma_m, \underline{\underline{\sigma}}, f\right) = \frac{3J_2}{\sigma_m^2} + 2q_1 f \cosh\left(\frac{q_2 I_1}{2\sigma_m}\right) - \left(1 + \left(q_3 f\right)^2\right) \qquad [2.50]$$

The coefficients q and f characterize the material.

Coupling results from the fact that the damage scalar variable is linked to porosity. At the initial state, the damage variable is not equal to zero, and can decrease in the event of compaction.

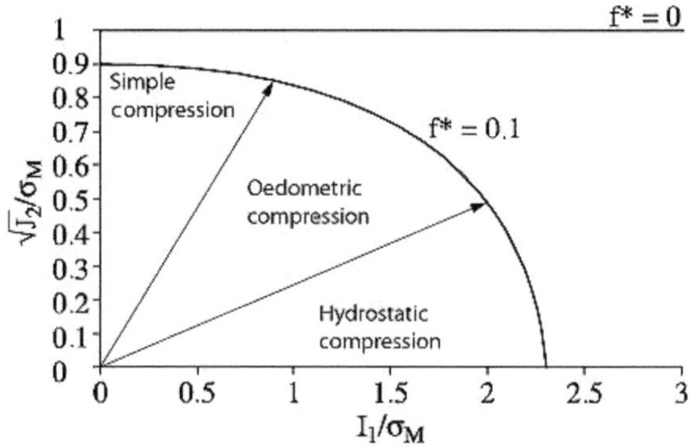

Figure 2.20. *Representation of Gurson's criterion in the plane of the first two invariants (from [BUR 97])*

2.6.6. *Model deriving from a hydrodynamic approach*

2.6.6.1. *"ARMOR"*

– Reference: [MAR 94].

– Principle: spherical and deviatoric separation of parts, perfect plasticity and damage.

– Application field: concretes and rocks under high pressure.

– Can model: compaction.

– Cannot model: negative strain hardening.

– Model data:

i) the volume mechanism is represented by a relationship between pressure and volume variation:

$$p = p(\mu) \quad p = \frac{1}{3} tr(\sigma) \quad \mu = \frac{\rho}{\rho_0} - 1 \qquad [2.51]$$

ii) the deviatoric mechanism is defined by a threshold of plasticity in the plane (p,q)

$$q^2 = \frac{3}{2} s_{ij}^2 = 3J_2 \qquad [2.52]$$

The material's threshold curve depends on a damage variable D:

$$S = S_s(1-D) + S_f(D) \qquad [2.53]$$

where S_S = threshold for the undamaged material and S_f = threshold for the damaged material (Figure 2.8).

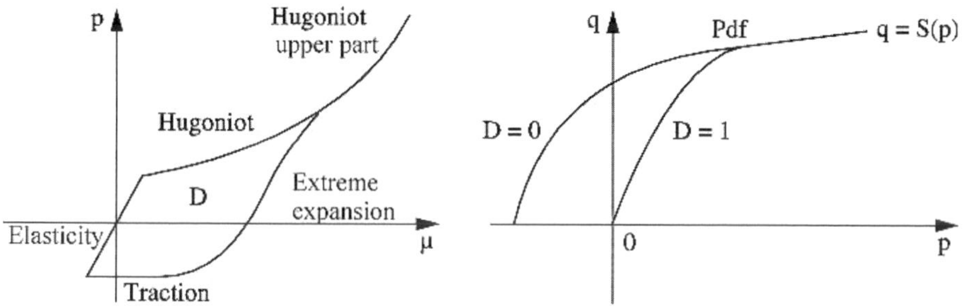

Figure 2.21. *"ARMOR" model, elastic field in the density-pressure plane and flow threshold surface in the p-q plane (from [MAR 94])*

2.6.6.2. *Holmquist*

– Reference: [HOL 93].

– Principle: spherical and deviatoric separation of parts, perfect plasticity and damage.

– Application fields: high stresses, high strain rates and high pressure.

– Can model: compaction, damage.

– Cannot model: negative strain hardening.

– Model data:

i) volume-pressure relationship, state equation;

elastic field: $P = K_{elastic} * \mu$ with $\mu = \dfrac{\rho}{\rho_0} - 1$ [2.54]

transition field: $P = \left[(1-F)K_{elastic} + F\ K_1\right]\mu$ [2.55]

"The Hugoniot's upper part":

$$P = K_1\bar{\mu} + K_2\bar{\mu}^2 + K_3\bar{\mu}^3 \text{ with } \bar{\mu} = \frac{\mu - \mu_{lock}}{1+\mu_{lock}} \quad [2.56]$$

ii) plasticity threshold $f(P, \dot{\varepsilon}, D)$ given by the effective stress:

$$\sigma^* = \left[A(1-D) + BP^{*N} 1 + C\ln\dot{\varepsilon}^*\right], \quad \sigma^* = \frac{\sigma}{f_c} \quad [2.57]$$

Figure 2.22 illustrates the different terms of the formulation.

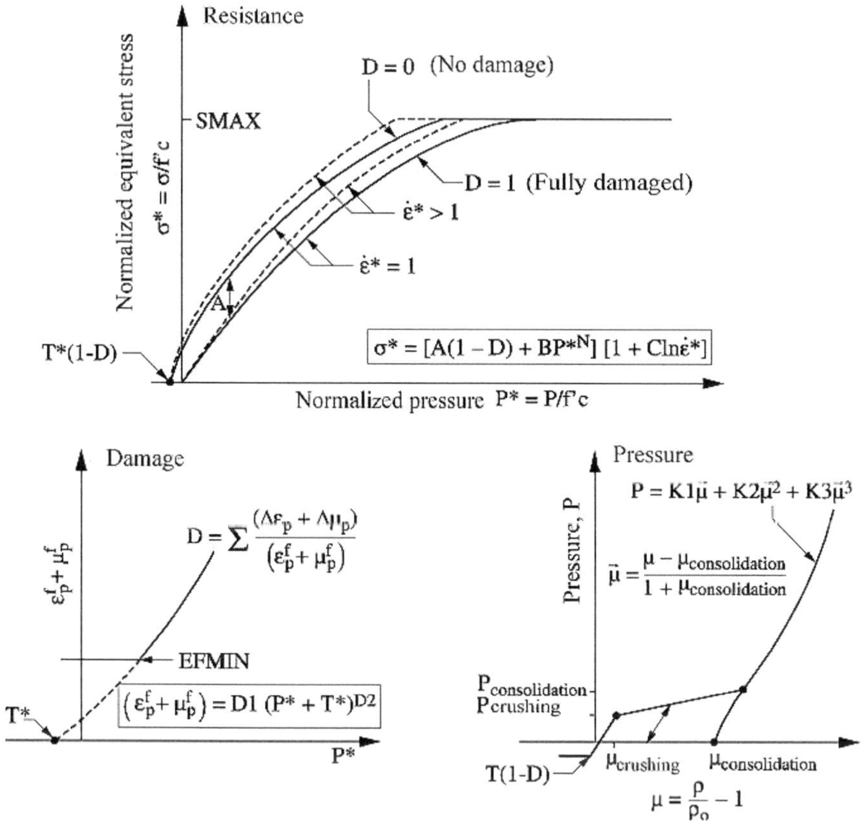

Figure 2.22. *Homlquist's model, threshold plasticity surface, damage-inelastic strain relationship, density-pressure relationship (from [HOL 93])*

2.6.7. *Endochronic models*

2.6.7.1. *Bazant and Bhat*

– Reference: [BAZ 76].

– Principle: similitude between a strain stress relationship and the evolution of strain in a viscous body model. The time variable is replaced by a variable depending on the accumulation of plastic strain.

– Application field: cyclic loadings.

– Can model: plastic dilatancy, strain hardening (positive or negative).

– Cannot model: compaction.

2.6.8. *Discrete element method*

Modeling a medium as an assembly of particles is an alternative to representation as a continuous medium. The concept has been applied to both fluid and solid mechanics, and to pulverulent media where there is an identity between the model's particle and the grain of the material. It has recently been extended to cohesive media. Two approaches co-exist in this frame:

– The medium is treated as undeformable particles with a mass (ED). The stresses and strains are represented by interactions between the particles (distance and binding force). Thus, behavior, both elastic and inelastic, is completely expressed by the linking conditions of EDs [CUN 97].

– The medium is considered as if composed of EDs which are deformable solids, the behavior of which is described through elasticity and plasticity in continuum mechanics. The links between EDs only express local phenomena, such as contact loss and friction [MOR 96].

2.6.8.1. *Camborde and Mariotti*

– Reference: [CAM 99].

– Type: interactions between particles.

– Each particle is representative of a mesoscopic scale consisting of a few aggregates, cement and empty space. Initially, each link that is of the cohesive type transmits compression and traction loads. Two types of elastic forces are present (Figure 2.23): the normal strength of stiffness Kn and the shear force of stiffness Ks.

The inelastic behavior involves a traction failure threshold T on F_n and a Mohr-Coulomb shear failure threshold on F_s, defined by the cohesion C and a friction angle ϕ. In a traction failure ($F_n < 0$), the link is simply suppressed. In a shear failure ($F_n > 0$), we change from a cohesive type law to a friction type law (2.24).

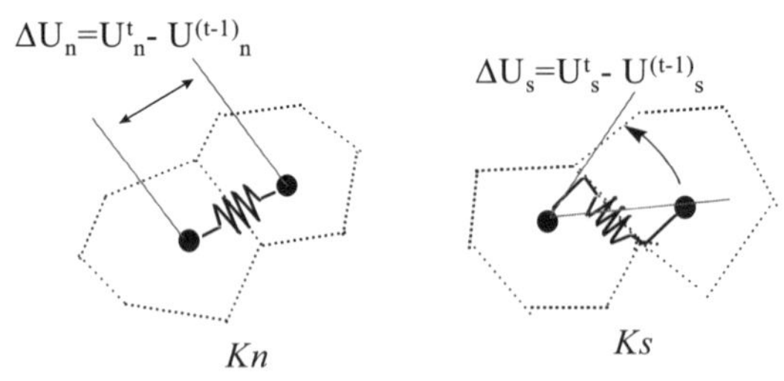

Figure 2.23. *Details of the particles' interaction ΔU_n normal displacement, ΔU_s tangential displacement (distortion) (from [CAM 99])*

Porosity is introduced not on the geometric level, but through an inner variable that controls the evolution of stiffness Kn, as is the case with the Holmquist and Johnson-type model.

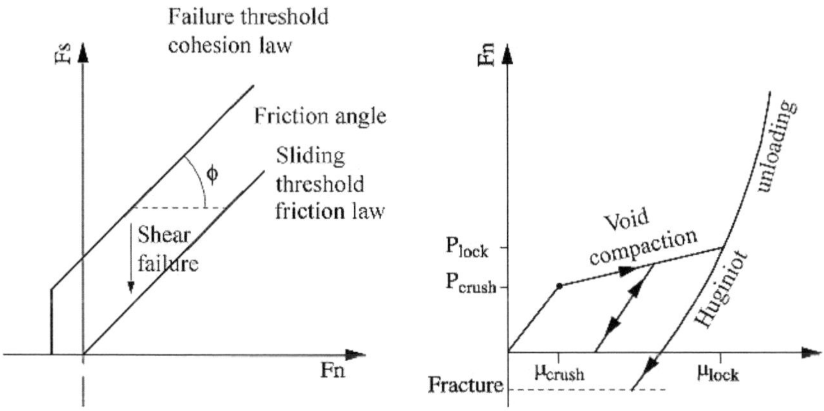

Figure 2.24. *Domain of the contact forces and state equation (from [CAM 99])*

2.6.8.2. *Donzé*

– Reference: [DON 98].

– Type: interaction between particles.

– One original feature of this model is definition of the interactions between particles. Two interacting elements are not necessarily in contact. Two elements a and b of radii R^a and R^b will be in interaction if $\gamma\ (R^a + R^b) > D^{a,b}$, with the interaction coefficient γ (> 1), $D^{a,b}$ being the distance between the centers of EDs. This definition of interacting elements is different from the one typically used with spherical EDs, where only the elements in contact interact ($\gamma = 1$). This definition of initial interactions allows us to increase the number of interactions for one element, and therefore to model the action of the matrix between the concrete aggregates. The interaction force between two EDs has a normal component F_n and a tangential component F_s. Figure 2.11 shows the cyclic loading response of the bonding between two EDs. A Mohr-Coulomb criterion is used to characterize the failure of the bonding between two EDs.

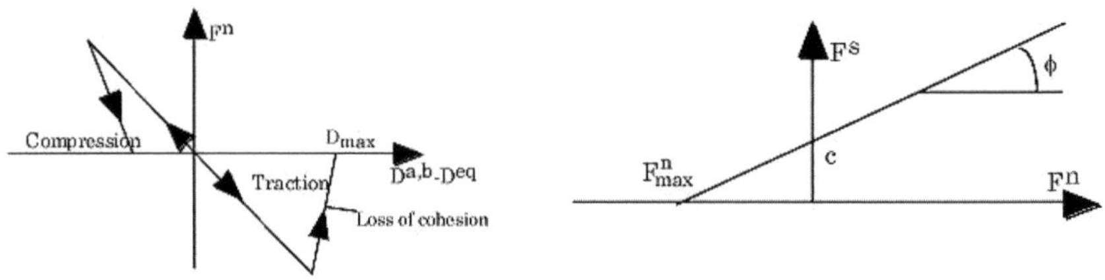

Figure 2.25. *Relative displacement-normal load answer of a bonding between two EDs and strength field limit (from [DON 98])*

2.6.8.3. *"LMGC" type*

– Reference: [MOR 96].

– Type: "dynamic contact".

The directing idea of this model is dealing separately with the elasticity of the medium and the adhesive contact laws where damage can appear. The laws linking the elements are of the graph type and describe the contact physical behavior. Between two elements, two physical principles exist: the existence of contact that is described using a Signorini condition, and the potential friction between elements. To this law, a new law is added to build a cohesive medium, namely adherence. Two media stand out, the micro-medium between discrete elements respecting the adhesive friction contact laws, and the macro-medium which describes the elastic continuous medium governed by the first principle of dynamics and the elements' elastic (or viscoelastic) strain laws.

2.7. Conclusion

2.7.1. *Main features of the models*

Most models are constructed within an elastoplastic framework, with a criterion determined by the first stress invariant. Taking the loading field into consideration, models generally include:

– a plastic flow due to shear;

– a plastic volume strain due to compaction;

– scalar damage due to cracking (usually).

Inevitably, whatever the model, the number of the parameters to be identified will be important, and complete identification will require several experiments. In practice, completing identification of all the parameters of a model is difficult. It requires having a concrete typology with the average values corresponding to each type for each model. Then the user only has to identify the type of concrete using a few simple tests, and can infer the whole set of corresponding parameters.

2.7.1.1. *Taking the dynamic aspect into account*

Should sensitivity to strain rate be taken into account in a concrete behavior model? When considering the models presented, it is easy to see that opinions are divided. The works of the "GEO" network [BAI 99] tried to answer this question. Two elements clearly emerge.

If dynamic compression tests (simple or with low confinement) are simulated using dynamic structure calculation carried out with models that are not sensitive to the strain rate, the apparent strength increase is quite well-rendered. Actually, the non-homogenity of the stress state and the "inertial" confinement effect has been established.

Unlike metal materials, concrete failure is accompanied by great dilatancy, which induces radial displacements slowed down by inertia. An increase of pressure within the test structure, and a higher failure threshold will follow [GAR 98a], [LEN 00].

Comparisons of static and dynamic compaction tests have rarely been conducted. Such comparisons are consistent if the stress path is the same in both cases. Experiments and simulations performed to that effect obviously show the importance of the strain rate on the compaction behavior [GAR 98b].

Rather paradoxically then, models integrating sensitivity to the strain rate do so in relation to the shear induced flow instead of the irreversible volume variation.

2.7.2. *Contribution of distinct elements*

This modeling type, initially designed for aggregate materials, is used to represent cohesive materials. Dynamic compression test simulations confirmed the conclusions of mechanics of continuous media studies with regard to the strain rate sensitivity.

Simulating structure tests with these methods allows us to describe post-ultimate phase aspects such as chip projection. This goal is difficult to achieve using the mechanics of continuous media and numerical finite element codes (except in the case of advanced codes, where finite elements meet EDs).

2.8. Bibliography

[BAI 99] BAILLY P., "Comportement des ouvrages en dynamique rapide, comportement dynamique du béton", final scientific report of the research group formed in the "GEO" laboratories, 1999.

[BAZ 76] BAZANT Z.P., BHAT P.D., "Endochronic theory of inelasticity and failure of concrete", *Journal of the Engineering Mechanics Division*, Vol. 102, N° EM4, August 1976.

[BIC 83] BICANIC N., ZIENKIEWICZ O.C., "Constitutive model for concrete under dynamic loading", *Earthquake Engineering and Structural dynamics*, Vol. 11, pp. 689–710, 1983.

[BIS 91] BISHOFF P.H., PERRY S.H., "Compressive behavior of concrete at high strain rates", *Materials and Structures*, Vol. 24, pp. 425–450, 1991.

[BUR 97] BURLION N., Compaction des bétons: éléments de modélisation et caractérisation expérimentale, PhD Thesis, Ecole Normale Supérieure de Cachan, 1997.

[BUZ 99] BUZAUD E., DON D., CHAPELLE S., GARY G., BAILLY P., "Perforation studies into MB50 concrete slabs", *9th International Symposium on Interaction of the Effect of Munitions with Structures*, Berlin, Germany, 1999.

[CAM 99] CAMBORDE F., Modélisation du comportement dynamique du béton, application aux problèmes d'impacts et d'explosions, Thesis, Ecole Centrale de Nantes, 1999.

[CED 77] CEDOLIN L., CRUTZEN R.J., POLI S.D., "Triaxial stress-strain relationship for concrete", *Journal of the Engineering Mech. Division*, Vol. 103, N° EM3, June 1977.

[CHE 75] CHEN ANDRIE C.T., CHEN WAI-FAH, "Constitutive relations for concrete", *Proceedings of the American Society of Civil Engineers*, Vol. 101, N° EM4, August 1975.

[COU 91] COUSSY O., *La Mécanique des Milieux Poreux*, Editions Technip, Paris, 1991.

[CUN 97] CUNDALL P.A., "Distinct element models of rock and soil structure", *Analytical and Computational Methods in Engineering Rock*, Allen & Unwin Eds, Ch. 4, pp. 129–163, London, 1997.

[DAR 90] DARVE F., "The expression of rheological laws in incremental form and the main classes of constitutives equations", *Geomaterials Constitutive Equations and Modeling*, Ed. Darve, Elseiver Applied Science, London, pp. 123–148, 1990.

[DON 98] DONZÉ F.V., MAGNIER S.A., "Numerical simulation of impacts using the discrete element method", *Mech. Cohes.-Frict. Mater.*, Vol. 3, pp. 257–276, 1998.

[DRU 52] DRÜCKER D.C., PRAGER W., "Soil mechenics and plastic analysis or limit design", *Quarterly of Applied Mathematics*, Vol. 10, pp. 157–175, 1952.

[DUB 94] DUBÉ J.F., Modélisation simplifiée et comportement visco-endommageable des structures en béton, PhD Thesis, University of Paris 6, 1994.

[ELW 79] ELWI A.A., MURRAY D.W., "A 3D hypoelastic concrete constitutive relationship", *Journal of the Engineering Mechanics Division*, Vol. 105, N° EM4, August 1979.

[GAR 98a] GARY G., BAILLY P., "Behavior of quasi-brittle material at high strain rate. Experiment and modeling", *European Journal of Mechanics, A/Solids,* 17, N° 3, pp. 403–420, 1998.

[GAR 98b] GARY G., BAILLY P., GATUINGT F. "Testing concrete at high strains and high rates of strain". *3rd International Symposium on Impact Engineering'98,* Singapore, Singapore, 7–9 December 1998.

[GUR 77] GURSON A.L., "Continuum theory of ductile rupture by void nucleation and growth: Part I – Yield criteria and flow rules for porous ductile media", *Engng. Materials and Technology*, Vol. 99, pp. 2–15, 1977.

[HAN 85] HAN D.J., CHEN W.F., "A non-uniform hardening plasticity model for concrete materials", *Mechanics of Materials,* 4, pp. 283–302, 1985.

[HER 69] HERMAN W., "Constitutive equation for the dynamic compaction of ductile porous materials", *Journal of Applied Physics*, Vol. 40, N° 6, May 1969.

[HOL 93] HOLMQUIST T.J., JOHNSON G.R., COOK W.H., "A computational constitutive model for concrete subjected to large strains, high strain rates, and high pressures", *14th International Symposium on Ballistics*, Quebec, Canada, 26–29 September 1993.

[LAB 91] LA BORDERIE C., Phénomènes unilatéraux dans un matériau endommageable: modélisation et application à l'analyse de structures en béton, PhD Thesis, University of Paris 6, 1991.

[LAS 88] LASRY D., BELYTSCHKO T.B., "Localization limiters in transient problems", *Int. J. Sol. Struct.*, Vol. 24, pp. 581–597, 1988.

[LEM 85] LEMAITRE J., CHABOCHE J.L., *Mécanique des matériaux solides*, Dunod, Paris, 1985.

[LEN 00] LE NARD H., BAILLY P., "Dynamic behavior of concrete: the structural effect of compressive dynamic strength increase", *Mechanics of Cohesive-Frictional Materials*, Vol. 5, pp. 491–510, 2000.

[LEV 98] LE VU O., Etude et modélisation du comportement du béton sous sollicitation de grande amplitude, application au problème de la pénétration dynamique, Thesis, Ecole Polytechnique, Palaiseau, France, 1998.

[LOR 87] LORET B., "Application de la théorie des multimécanismes au comportement des sols", *Manuel de rhéologie de géomatériaux*, presses de l'ENPC, 1987.

[MAR 94] MARIOTTI C., THOMAS F., "Loi de comportement ARMOR pour géomatériaux sous chargement dynamique", *Journal de Physique IV*, Vol. 4, September 1994.

[MAZ 84] MAZARS J., Application de la mécanique de l'endommagement au comportement non linéaire et à la rupture du béton de structure, PhD Thesis, University of Paris 6, 1984.

[MCQ 70] MC QUEEN R.G., MARSH S.P., TAYLOR J.W., FRITZ J.N., CARTER W.J., "The equation of state of solids from shock wave studies", in *High Velocity Impact Phenomena*, R. Kinslow (Ed.), Academic Press, New York, USA, 1970.

[MOR 96] MOREAU J.J., JEAN M., "Numerical treatment of contact and friction: the dynamic contact method", *Proc. Eng. Systeme Design and Analysis Conference, ASME*, New York, USA, Vol. 4, pp. 201–208, 1996.

[PER 63] PERZYNA P., "The constitutive equations for rate sensitive plastic materials", *Quaterly Applied Mechanics*, Vol. 20, N° 4, pp. 312–332, 1963.

[PIE 88] PIETRUSZCZAKS, JLANG J., MIRZA F.A., "An elastoplastic constitutive model for concrete", *Int. J. Solids Structures*, Vol. 24, N° 7, pp. 705–722, 1988.

[PIJ 87] PIJAUDIER-CABOT G., BAŽANt Z., "Non-local damage theory", *J. Eng. Mech., ASCE*, Vol. 113, pp. 1512–1533, 1987.

[PON 95] PONTIROLI C., Comportement au souffle des structures en béton armé – analyse expérimentale et modélisation, PhD Thesis, ENS Cachan, 1995

[SAN 76] SANDLER I.S., DIMAGGIO F.L., BALADI G.Y., "Generalized CAP model for geological materials", *Journal of the Geotechnical Engineering Division*, Vol. 102, N° GT7, July 1976.

[SUA 84] SUARIS W., SHAH S.P., "Rate sensitive damage theory for brittle solids", *Journal of Engineering Mechanics*, Vol. 110, N° 6, June 1984.

[ULM 93] ULM F.J., "Un modèle d'endommagement plastique: application aux bétons de structure", Report from a PhD Thesis, Ecole Nationale des Ponts et Chaussées, 1993.

[ZAO 93] ZAOUI A., PINEAU A., FRANÇOIS D., *Comportement mécanique des matériaux*, Hermes Science Publications, Paris, 1993.

Chapter 3

Seismic Ground Motion

3.1. Introduction

Seisms are sudden fracture failures, with aftershock that can reach several meters, on areas that can exceed 10,000 km². The sudden sliding emits seismic waves that propagate in the ground, causing multi-directional vibratory movements u_i (x,y,z; t) in horizontal and vertical directions that generate vibrations on the surface of existing structures. Depending on their amplitude, frequency and duration characteristics, these vibrations can damage buildings or even cause their complete collapse. Seisms not only affect man-made structures, but also lead to the damage or collapse of "natural structures" like soils, causing liquefaction, settlement or the formation of unstable slopes.

The main aim of "seismology engineering" is either to estimate *a priori* the main features of such vibrations so that surface structures can be dimensioned safely, or to estimate the potential damage that could accrue on pre-existing natural or man-made structures. Whilst, on the one hand, seismology attempts to analyze the space and time repartition of vibrations (their location, depth, size and frequency of earthquakes), in close connection with "conventional" geologists and seismologists, it will also attempt to quantify the associated vibrating motions, and characterize them using parameters representing their "damaging power". The aim is to understand the phenomena that cause vibrations and the physical values that control them as well as possible, so that future seismological events can be predicted.

Chapter written by Pierre-Yves BARD.

Only those aspects linked with seismic motion itself will be dealt with here: readers interested in seismology can read [BER 03], [LAM 96], [LAM 97] and [MAD 91].

This chapter briefly outlines three areas:

i) techniques for the measurement of seismic motions for engineering purposes;

ii) qualitative characterizations, with a view to para-seismic dimensioning; and

iii) physical descriptions of the phenomena affecting the characteristics of such motions, including their origin, propagation and site effects.

3.2. Measuring seismic motions

3.2.1. *Differences between seismological and accelerometer networks*

The first seismological instruments were built almost 2,000 years ago by the Chinese, who developed a device for locating the epicenter of an earthquake. The first "modern" instruments, which date back to the end of the 19th century, aimed at detecting and recording motions that were imperceptible to man. Since then, a vast range of seismological instrumentation has come online, with increased sensitivities and frequency responses, designed to locate quake event maxima (whether local, regional or global) and to provide information on the inner structure of the terrestrial globe.

For technological reasons, these very sensitive instruments were initially intended to monitor violent events in their immediate vicinity. This is why Californian engineers since the 1930s have tended to design low gain instruments. The first accelerometers (called "strong motion" instruments) were low gain devices because these make it easier to record acceleration directly. Even now, when the velocimeters available have wide recording dynamics, the "strong motion" ground acceleration has endured as a testing parameter. It has also been kept alive by the fact that most of the instrument networks set up globally to monitor violent motions still use accelerometers. The logic behind the siting of stations remains quite different; "strong motion" instruments are typically installed in high economic stake areas (city areas, important works). The ways in which seismological networks use data on strong motions also remain quite different. Besides "traditional" seismology, a more practical "violent motion seismology" has developed, which does not necessarily leave aside relatively sophisticated treatment or modeling: this forms the subject matter of this chapter.

This chapter deals with the characterization of seismic stresses using values suited to para-seismic design, and details the ways in which these values are estimated and the way they are used to predict future earthquakes using either empirical or more physical models.

3.2.2. *Accelerometer networks*

Several thousand accelerometers are in operation all over the world. Organized as regional networks, they are sometimes installed as low span very dense local networks, or installed inside a wide range of civil engineering works (buildings, dams, nuclear power stations and geo-technical works). Generally, there are two kinds of instrument: "open field" instruments, intended for recording the ground motion (or seismic action); and "structure" instruments, used for recording the actual seismic behavior of the monitored structure.

Until the last decade, the only purpose of the latter instrumentation was to improve para-seismic knowledge and safety regulation. Telecommunications advances have, however, allowed two other uses in quasi-real time: real time alert generation in the case of big distant and remote earthquakes (like the Pacific subduction area earthquakes and their effects in Mexico) and crisis management to make a chart of the most severely hit areas in a very short time, for co-ordination and decision-making in relief efforts (for example, gas network management in Taiwan in 1999 – Tokyo Gas manages over 3,000 stations on its own behalf, or high-speed trains – Shinkansen in Japan and Mediterranean TGV in France).

The most important open field networks, as far as volume is concerned, have been installed in the Pacific area, notably Japan, Taiwan, California and Mexico. Europe and countries around the Mediterranean with a few exceptions are somewhat behind, but efforts have been made to create new networks (such as the RAP in France) or to improve them (as in Italy).

Structure instrumentation is far less widespread, and the data obtained is typically under-used, for reasons probably arising from "community cultures" that are more or less modeling confident. Whilst glaring gaps in the networks still exist in Europe, the USA and Japan have started their own consistent and sustained instrumentation programmes. Their results are not always easy to "import" into Europe because of differences in building processes, particularly when modeling ancient buildings.

3.2.3. *Accelerometer data banks*

The rate at which new recordings are collected is constantly increasing, and a cursory inspection of recent strong motion seismological data demonstrates they have all been caused by "astonishing" recordings, the analysis and interpretation of which has forced engineers to re-evaluate certain hypotheses in the field.

Whereas this data was spread confidentially until the 1980s, the advent of the Web has forced a state of mind change thanks to the volume now available online. Several sites are listed at the end of this chapter, but the system evolves so rapidly that many of these may already be obsolete.

Two Internet site categories are given, some presenting recordings from specific networks and others gathering data from quite varied sources to compile as databases.

3.3. Quantitative characterization of seismic movements

Real accelerograms possess different forms of time, amplitude and frequency content. In most cases, para-seismic dimensioning is not carried out on complete accelerograms (used only in non-linear dynamic analysis cases), but is performed using a restricted number of reducing quantities which are representative of the damage a structure might receive.

Only those most commonly used are referred to; more elaborate descriptions are given in [BET 03], [JEN 83] and [KRA 96].

3.3.1. *Time maximum values*

The most direct and simple quantity is the acceleration maximum, often referred to as the "PGA" (peak ground acceleration). For a long time, when only analog recordings were available, this was the only quantity available. As we will see later, it remains in use because it links directly to the spectral characterization that para-seismic engineers prefer, i.e. the acceleration response spectra.

For destructive earthquakes, PGA data has sensitivity better than 1 m/s^2. Even if until the 1970s it was generally thought that ground acceleration could never be more than 1 g, observations have since indicated values in excess of 20 m/s^2, even along the vertical component (which may explain some reported discrepancies in construction works with spans resting on their supports).

Obviously, the relationship of this quantity to the damage a seismic signal can cause is somewhat obscure. It gives no information on the duration or spectral content of the stress, both essential parameters for modeling non-linear oscillators with several degrees of freedom, which lie at the root of paraseismic calculations.

Nevertheless PGA-macroseismic connections are often used quantitatively. Everybody has to be aware that qualitative relationships are marred by huge uncertainties and should never be used without explicitly taking into account the associated standard deviation (a minimum factor 3). Most existing correlations show that whenever intensity rises by 1 degree (MM or MSK), the PGA is multiplied by a factor of 2 or 3.

To be able to take spectral content into account in an approximate way whilst keeping the simple concept of a maximum value, we consider the ground velocity ("PGV") and displacement ("PGD") maxima as well. Whereas the maximum acceleration is mainly associated with the high-frequency spectral content (beyond 5 Hz), the maximum speed is associated with the intermediate frequencies (between 5 and a few Hz), and the maximum displacement is associated with low frequencies (below 1 Hz).

For a destructive earthquake, values of PGV/PGD typically range from a few cm/s to over 1 m/s, and from a few millimeters to a few meters. They are far more sensitive to the size of the shock than the maximum acceleration, because size influences the low frequency content far more than the high frequency content.

Nevertheless, neither of these two values shows a better correlation with macro-seismic intensity. Furthermore, their estimation is less direct than that of the PGA, since acceleration histories have to be integrated numerically. For analog or first generation digital instruments, it is also less reliable because of the greater sensitivity of the numerical process to low frequency noise.

However, it should be noted that knowledge of all three values gives a good idea of the frequency level and content of a given accelerogram. Furthermore, the non-dimensional ratio, PGA x PGD/(PGV)^2, which relates to the width of the excited frequency response, generally varies little from one recording to the next, with values typically ranging from 1 to 10, and even from 2.5 to 7.5 [BET 03].

3.3.2. *Spectral characterizations*

Whilst characterizations that are more complex are often used in seismic engineering, spectral characterizations convey information that is richer and more consistent than mere maximum values. Unfortunately, two different spectral characterizations are used by two distinct communities: the "producers" and "users". Seismologists prefer using Fourier spectra, due to mathematical practicability and the possibility of establishing useful relationships with wave emission and propagation physics. On the other hand, "users" like civil engineers (structure, geotechnology) have become accustomed to reasoning in terms of response spectra, as they are easily adapted to the simplified modeling of civil engineering structures.

3.3.2.1. *Response spectra*

The origin and interest in response spectra lie in the fact that they reduce (as a rough approximation) the seismic behavior of a building to a simple oscillator with 1 degree of freedom. Representation as a response spectrum directly accesses the motions of a structure's center of gravity.

Let us consider a 1-dof linear viscoelastic oscillator, characterized by its frequency f and damping ζ. Due to an earthquake characterized by an acceleration a(t), the oscillator will undergo a relative displacement x(t), and an absolute acceleration $x''(t) + a(t)$. Response spectra are defined as the time maximum responses of the oscillator for a *relative* displacement, *relative* speed and *absolute* acceleration:

$$Sd\,(f,\xi) = \text{Max}_t\ \{x(t)\}$$

$$S_v\ (f,\xi) = \text{Max}_t\ \{x'(t)\}$$

$$S_a\ (f,\xi) = \text{Max t.}\{x''(t) \circ + \circ\, a(t)\}$$

On varying the frequency (f) of the oscillator with a constant damping (ζ), we obtain three curves, Sd, S_v, S_a defining respectively the response spectra for displacement, speed and acceleration. These are usually calculated for damping discrete values: 0%, 2%, 5%, 10%, and 20%, with 5% as the most frequent value.

With the following expression for the basic equation of a mass (m), stiffness (k), and damping (c) of a simple oscillator with 1 degree of freedom:

$$m\,[\,a(t) \circ + \circ\, x''(t)\,] = -\xi\ c\ x'$$

$$\omega_0 = \sqrt{(k/m)};\xi = \tfrac{1}{2}\ c/\sqrt{(km)}$$

$$\omega_1 = \omega 0 \sqrt{\left(1-\xi^2\right)}$$

The expression for x(t) conventionally derives from Duhamel's integrals:

$$X(t) = -1/\omega_1 \int_0^t a(\tau)\ \exp\left[-\xi\omega_0 (t-\tau)\right] \sin\left[\omega_1 (t-\tau)\right] \sin\left[\omega_1 (t-\tau)\right] d\tau$$

However, it can also be derived fast and simply using the Fourier transform.

According to these formulae, we can then see the following properties:

$$S_a\ (\infty,\xi) = \text{Max t.}\{a\ (t)\} = a_{max} = \text{PGA} \qquad \forall\ \zeta$$

$$S_d(0,\xi) = \text{Max t.}\{-\ d\ (t)\} = d_{max} = \text{PGD} \qquad \forall\ \zeta$$

$$S_a\ (f,\ 0) = 4\pi^2\ f^2\ .\ S_d(f,\ 0)$$

$$S_v\ (f,\xi) \approx 2\pi\ f\ .\ S_d(f,\xi)$$

$$S_a\ (f,\xi) \approx 4\pi^2\ f^2\ .\ S_d(f,\xi)$$

The last two formulae are only approximate with non-zero damping; two new quantities have been introduced, called speed (S'_v) and acceleration (S'_a) pseudo-spectra, which can be defined as:

$$S'_v(f,\xi) = 2\pi\ f\ .\ S_d(f,\xi)$$

$$S'_a(f,\xi) = 4\pi^2 f^2 . S_d(f,\xi) = 2\pi\ f\ .\ S'_v(f,\xi)$$

The latter relationships allow a quadri-logarithmic representation of the pseudo-response spectra, with the logarithm of the frequency (or the period T) on the abscissa, and the logarithm of S'_v on the ordinate. S'_a and S_d values can then be read at once in relation to the diagonal axes, since:

$$\log\ S'_a = \log S'_v + \log(2\pi) + \log\,(f) = \log S'_v + \log(2\pi) - \log\,(T)$$

and as a correlation:

$$\log S_d = \log S'_v - \log(2\pi) - \log\ (f) = \log S'_v - \log(2\pi) + \log\,(T)$$

Thus, it is quite easy, from a given accelerogram, to derive response spectra. However, a given response spectrum (for a given damping value) corresponds to an infinite number of time histories with the same acceleration and displacement maxima but different durations and phases. The repartition of the energy arrival in time is not constrained by the response spectrum data. The absence of equivalence constitutes a great handicap when non-linear time studies have to be led, and using even adapted accelerograms to a given response spectrum is a difficult task which can only be performed by seismologists if they are to appear realistic.

Elastic response spectra are often used in paraseismic engineering because, as a first estimate, simple structures can be compared to a 1-dof oscillator with a known period and damping. Caution must be urged in this regard, as the conventional formula T = N/10 (with N being the number of floors) has been imported from the USA where most buildings are framework constructions and are not applicable in France, where the prevailing structures have stiffer walls for which the formula T = N/25 is more appropriate. The motions of the center of gravity can be reasonably well estimated, as long as the structure is assumed to have a linear elastic behavior.

This last hypothesis is obviously not true when the structure has been seriously damaged: that is why other non-linear spectra have been developed, among which the simplest correspond to perfect elasto-plastic behavior. They involve the introduction of an additional parameter, the ductility demand, (μ), which defines the relationship between the maximum displacement of the elasto-plastic structure and that of the associated elastic structure (with the same low acceleration stiffness and damping).

As estimating the dynamic response of non-linear systems is difficult, it has recently been proposed that dynamic non-linear analysis is replaced by a static non-linear analysis, called "push-over analysis". The gross result of this is a curve that links the applied horizontal force to the displacement. Procedures have also been proposed for converting elastic response spectra into "stress-strain" curves, or to transform the latter into "S_a-S_d" curves, linking force to acceleration by means of the modal mass.

3.3.2.2. *Fourier spectra*

For seismology, the most commonly used spectral representation resorts to Fourier transform, an algorithm well-assimilated and quick to use from a numerical point of view. The Fourier acceleration spectrum is defined by:

$$F_a(f) = A(f) . \exp(i\varphi(f)) = \int_{-\infty}^{+\infty} a(t)\exp(-i2\pi ft)dt = \int_0^D a(t)\exp(-i2\pi ft)\,dt$$

where A refers to the modulus, φ is the phase and D is the total duration of the recording.

Similar expressions exist for speed $F_v(f)$ and displacement $F_d(f)$ spectra, and the conventional relations of Fourier transform are quite convenient and useful:

$$a(t) = \int_{-\infty}^{+\infty} F_a(f) . \exp(+i2\pi ft)\,df = 2\int_0^{f_N} F_a(f) . \exp(+i2\pi ft)\,df$$

where f_N is the maximum frequency which can be calculated taking the sampling time step Δt: $f_N = 0.5/\Delta t$.

$$F_a(f) = i.\, 2\pi f.\, F_v(f) = -4\,\pi^2 f^2$$

$$Fd(f) = \int_0^{f_N} \left|F_a(f)\right|\,df = 1/2 \int_0^D a^2(t)\,dt$$

This last relationship (Parseval's theorem) allows us to link time descriptions with spectral data: it is particularly used in many "spectral" models that allow the quantitative parameters of seismic movement to be linked not only to the physical properties that characterize the seismic source (emitting waves), but also to their crust (i.e. very deep) propagation and site effects. However, it should be noted that reconstructing the time signal from its Fourier spectra is only possible if both the modulus A and the phase φ are known. In general terms, only relationships between the Fourier modulus and various parameters describing both origin and propagation are well-established.

Therefore, methods and practices for describing the non-stationary time state of a seismic signal and its consequences for the strength of building works are still evolving. Some quite interesting tracks have been opened using "group delay time" or the phase φ derivative in relation to frequency f.

3.3.2.3. *Relationships between response and Fourier spectra*

A qualitative correspondence exists between these two spectral representations, which both express the frequency repartition of energy of the seismic signal. However, no strict mathematical relation between the elastic response spectra and the Fourier spectra exists. The only relationship that can be demonstrated is an inequality, given by:

$$S_v(f,0) \geq |F_a(f)|$$

3.3.2.4. *Generic spectral shapes*

These spectra are characterized by simple shapes (Figure 3.1): once smoothed, a Fourier spectrum A(f) is characterized by a plateau shape between two frequencies f_c and f_{max}. The first (the corner frequency) is inversely proportional to the size of the failure area, but it is also linked to the stress drop caused by the earthquake and to the directivity process, that is, to the position of the receptor at the front or at the back of the failure propagation direction. For destructive earthquakes, it is generally less than 1 Hz, and can drop well below 0.1 Hz for very high magnitude earthquakes. The second frequency is linked to inelastic attenuation phenomena which affect the waves during their in-depth and surface propagation, and to processes that govern failure dynamics on the crack. This frequency is generally higher than 4-5 Hz, and can sometimes exceed 20 Hz. Below f_c, the acceleration Fourier spectrum modulus varies as f^2, which corresponds to a plateau for the displacement Fourier spectrum, usually written Ω_0, and is linked to the properties of the source by the following relationship:

$$\Omega_0 = M_0.R_{\theta\varphi} / (4\pi\rho\, R\, c^3)$$

where M_0 is the seismic moment $\mu .D_0. S$, μ is the shear stiffness at focal depth, D_0 is the average reactivation on the fault, S is the total failure area one the fault plane, R is the focal depth, and c is the wave propagation speed (and thus generally that of the S waves if we are considering horizontal component forces).

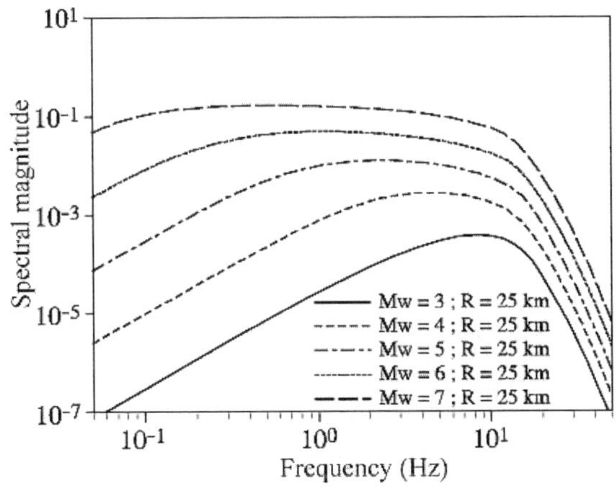

Figure 3.1. *Generic shapes of acceleration Fourier spectrum, for increasing magnitudes ranging from 3 to 7 (these curves were established on the basis of scale laws specified in section 3.4.1)*

Beyond f_{max}, the acceleration Fourier spectrum modulus decreases in $f^{-\alpha}$, but with an exponent α that can vary according to the site and the event.

As shown in Figure 3.2, the acceleration response spectrum, represented as a function of the period, has a characteristic shape, with an ordinate equal to the PGA at the origin. It has a plateau between two periods depending on the site and the size of the event and a level typically between 2 and 3 PGA, followed by a branch decreasing first as 1/T, and then as $1/T^2$.

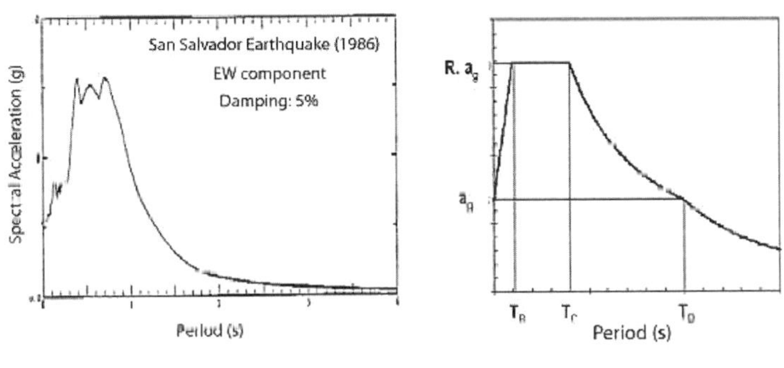

Figure 3.2. *Example of a real response spectrum for a 5.5 magnitude earthquake (San Salvador 1986), and generic shape of the pseudo-acceleration response spectrum, as defined in most para-seismic regulations (EC8 in this case)*

3.3.3. *Features of hybrid characterizations*

Other representative values are used in the field: from a physical point of view they are more significant than peak values (PGA, PGD and PGV) and simpler than spectral representations. Many proposals have been made; here we will only refer to those that have been accepted by the majority of the para-seismic community.

3.3.3.1. *Spectral intensity*

This value was introduced initially by Housner to characterize the stress average level of the most "common" structures: he considered the response spectrum integral on periods ranging from 0.1 to 2.4 seconds:

$$SI\,(\xi) \approx \int_{T=0.1}^{T=25} Sv(T,\xi)\,dT.$$

As a rule, SI is calculated for high damping values (ζ: typically 20%), in order to integrate the already highly smoothed spectrum.

Other values can be defined by changing the integration bounds, altering the mute variable (period or frequency), or by carrying out the integration on a logarithmic axis. None really stands out as superior to any other in practice.

However, the notion of spectral intensity has come back in force in Japan over the last two decades, namely IS SI(0.2)/2.4 s (then associated with the speed response spectrum average level on the 0.1 – 2.5 s range). The reason for this is the excellent correlation observed with the intensity of damage, and with a significant damage apparition threshold for IS ≥ 30 cm/s. The correlation has proved good enough for Tokyo-Gas to use it in Tokyo and its suburbs, where over 3,000 instruments calculate it directly and transmit it in real time to a monitoring center, for the purpose of mapping the damage estimated layout to decide which safety floodgates should be activated.

3.3.3.2. *Duration, average quadratic acceleration and the Arias intensity*

Many characterizations are based on the increasing monotonous function W(t), defined by:

$$W(t) = \int_0^t a^2(\tau)\,d\tau$$

Its final value W_∞, is directly linked to the energy contained in the signal a(t). It can also be connected to the spectral content using Parseval's relation (see above). For this reason, Arias used it to define the "Arias intensity", viz: $I_A = \pi/2g.\ W_\infty$.

This is also used to define a "high phase duration", D (ε_1, ε_2), given by:

$$D(\varepsilon_1, \varepsilon_2) = T(\varepsilon_2) - T(\varepsilon_1)$$

where $T(\varepsilon_i)$ is defined by: $W(T(\varepsilon_i)) = \varepsilon_i.\ W_\infty$.

If ε_1 is nearly always chosen as equal to 5%, the values of ε_2 are either 75% (which involves quite short a duration) or 95% (which, on the contrary, leads to a rather long duration).

Correlatively, as soon as duration has been determined, we can define an "average quadratic acceleration", a_{rms} (ε_1, ε_2) for this duration:

$$a_{rms}(\varepsilon_1, \varepsilon_2) = \sqrt{\left[(\varepsilon_2 - \varepsilon_1)\ W\infty / D(\varepsilon_1, \varepsilon_2)\right]}$$

Thus, once a duration has been defined, we can estimate time parameters owing to the Fourier spectrum.

3.3.4. *Caveats regarding differential motions*

In addition to considering pure translation motions, the rapid space variations shown by seismic motions should not be neglected either. These variations have multiple origins, including the differences due to wave propagation, decorrelations related to heterogenity or the complex wave field and local site effects.

Thus, the origin of spatial variability of seismic motions is multiple.

– whenever the angle of incidence is oblique ($\theta \neq 0°$), their horizontal propagation speed is finite and equal to $c/\sin\theta$. From then on, at two sites separated by a distance d (projected along the propagation direction), the signal (despite its shape and amplitude being identical for both sites) will be delayed by $\tau = d \sin\theta /c$. The delay has to be compared with the predominant period of the signal $T = 1/f$ and can induce significant effects when the ratio $\tau = d \sin\theta /c$ exceeds a threshold of 0.1. The "horizontal propagation" effect will become increasingly significant when the propagation speed c is low (therefore for weak stiffness grounds), the angle of incidence is great (therefore for surface waves), and the frequency is high;

– in reality, as the propagation medium is not homogenous (especially near the surface, and the incident wave field is not limited to one single wave), an infinity of them is generated on each point of the failure area, and they therefore arrive with different azimuths, different angles of incidence, different polarizations and different frequency contents. This effect is particularly marked in a near field; they give rise to multiples whenever they meet a heterogenity (e.g. reflection phenomena, refraction, geometric diffraction) whilst propagating. As the propagation delays associated with each individual wave vary from one wave to the other, the resulting signal is greatly modified from one site to the other, even if its overall spectral characteristics remain unaltered. These signal modifications are usually measured by the coherence;

– for structures with a large basemat into the ground (e.g. bridges and dams), the geological and geotechnical conditions can vary greatly along the foundations and involve significant variations of frequency content, as well as the phase modifications mentioned above.

Besides transient "dynamic" differential motions, residual static motions associated with the sudden or continuous *creep* of active faults or sliding activated or re-activated by earthquakes can exist. For example, for the new Rion-Antirion Bridge, it was necessary to take into account the "continuous" opening of the Gulf of Corinth, at a rhythm of about 2 cm a year, and the potentiality of a major earthquake that might bring about sudden spreading over 10 cm. Such imposed differential displacements should be taken into account in town planning and design, but it should be remembered that the damage associated with surface failure accounts for only about 1% of the total damage earthquakes cause.

3.4. Factors affecting seismic motions

The physics of seismic wave emission on fault propagation within the Earth's crust involves diverse phenomena associated with an extensive literature, and in some cases, are not well understood themselves. For this reason, the indications presented below are deliberately simplified: any reader interested in looking deeper into the subjects could profitably consult the works of [AKI 80], [BER 03], [BET 03], [KRA 96], [MAD 91] and [PEC 85].

3.4.1. *Spectral signature of the seismic source*

The term "seismic source" refers to wave emission caused by the creep of a finite sized fault area; this creep is neither instantaneous nor homogenous on the failure zone, and the emission process is so complex that it forms a research field in its own right.

When simplifying to extremes, the source of an earthquake (Figure 3.3) can be viewed as a failure starting at a focus F (given by the (ξ_0,η_0) coordinate hypocenter) which then propagates along the fault plan (ξ,η) at a failure speed V_R, causing at each point a sliding $\Delta u\ (\xi,\eta;\ t)$. At the end of each earthquake, sliding at these points reaches a final value $\Delta U_f(\xi,\eta)$.The source is completely determined by the space-time sliding function Δu and the waves emitted at point (ξ,η) have a displacement signal proportional to $\partial/\partial t\ [\Delta u\ (\xi,\eta;\ t)]$. At present, we do not know how to anticipate future earthquakes; nevertheless, with good instrument conditions, we can discover its characteristics for high wavelengths (several kilometers) and low frequencies (generally lower than 1 Hz) *a posteriori*. We are also beginning to reconstruct a stochastic description of the short wavelength and high frequency part.

However, without going into all the relevant details, it is now possible to explain certain general aspects of the spectral content of emitted waves, thanks to a few overall features of the seismic source. These overall features together with their effects are described below (Figure 3.3):

– the $\Delta U_f\ (\xi,\eta)$ quantity is reduced to the surface F of the failure zone (LW for a rectangular fault, πa^2 for a circular fault), and to the average sliding D_0. This enables the definition of a fundamental quantity indicating the importance of the earthquake, namely the "seismic moment", $M_0 = \mu\ S\ D_0$, where μ is the shear stiffness of the Earth's crust at the level of the fault. This quantity controls the low frequency level of the emitted waves, expressed in N.m;

– the stress drop $\Delta\sigma$ expresses the shear stress relaxation between the state immediately prior to failure, σ_0, and the state immediately following failure end, σ_1. Elastic dimensional analysis allows the following rough estimate to be obtained. $\Delta\sigma = C\ \mu\ D_0/L_c$ where L_c is the dimension characteristic of the fault (W for a rectangular fault, a for a circular fault), and C is a shape coefficient close to 1;

– the energy released during an earthquake corresponds to the work of shear stresses. It can be roughly approximated by the work of the average stress $E = S.\ [(\sigma_0 + \sigma_1)/2].\ D_0$. Thus, we can write $E = M_0.\ [(\sigma_0 + \sigma_1)/2\mu]$. Because, deep into the ground, the shear stiffness μ and the average stress $(\sigma_0 + \sigma_1)/2$ do not differ significantly, we can see that the seismic moment is a good indicator of the total energy released;

– the failure speed typically ranges from 2 to 3.5 km/s, but it can sometimes exceed the speed of the waves S. In such cases, the failure is termed "supersonic" or "super shear". Together with the dimensions of the failure area, this speed controls the wave emission time and the length of near-field strong motions. Moreover, it plays a part in directivity phenomena which are similar to the well known Doppler effect: a receiver located at the front of the failure propagation direction will receive waves emitted for a shorter time the nearer the failure speed is to wave propagation speed (S), which involves more intense motions. The opposite is true for a site located at the rear of the failure front;

– the final "global" parameter used is the "rising time" (τ). This is the time necessary for sliding at a specific point to pass from 0 (pre-failure) to its final value ΔU_f. In reality, τ varies from one point to another on the failure area. Nevertheless, for simple models, we assume the sliding function is identical on all points, i.e. that $\Delta u(\xi,\eta; t) = U_f(\xi,\eta) . S(t - d/V_R)$, S(t) being a function the derivative which has a bounded support [0, τ], and d being the distance between the current point (ξ,η) and the hypocenter (ξ_0,η_0). The simplest function is a linear ramp function (see Figure 3.3).

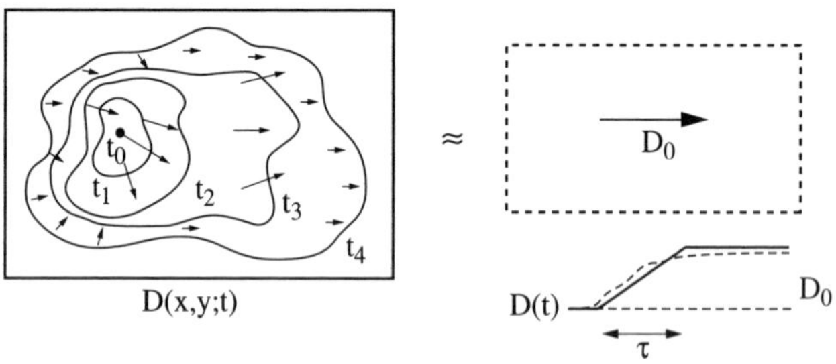

Figure 3.3. *Schematic representation of the failure process. The diagram on the left represents the real history of the sliding field on the fault D (x,y;t) (isochrones of the failure front at successive instants), whereas the diagram on the right illustrates its simplified representation as a (L,W) dimensioned rectangular fault, in which the dislocation function D(t) follows the same time history on each point (ramp function characterized by its rising time τ and its uniform final dislocation D_0)*

The analysis of these parameters for numerous earthquakes demonstrates they do not vary from one another, and therefore that the following three "scaling laws" exist:

– a "geometric" law; $W/L \approx c_1$, with $c_1 \approx 0.5$ (except for important strike-slip earthquakes, where width W is limited by the thickness of the brittle part of the Earth's crust, i.e. 15-20 km);

– a law associated with a relative constancy of the stress drop: $D_{0/}L_c \approx c_2$, with $c_2 \approx 10^{-5}$ and 10^{-4} (which gives $\Delta\sigma = C\ \mu\ D_{0/}L_c \approx \mu\ c_2 \approx 3.\ 10^5$ to 10^6, i.e. 0.3 at 3 MPa);

– a "dynamic" law: $\tau\ .\ V_{R/}L \approx c_3$, with $c_2 \approx 0.2$ (the rising time is proportional to the total length of failure propagation).

By integrating these empirical scaling laws into simple models, we can show that:

– the magnitude of the moment $M_W = 2/3 \log M_0 - 6$ varies like the logarithm of the fault's surface;

– the seismic moment, and the very low frequency content vary as the cubed fault's dimension and as $10^{1.5\ M_W}$;

– the corner frequency (which is linked to the emission length of the waves on the failure area), varies as the inverse of the fault's dimension, and then as $10^{-0.5\ M_W}$;

– from then on, the level of the plateau of the acceleration Fourier spectrum is roughly proportional to the fault's dimension, and then it varies as $10^{0.5\ M_W}$.

3.4.2. *Effects of propagation in the Earth's crust*

Once emitted, seismic waves propagate inside the Earth's crust: these deep propagation effects are characterized by three main phenomena, which are discussed in the following sections.

3.4.2.1. *Geometric expansion*

This is simply the decrease in wave amplitude due to distance, in close contact with energy preservation.

For "volume" waves, the wave front is spherical, and the conservation of energy (which is proportional to the squared amplitude on this surface) involves using a geometric expansion term varying in $1/R$.

For surface waves in which the energy concentrates in the direct vicinity of the surface, the wave front is cylindrical and the geometric expansion term varies in $1/R^{0.5}$.

These theoretical dependencies are of course, only valid for far fields, and for a perfectly homogenous space (or semi-space). In practice, the existence of near-field terms, the non-homogenity of the propagation medium and the multiplicity of the waves bring about geometric expansions with terms which are markedly different from the exponents – 1 and – 0.5.

3.4.2.2. *Multiple pathways*

As it meets heterogenities in the crust during propagation, the wave emitted from the source will be partially transmitted, partially reflected, and possibly partially diffracted. There are also possible conversions between radial P and S waves. Overall, this results in the existence of many additional waves with different paths that will prolong the signal, which decrease faster with the distance of direct waves.

For certain strong earthquakes, energy phases corresponding to reflections of direct waves at the base of the crust appear at distances ranging from 50 to 100 km This was evident in San Francisco in 1989 during the Loma Prieta earthquake (located 80–100 km away).

Effects like these are almost impossible to predict with simple models. However, their numerical modeling is relatively simple, providing the structure to the Earth's crust is known (location and characteristics of the main discontinuities).

3.4.2.3. *Inelastic attenuation*

As is the case in any real material, these waves undergo some energy loss (mainly via aggregate friction), which is characterized in seismology by a quantity factor Q that is related to the damping coefficient ζ by the relationship $Q = 0.5/\zeta$. The definition of Q is linked to the relative energy loss on a wavelength λ, chosen to be equal to $\Delta E/E = -2\pi/Q$. For a frequency f, plane waves propagating in the R direction with a speed c, have an associated spatial dependency given by:

$$A(R) = A_0\ (R).\ \exp\left(-i\ 2\pi f\ t\right) \cdot \exp\left(-\pi f\ r/Q\ c\right)$$

for which A_0 (R) corresponds to the geometric expansion decrease:

$$\left[A\left(R+n\ \lambda\right) = A(R) \cdot \exp\ \left(-\pi f\ n\ \lambda/Q\ c\right) = A\left(R\right) \cdot \exp\left(-\pi\ n/Q\right)\right]$$

This expression shows that for a given distance R, higher frequencies will be attenuated more than lower frequencies, assuming Q has the same value at high and low frequencies. This effect is often used to account for the high frequency decrease of the Fourier spectrum.

Typically, inside the Earth's crust, Q values are a few hundred (generally higher for P waves than for S waves), and this effect only becomes evident when the distance corresponds to a number n of value of Q wavelengths. However, in the most superficial materials, Q can reach values equal to only a few dozen: if we take a Q of 20 on a 200 m deep layer, with 200 m/s speed values, we calculate loss factors of exp (– π /20) at 1 Hz (negligible), and of exp (– π) at 20 Hz (significant!). This indicates that the high frequency decrease effect is due either to propagation over long distances within poorly attenuating media or to a very superficial effect within highly attenuating materials. We note that the values of Q found in the literature generally reflect two phenomena: on the one hand, the intrinsic inelastic attenuation, and on the other, energy losses through diffusion on short wavelength heterogenities – this often results in a frequency dependency of Q that is thought to be linked to the size of the heterogenities. This dependency is expressed generically as:

$$Q = Q_0 f^{\alpha} \text{ with } 0 \leq \alpha \leq 1$$

3.4.3. *Site effects*

Seismic motions can also be noticeably modified by close surface geological conditions. This is evident by comparing their wavelengths (λ = c/f, for c varying from 100 m/s to 2 km/s, and f varying from 0.5 to 10 Hz, i.e. for λ varying from 10 m to 4 km), with the dimensions of surface heterogenities that are in the same range. Because such heterogenities can be quite important, both mechanically and geometrically, the interferences between the incident wave fields can be very intense and cause modifications of the spectral characteristics of seismic motions.

Seismic effects are thus wave propagation effects leading to amplifications (or de-amplifications) located in space which can reach very high factors (greater than 10 in some extreme cases). They mainly affect topographic reliefs and sedimentary fillings. In the next section, we briefly explain the basic physics for these typical configurations. The curious reader will find a more comprehensive description in [BAR 99], which discusses the remaining unsolved questions, and gives an account of the diverse estimation methods used.

We also feel it is important to mention an additional site effect category – those characterized by ground located failures caused by intense vibrations. These include liquefaction affecting saturated sandy soils (causing subsidence) and slope instability (landslips, sliding, both superficial and deep). If the latter arise in static conditions, the earthquake is only a starting element, but the former are specific to earthquakes. Taking both of these phenomena into account depends more on particular dispositions (especially for foundations or ground improvement) than structure calculation, so they are not covered in detail here.

3.4.3.1. *Effects of topographic relief*

After destructive earthquakes, the damage reported is significantly worse at the hilltop than the base. The most outstanding French examples took place in Rognes and Vernegues (1909 earthquake) and in Castillon (1887 earthquake). This "qualitative" observation has been confirmed by numerous instrument measurements, which revealed amplitude relationships between the summit of a relief and its foot which sometimes reach a 3 to 4 factor difference in their time maximum (PGA, PGV) and a factor higher than 10 in the spectral field (Figure 3.4). The number of experimental studies on the subject is extremely low compared with the studies dealing with the underground amplification, so it is not yet possible to derive statistically significant empirical laws. Theoretical and numerical models also predict a seismic motion systematic amplification perpendicular to the convex parts of a topographical relief (cliff edge, hilltop), in addition to sensitivity to the field characteristics of the incident wave (wave types, azimuth, incidence). The sensitivity of seismic motions to the topographic relief seems to be linked to three physical phenomena:

– sensitivity of surface motion to the incidence angle, especially around the limit angle for SV waves;

– focusing of seismic waves in a convex relief (summit) and defocusing in a concave relief (foot);

– volume and surface wave diffraction over all the surface irregularities, which generates interferences (constructive and/or destructive depending on the position and the frequency).

The known facts on this topic can be summarized as follows:

– as far as quality is concerned, agreement between theory and observations is quite satisfactory;

– a convex relief generates amplification whereas a concave relief brings about de-amplification;

– the amplification is generally higher for horizontal component forces than for vertical component force. In the case of 2D reliefs (ridges), the horizontal component perpendicular to the crest's axis is often more amplified, and the transverse stiffness is weaker than the longitudinal stiffness;

– amplification level is linked to the topographic slenderness (height/width): the stiffer the average slope, the higher the summit amplification;

– summit amplification (as well as base de-amplification) shows a strong frequency dependency. The maximum effects correspond to wavelengths similar to the relief's horizontal dimensions, favoring an interpretation of the topographic effect by means of lateral interferences of the diffracted waves on all the surface irregularities;

– qualitatively, the situation is rather confused. Besides a few well documented cases for which the *in situ* instrument observations reveal moderate amplifications (crest/base spectral amplification between 2 and 3), in good accordance with the numerical modeling, in many cases the observed amplifications are far more important than the theoretical previsions obtained from sometimes sophisticated models, which are either two- or three-dimensional. Thus, an important proportion of the instrumented sites showed spectral amplifications around 10, whereas this value was reached by only two simulations, and these resorted to other effects rather than geometry alone (e.g. mechanical contrast).

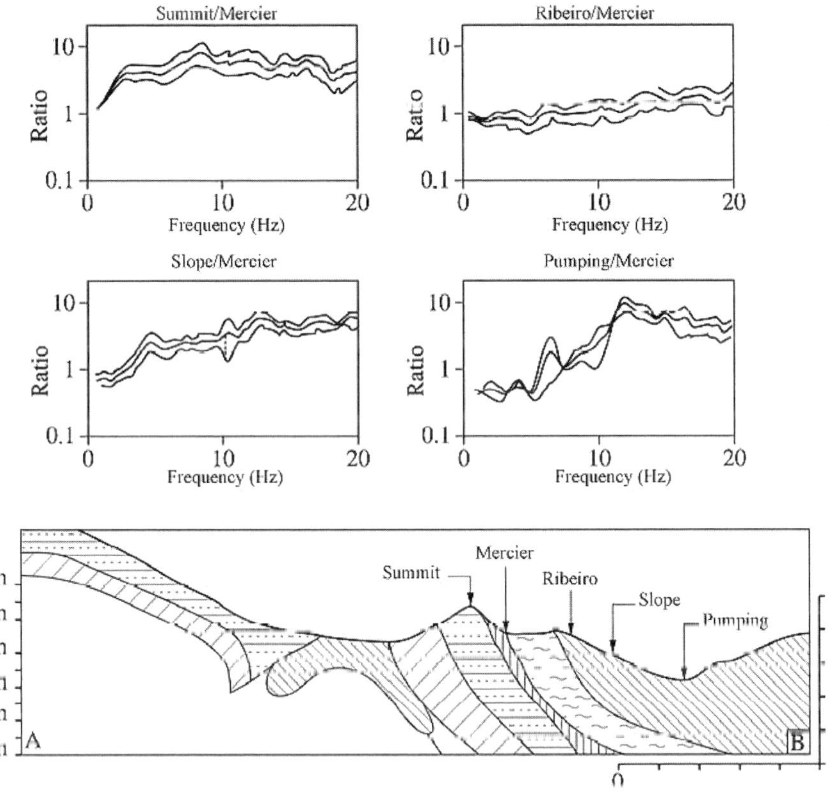

Figure 3.4. *Example of topographic amplification in the village of Castillon (maritime Alps), severely damaged during the 1887 Ligure earthquake. The instrument transfer functions (on top) show the amplification on the "summit" site compared with the "Mercier" site (from [NEC 95])*

To conclude this section, whilst theoretical convex relief focusing certainly plays a significant part, it does not seem to be the only physical phenomenon involved. The fact that only this phenomenon has been incorporated into the French national para-seismic regulations is quite justified. We should, however, be aware that more important effects that are not well delineated at present do take place, but can only be grasped by measurement.

3.4.3.2. *Effects of the sedimentary or alluvial cover*

3.4.3.2.1. Observations

Earthquake damage is generally more significant in sediment-filled areas than on rocky outcrops. The intensity local increments observed commonly reach 1 or 2 degrees in intensity (MM or MSK scale) and have sometimes exceeded three degrees (Mexico and San Francisco for instance), indicating in certain cases the complete control of damage distribution on the nature of the ground. These observations have given rise to large numbers of instrumental studies that have confirmed the existence of amplification phenomena, and to as many theoretical and numerical works, the results of which are summarized below. The curious reader is referred to the different synthesis articles in the bibliography.

3.4.3.2.2. Physical explanations for amplification

The basic phenomenon responsible for amplification is the trapping of seismic waves – particularly the S waves – into low mechanical (shear) stiffness superficial formations. For horizontally-stratified formations ("1D" structures), this only affects the incident volume waves that travel vertically back and forth between the surface and the substratum-sediment interface (Figure 3.5 left). When these are not very stiff, superficial formations present thickness side variations whether they are 2D (valleys) or 3D (basins). We can also discern generation of local surface waves (through diffraction on lateral heterogenities) that can reflect between the valley or basin sides.

The interference between trapped waves causes resonance phenomena, the features of which (frequencies, natural modes, amplification) depend on the geometric and mechanical characteristics of the structure. Diffraction and trapping phenomena also have other consequences apart from amplifying certain frequencies.

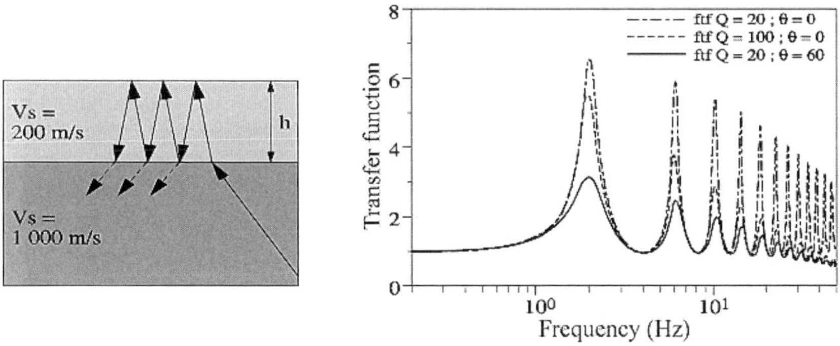

Figure 3.5. *Principle of site effects associated to the alluvial cover: the incident waves are trapped therein and their vertical reflections (figure on the left) cause a resonance phenomenon characterized by amplification peaks at certain frequencies (fundamental and harmonic). The curves on the right illustrate the variability of the amplification according to damping (full line: $\zeta = 2.5\%$, $\theta = 0°$; dotted line: $\zeta = 0.5\%$, $\theta = 0°$) and the angle of incidence (dashes: z = 2.5%, $\theta = 60°$)*

3.4.3.2.3. Spectral signature

Within the frequency field, trapping effects are expressed through the strong frequency dependency of surface amplification (or "transfer function" H(f), and the relationship of the surface Fourier spectrum As(f) with that of the depth incident motion A0(f). These transfer functions can be calculated with exact methods for a limited number of configurations (1D stratified media, "canonical" valleys or basins perfectly semicircular, elliptical or spherical geometrically speaking). Here we only give results for a simple model with only one sediment layer (medium 1) resting on a rocky substratum (medium 2) by means of a horizontal interface.

In this case, the complex transfer function for vertically incident S waves is given by the relationship:

$$H(f) = 2\,C/\left[C \cdot \cos\left(2\pi\, f\, h/\beta_1\right) + i \cdot \sin\left(2\pi\, f\, h/\beta_1\right)\right]$$

where h is the depth of the layer, f the frequency, C the mechanical impedance contrast $\rho_2\beta_2 / \rho_1\beta_1$, ρ_i the density of medium i and β_i the S wave speed of medium i.

This general formula gives access to resonance frequencies f_n and to the corresponding amplifications (Figure 3.5 right):

– resonance frequencies;

$$f_0 = \beta_1 / 4\,h \qquad \text{(fundamental mode)}$$

$$f_n = (2n+1)\,f_0 \qquad \text{(harmonics)}$$

where n is a positive integer;

– amplification: in the absence of any inelastic damping, the resonance amplification – apart from free surface effect – is equal to the mechanical impedance contrast:

$$A_0 = \left|H\left(f_0\right)/\,2\right| = C = \rho_2\ \beta_2 \,/\, \rho_1\ \beta_1 \qquad \text{(all modes)}$$

– in the presence of damping, the amplifications are reduced, in a more and more marked way for higher harmonics. As for the fundamental mode, the formula is still simple:

$$A_0 = C/\left(1 + 0.5\ \pi \xi_l C\right)$$

where ζ_l is the damping inside the surface layer.

Whilst the value of the fundamental frequency is rarely modified, the same is not true of amplitudes (except when we are interested in the vertical component force and when the S wave speed β_1 has to be replaced by the P wave speed α_1). Certain numerical calculations for deep valleys have shown over-amplification factors up to 4, with very deep modifications of transfer functions beyond the fundamental frequency f_0. In addition, lateral reflections generate other harmonics that can combine to create wide band amplification areas.

3.4.3.2.4. Time signature

In the time field, these effects alter maximum amplitudes, wave shapes and motion durations, especially for 2D or 3D structures.

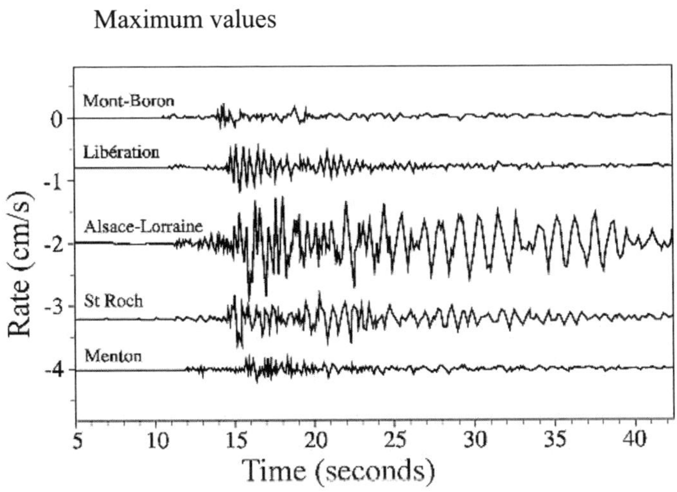

Figure 3.6. *Example of the appearance of site effects in the time field. RAP recordings of 25 February 2001 (M =4.7) earthquake in the city of Nice. The "Mont-Boron" station is on rock, whereas the four others are on the Paillon's alluvial deposits; the "Alsace-Lorraine" station is the thickest*

It is almost impossible to give general indications, as the effects depend too much on the position of the frequency band amplified in comparison with the incident spectral content. For example, a site that is not very thick (high f_0) may give rise to strong PGA amplification during a moderate earthquake nearby, but no particular effect during a wide amplitude earthquake far away – even if the absolute incident PGA levels are identical! Conversely, very thick sites may not affect, or even reduce, the PGA during moderate earthquakes nearby, yet amplify it considerably for those far away, while the transfer functions remain unchanged.

3.4.3.2.5. Duration

Because motion duration is rarely taken into account in paraseismic design, few attempt at quantifying the consequences of resonance and trappings on duration have been published. From a qualititative point of view, however, valley and basin configurations seem to prolong motions in a significant way ([BEA 03], [COR 03], [PAR 03]). The study of prolongation may even constitute a good method for detecting the presence of 2D or 3D effects (Figure 3.6).

3.4.3.2.6. Modifications due to non-linearity

As with most other materials, grounds deteriorate when subjected to high mechanical stresses: as shown by [SEE 69]'s pioneering works, the secant stiffness modulus decreases and the damping increases when the imposed strain increases: $G = G_{max}\ 1./(1+\gamma/\gamma_r)$. Such behavior might be expected to lead to a decrease of both frequencies and resonance associated amplifications, as well as reductions in the high-frequency content (maximum acceleration, especially).

The question is knowing where the strains imposed by an earthquake are located (they can quite often exceed 10^{-3}, and sometimes reach 10^{-2}) with regard to the critical strain values (γ_r). Seismologists and geotechnicians disagree on this point, the former believing that accelerometer observations were best explained by reference to linear visco-elastic behaviors, whilst the latter measured (γ_r) values under 10^{-3} under laboratory conditions. In the last two decades, observations have tended to reconcile both points of view, owing to the simulation of far less non-linear behaviors under lab conditions, notably for very plastic grounds [VUC 91], and the observation of non-linear *in situ* effects, especially in sand layers. Some quantitative disagreements still exist, since even the best accelerometer data seems to reveal a slightly to appreciably less non-linear behavior than that predicted by numerical models derived from laboratory measurements [BON 03].

It is important to grasp the reason for inconsistencies between *in situ* and laboratory observations, particularly for seismic zones such as France. Currently, French regulation (PS92) authorizes a 20% decrease of the high frequency content in soft ground, implicitly assuming highly non-linear grounds, and has proved to be rather conservative. The EC8 recommendations allow high frequency amplifications of up to 80%, whilst the latest American propositions reach 200%.

3.5. Conclusions

Strong motion seismology is a relatively new subject which has evolved a great deal over the last decades, advances often being due to the questions raised by "abnormal" damage and intensity observations made during destructive earthquakes, generally via accelerometer recordings. Whilst estimating seismic motions was mainly empirical in the last century, many models that were satisfactory from a physical point of view emerged in the following decades, and have been progressively adjusted and calibrated to real time instrument measurements.

The percolation time between "normal observations" and their use in seismic engineering practice remains quite long (about a decade and even more for conventional regulations) for the following reasons: if we simplify in the extreme, the seismic force F which a building will have to withstand is proportional to three terms: $F \propto a_z.\ S'_a(T,\zeta)/Qs$, where:

– a_z is the rated acceleration mainly depending on the seismic area (regional zoning);

– $S'_a(T,\zeta)$ is a spectral shape standardized with regard to the rock, taking the magnitude of the most frequent earthquakes into account, but which mostly depends on the surface geological conditions;

– *Qs* (behavior coefficient) is a reduction factor which takes into account the resistance reserves of the structure, linked to the more or less ductile behavior of its constitutive materials and their assembling mode: this coefficient can vary from 1 to 1.5 for very brittle structures (non-reinforced masonry), and up to 8 for very ductile structures (steel).

Logically, evaluation of all these terms should be performed in separate and independent steps. This is rarely the case, as so many uncertainties exist when evaluating the unknown factors. This is coupled to the inertia of building practices in different countries, some of whom do not think it is possible to clearly distinguish unknown parameters from design parameters. Final regulations for a given country are usually a compromise reached when it is difficult to separate the safety coefficients inherent to each step, and there is a reluctance to modify any unknown factor values (a_z for example), without modifying the whole regulations. Nevertheless, such weaknesses will be corrected eventually, which will enhance the weak links, and assist in fields where uncertainties are the highest and greater efforts have to be made.

As far as unknown factors and seismic motions are concerned, despite improvements made in the last decades, many uncertain areas (motions very close to the source, importance of non-linearities, possible maximum motions) remain. Experience shows that only instrument observations can lead to further progress in practice. It is therefore absolutely essential to support the installation of more instrumentation on open field sites, as well as inside structures, particularly since the cost of an accelerometer (between €3,000 and €10,000) is derisory if compared with the cost of rebuilding a bridge, a dam, a nuclear power station or a 10-floor building; moreover, because of recent advances in telecommunications techniques, accelerometer maintenance costs are now comparatively modest.

3.6. Bibliography

[AKI 80] AKI K., RICHARDS P.G., *Quantitative Seismology. Theory and Methods*, Freeman and Co, San Francisco, USA, 2 Volumes, 932 pages (re-edited into a single volume in 2002), 1980.

[BAR 99] BARD P.-Y., RIEPL-THOMAS J., "Wave propagation in complex geological structures and local effects on strong ground motion", in *Wave Motion in Earthquake Engineering* (Chapter 2), E. Kausel and G.D. Manolis (Eds.), WIT Press ("Advances in Earthquake Engineering" Series), ISBN 1-85312-744-2, pp. 38–95, 1999.

[BEA 03] BEAUVAL C., BARD P.-Y., MOCZO P., KRISTEK J., "Quantification of frequency-dependent lengthening of seismic ground motion duration due to local geology: applications to the Volvi area (Greece)", *Bull. Seism. Soc. Am.,* 93-1, pp. 371–385, 2003.

[BER 03] BERNARD P., *Qu'est-ce qui fait trembler la terre ? A l'origine des catastrophes sismiques,* EDP Sciences, collection "Bulles de Sciences", 287 pages, 2003.

[BET 03] BETBEDER-MATIBET J., *Génie parasismique*, 3 volumes, Hermes, Paris, 2003.

[BON 03] BONILLA L.F., COTTON F., ARCHULETA R.J., "Quelques renseignements sur les effets de site non-linéaires en utilisant des données de forage: la base de mouvements forts Kik-net au Japon", *VIe Colloque National de l'AFPS, Ecole Polytechnique*, 1–3 July 2003, Vol. I, pp. 181–192, 2003.

[COR 03] CORNOU C., BARD P.-Y., DIETRICH M., "Contribution of dense array analysis to basin-edge-induced waves identification and quantification. Application to Grenoble basin, French Alps (II)", *Bull. Seism. Soc. Am.*, 93-6, pp. 2624–2648. 2003.

[JEN 83] JENNINGS P.C., "Engineering seismology", in *Earthquakes: Observation, Theory and Interpretation,* H. Kanamori and E. Boschi (ed.), North-Holland, pp. 138–173, 1983.

[KRA 96] KRAMER L.K., *Geotechnical Earthquake Engineering*, Prentice Hall International Series in Civil Engineering Mechanics, Upper Saddle River, 1996.

[LAM 96] LAMBERT J., LEVRET-ALBARET A. (Ed.), *Mille ans de séismes en France, Catalogue d'épicentres paramètres et références*, Ouest Editions, Presses académiques, 80 pages, 1996.

[LAM 97] LAMBERT J. (Ed.). *Les tremblements de terre en France*, Editions BRGM, 196 pages, 1997.

[MAD 91] MADARIAGA R., PERRIER G., *Les tremblements de terre,* Presses du CNRS, 211 pages, 1991.

[NEC 95] NECHTSCHEIN S., BARD P.-Y., GARIEL J.-C., MÈNEROUD J.-P., DERVIN P., CUSHING M., GAUBERT C., VIDAL S., DUVAL A.-M., "A topographic effect study in the Nice region", *Proceedings of the Fifth International Conference on Seismic Zonation*, Nice, France, 17–19 October, II, pp. 1067–1074, 1995.

[PAR 03] PAROLAI S., BARD P.-Y., "Evaluation of amplification and frequency-dependent lengthening of duration of seismic ground motion due to local effects by means of joint analysis of sonogram and standard spectral ratio", *Journal of Seismology*, 7, pp. 479–492, 2003.

[PEC 85] PECKER A., *La dynamique des sols appliquée aux séismes*, Presses des Ponts et Chaussées, 1985.

[SEE 69] SEED H.B., IDRISS I.M., "Influence of soil conditions on ground motions during earthquakes", J. *Soil Mech. and Found. Div.*, ASCE, 95 (SM1), 1969.

[VUC 91] VUCETIC M., DOBRY R., "Effect of soil plasticity on cyclic response", *ASC Journal of Geotechnical Engineering*, 117(1), pp. 89–109, 1991.

List of accelerometry websites

European accelerometry data

European accelerometer data: http://www.isesd.cv.ic.ac.uk/ESD/frameset.htm

France: http://www-rap.obs.ujf-grenoble.fr

Greece: http://www.itsak.gr

Italy: http://www.serviziosismico.it, http://www.dstn.pcm.it

Switzerland: http://seismo.ethz.ch

Turkey: http://angora.deprem.gov.tr

Japanese accelerometry data

K-NET network (1,000 stations with geotechnical information): http://www.k-net.bosai.go.jp/

KiK-NET network (500 coupled surface-depth stations): http://www.kik.bosai.go.jp/

Californian accelerometry data

http://quake.crustal.ucsb.edu/scec/smdb

http://www.consrv.ca.gov/cgs/smip

World accelerometry data

http://agram.wr.usgs.gov/docs

http://mceer.buffalo.edu

http://perun.wdcb.rssi.ru/SMDB/

http://www.ednes.org/CGDS/

http://www-socal.wr.usgs.gov/smdata.html

National Geophysical Data Center (NGDC):
htpp://www.ngdc.noaa.gov/seg/hazard/strong.html

Chapter 4

Soil Behavior: Dynamic Soil-Structure Interactions

Introduction

As suggested by the name, soil-structure interaction aims at assessing the response of a structure resting on the ground and subjected to any stimulation, whilst taking into account coupling with the support medium and the soil, which has its own deformability and even resistance characteristics. This interaction is expressed through modifications of the motion of the soil near the structure as compared with the open field – without any structure configuration; in the same way, the motion of the structure is different from that which would result from the hypothesis of an infinitely stiff soil: a structure "embedded" at its base corresponding to the hypothesis taken into account in paraseismic building codes. This interaction is of variable importance, depending on the nature of the soil, the characteristics of the structure and its type of foundations. For light structures with superficial foundations, it can be almost negligible. However, whenever it is necessary to study the seismic response of a structure, and to consider it as an integral part of a whole system consisting of the ground, foundations and structure, soil-structure interaction analysis becomes increasingly important when building structures like dams, nuclear facilities, liquefied natural gas tanks and very high buildings.

Chapter written by Alain PECKER.

Many factors need to be considered to completely solve a soil-structure interaction problem. They include: the definition of the seismic unknown factor and the resulting motion, soil behavior under cyclic stress, an estimate of soil response in the open field and the response of structures under dynamic loading. In this chapter, only those aspects dealing with the behavior of soils under cyclic loading will be tackled, together with soil-structure interaction modeling.

The extensive literature dealing with the soil-structure interaction reflects both the complexity of the phenomenon as well as the interest in the subject shown by the scientific community. Two general syntheses have proposed a classification of the methods used to take soil-structure into account ([LYS 78] and [IDR 80]). Both publications stressed at the time that the study of interaction phenomena was essentially limited to cases involving linear problems, approaches for non-linear problems still being rare. However, in the last few years, important advances have been made as far as our understanding of non-linear phenomena is concerned, which makes broaching these problems possible.

4.1. Behavior of soils under seismic loading

4.1.1. *Influence of the nature of soils on seismic movements*

Observations made on sites during real earthquakes clearly show the influence of the nature of surface layers on the seismic motion recorded. This fact has been acknowledged for about 20 years and has led to response spectra being interpreted according to the nature of the soil.

The recent major earthquakes that have affected the world (Mexico, 1985; Loma Prieta, 1989; Northridge, 1994; Kobe, 1995) confirmed the following facts: alluvial soils tend to amplify incident motion, especially at low frequency. As an illustration, recordings of the Loma Prieta earthquake obtained on rocky sites around San Francisco showed a maximum acceleration of about 0.10 g; recordings of the same earthquake made on alluvial sites showed maximum accelerations two to three times as high (Table 4.1 [COL 90]), with spectra presenting important low frequency peaks. As the epicenter was far from the recording sites (about 85 km), this could not be a local effect due to the source (directivity). By the same token, as the recording sites were near to one another (within a radius of a few kilometers), it could not be an effect due to propagation between the source and the site inside the Earth's crust. The only parameter that could have affected the nature of the recorded motion was the geological nature of the sites, i.e. the mechanical characteristics of the soil near the surface.

However, we should not deduce from the previous observations (or other similar ones made in Mexico) that alluvial soils systematically amplify seismic accelerations. If we again consider the San Francisco sites, it seems that at the time of the 1957 earthquake, which also originated on San Andreas's fault but nearer the city (between 15 and 20 km), the accelerations registered on rocky sites were also about 0.10 g, as the earthquake had a lower magnitude (5.3 instead of 7.1). Yet at the surface of the alluvial sites, the recorded accelerations for the earthquake were 1.5 to 2 lower than those on the rock (from 0.05 g to 0.07 g).

To be able to estimate such differences, it is necessary to have a thorough knowledge of soil behavior under cyclic loading so that it can be integrated into elaborate calculation models. At the present time, even if many aspects remain to be clarified, our knowledge of soil behavior has progressed and the calculations models have developed sufficiently to allow an evaluation of the phenomena that will satisfy engineers.

Station	Stratigraphy	Ground maximum acceleration	
		1957	1989
Golden Gate Park	Rock	0.13	
Market/Guerrero St	Rock	0.12	
State Building	Sand + clayed sand (60 m)	0.10	
Mason/Pine St	Rock	0.10	
Alexander Building	Clayed silt + sand (45 m)	0.07	0.17
Southern Pacific B.	Soft clay	0.05	0.20
Rincon Hill	Rock	0.10	0.09
Oakland City Hall	Clay, sand (30 m) + stiff clay (270 m)	0.04	0.26

Table 4.1. *Maximum acceleration in San Francisco (from [COL 90])*

On the basis of these observations, the various paraseismic building codes acknowledge the necessity of taking into account the geological nature of the soil in the definition of seismic stresses. This is expressed in the way response spectra are expressed differently according to the nature of the ground, which is characterized by the average propagation speed of shear waves in the 30 uppermost meters of the ground layer [COL 02].

4.1.2. *Experimental description of soil behavior*

Both the dynamic response calculations of a soil profile and soil-structure interaction problems usually consider seismic motion are caused by a shear wave propagating vertically from the substratum. Under such conditions, a soil element taken from the soil layer is subjected to the stress cycles represented in Figure 4.1.

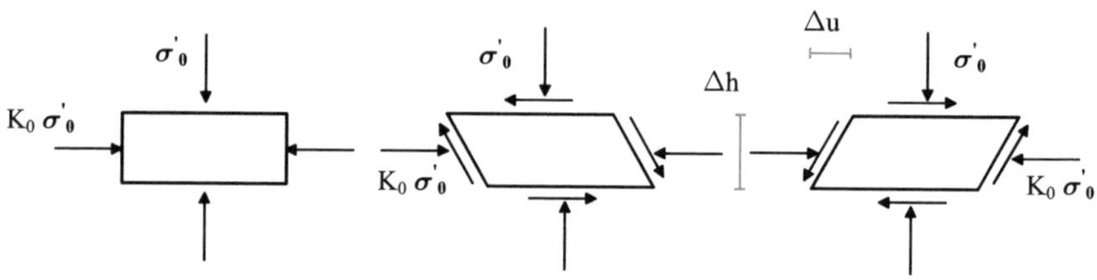

Figure 4.1. *Idealized loading sequence*

Initially, in the case of a horizontal soil profile, the element is balanced under the real vertical stress σ'_v and the real horizontal stress $K_0\ \sigma'_v$, where K_0 is the thrust coefficient of the ground at rest. The passing shear wave is discerned by the application of a shear stress $\tau(t)$ on the horizontal sides, and then on the vertical sides of the soil element (to preserve equilibrium conditions). Under the effect of the stress, the sample is subjected to a simple shear strain, which, for an elastic behavior material, involves a volume variation equal to zero. The shear strain, also called distortion, is described in Figure 4.2.

$$\gamma = \frac{\Delta u}{\Delta h} \qquad [4.1]$$

If in a laboratory we reproduce symmetric and constant amplitude strain cycles similar to those in Figure 4.1, we obtain the curves shown in Figure 4.2. The latter shows that in the plane (τ, γ), soil behavior is characterized by a hysteresis loop, the area and inclination of which is determined by the amplitude of the strain. The wider the latter, the more important loop area is, and the less horizontally inclined it becomes. In addition, it shows that experimentally, the shape of the hysteresis loop is unaffected by the loading rate.

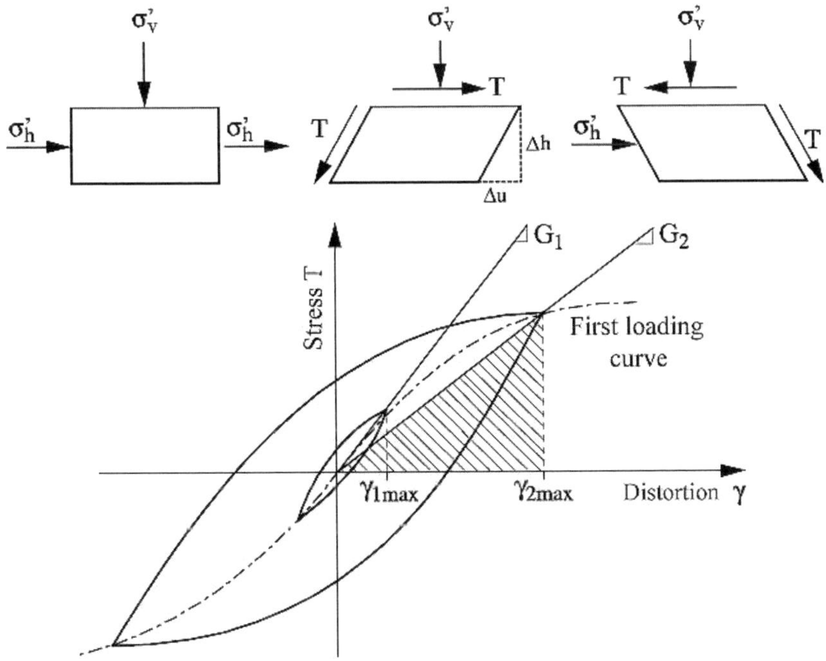

Figure 4.2. *Cyclic strain-stress curve*

When cyclic loading is not closed, the behavior becomes more complicated to describe. An example of this is given in Figure 4.3. Up to b, the path followed is identical to the one in Figure 4.2 (first loading curve then discharge curve). At point b, where $\sigma_b < \sigma_a$, the loading sign gets inverted again; the path followed is given by curve bc, then possibly by $c\varepsilon_c$ if the loading changes its sign again in c. Yet if the loading is continued beyond c, the path is represented by ca, then beyond c it follows the first loading curve anew.

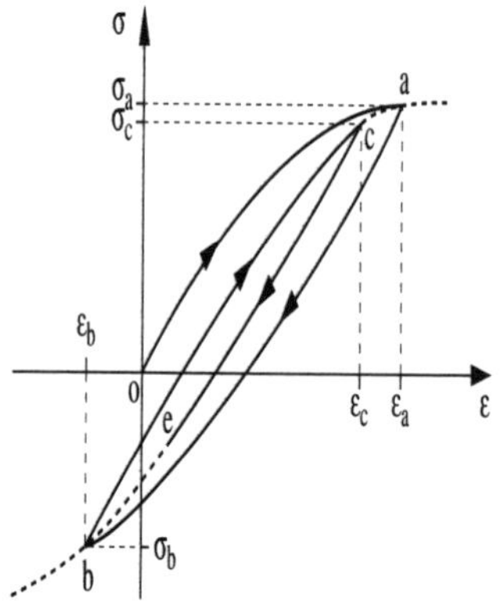

Figure 4.3. *Any cyclic loading*

Associated with these shear strains, volumic strains also appear (Figure 4.4). These irreversible volumic strains express the fact that soil behavior is non-linear. These irreversible volumic strains cause hardening in a dry material. Thus, even for closed cycles that are symmetric and have identical strain amplitudes, the hysteresis loop obtained during the fourth loading cycle is different from that obtained during the first cycle. The latter loop is less inclined on the horizontal and has a smaller area. For a saturated, low-permeability soil, the strains occur at constant volume, because the interstitial water does not have time to drain off the skeleton. However, the tendency for volume variation exists, and involves an increase of interstitial pressure, and therefore a decrease in effective strain.

The few simplified examples above show the complexity of the soil behavior, which is highly non-linear and inelastic.

The modeling adopted in practice and described below only refers to the deviatoric behavior of the soil (Figure 4.2). The volume variation aspects are often neglected, except in cases where real elasto-plastic behavior laws are used.

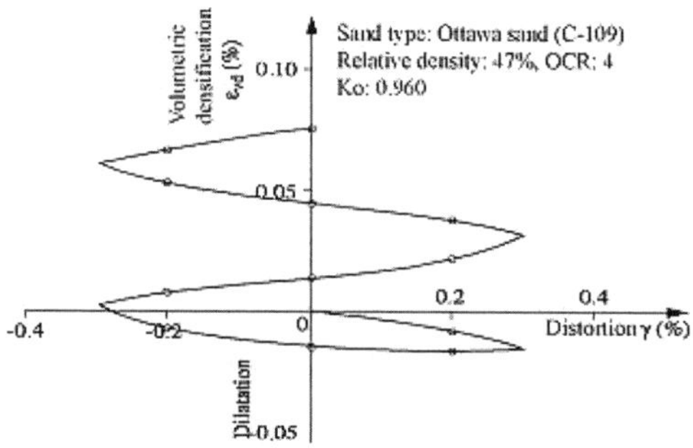

Figure 4.4. *Volume variations under cyclic loading*

4.2. Modeling soil behavior

An exhaustive description of soil behavior can be obtained from a state of balance, characterized by a stress field $\underset{\sim}{\sigma}$ and a strain state (characterized by tensor $\underset{\sim}{\varepsilon}$), if it is possible to determine the stress field (or the new strain state) corresponding to the new state of balance after applying strain increment d $\underset{\sim}{\varepsilon}$ (or respectively of a stress increment d $\underset{\sim}{\sigma}$). In most cases, time also plays a part in the expression of behavior law, but as a rule, for most soils, this parameter can be neglected. Establishing the behavior law is the ultimate goal of the description of the soil behavior. Nevertheless, because of the complexity of the experimental behavior of the soil, establishing a true behavior law is both delicate and costly. At present, no universal behavior law exists: each law available in literature has advantages as well as drawbacks and limitations.

When confronted with these difficulties, geotechnicians often prefer to use procedures commonly employed in soil mechanics, which involve anticipating the stress mode to which an element isolated inside the soil layer will be subjected and reproducing the corresponding loading path, either inside a sample or *in situ*. The parameters measured during the test can then be used directly for the calculations. For example, a similar approach is used in soil mechanics to study the subsidence of a compressible clay layer due to the weight of a backfill, the extension of which is important relative to the layer thickness: the test used would be an oedometric test with no radial strain.

We have to realize that such an approach is not the same as determining a behavior law, even if the stress-strain curves obtained are represented by mathematical equations. This modeling of soil behavior is only valid for loading paths close to those used to establish the model; extrapolation to different paths would be erroneous and invalid. Furthermore, quite often, the modeling obtained in this way only partly represents the physical phenomena; thus, the equivalent linear visco-elastic modeling (which will be described later) does not describe or incorporate volume variations (subsidence) under purely deviatoric loadings. Moreover, the stress paths represented in tests are only rudimentary idealizations of real stresses. This type of approach represents a compromise between the phenomenon to be modeled and ease of implementation. If it is used with good judgment, however, it is quite a powerful tool.

Before attempting an experimental description of the phenomena to be modeled by mathematical representation, it is important to realize that on the time scales involved in seismic stresses, most soils exhibit non-drained behavior during the stress. Soil permeability is not sufficient (compared to the loading rate) to permit drainage. As a consequence, in the approach described above, we reason in terms of total stresses. Once again, this is a simplification, as real soil behavior is controlled by the actual stresses.

The rest of this text will be limited to examining soil behavior before failure. The study of soil behavior on failure gives rise to different approaches made necessary by the adopted schematization. If we had a real behavior law at our disposal, such a distinction would not be necessary; in fact, the behavior law would allow us to reproduce soil behavior from the smallest strains (quasi-elastic strains) to the very high strains associated with failure.

For further descriptions of soil behavior, see [HAR 78], [PEC 8] and [PRE 78].

What emerges from the experimental statements in section 4.1.2 is that the soil cannot be represented by an elastic model, at least as soon as the strains become significant.

The non-linearity appearance thresholds generally correspond to low strains (in the range 10^{-4} to 10^{-6}). However, we must distinguish between reversible (or quasi-reversible) and irreversible non-linearities, as the appearance thresholds of the latter are higher (10^{-4} to 10^{-3}).

		Behavior linearity	Elasticity and plasticity	Cyclic degradation for saturated soils	Analysis method
Very low	$0 \leq \gamma \leq \gamma_s$	Practically linear	Practically elastic	Non-degradable	Linear
Low	$\gamma_s \leq \gamma \leq \gamma_v$	Non-linear	Weakly Elasto-plastic	Practically non-degradable	Linear equivalent
Average to large	$\gamma_v \leq \gamma$	Non-linear	Elasto-plastic	Degradable	Non-linear

Table 4.2. *Behavior field for cyclic stresses*

The appearance thresholds for these non-linearities, respectively named γ_s and γ_v, depend on the nature of the material, which is roughly characterized by its plasticity index IP.

Table 4.2 and Figure 4.5 allow us to characterize each of the behavior fields as well as the analysis methods used to express their behavior in digital studies.

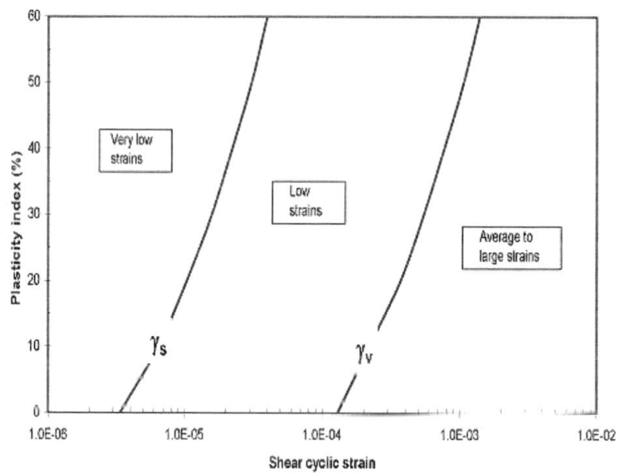

Figure 4.5. *Shear cyclic strain*

The stresses generated by major earthquakes expected in the French context could induce strains causing significant linearity losses, or even irreversible strains in the highest cases (where $\gamma \geq \gamma_s$ or γ_v).

As indicated in the previous table, each domain corresponds to a behavior type, the characterization of which involves measuring specific parameters.

4.2.1. *An experimental description of elastic soil behavior ($\gamma \leq \gamma_s$)*

For strains lower than 10^{-6} to 10^{-5}, soil behavior is typically linear elastic. Some saturated materials can present a slight damping of viscous origin (a few percent), in which case soil behavior can be characterized by an elastic or possibly visco-elastic, linear type of behavior law. In the case of an isotropic material, the modulus of shear elasticity G (similar to Lame's modulus μ in continuous medium mechanics) and a volumic bulk modulus B allow the complete characterization of the behavior. Alternately, it is possible to use the propagation rate of the elastic waves V_s (shear waves) and V_p (compression waves) linked to the previous values by:

$$G = \rho\, V_s^2 \quad B = \rho\left(V_p^2 - \frac{4}{3}V_s^2\right) \qquad [4.2]$$

where ρ is the density of the material.

Measuring these values can be either done *in situ* (V_s, V_p) or under laboratory conditions on intact samples (G, B).

The measurements correspond to very specific techniques. Among the *in situ* techniques used, we include the logging suspension test, spectral analysis of SASW surface waves, or any other drilling measure technique (crosshole or downhole test) as examples; in the laboratory, the most appropriate test is the resonant column test.

The study of soil behavior within its elastic field is important, and a whole range of problems exist for which the models are valid: cases of vibrations of well conditioned machine blocks, low amplitude seismic stresses like those caused by geophysical tests are examples.

4.2.2. *Linear visco-elastic models for medium strain domains where $\gamma_s \leq \gamma \leq \gamma_v$*

In this strain field, more or less marked non-linearities appear in the stress-strain curve.

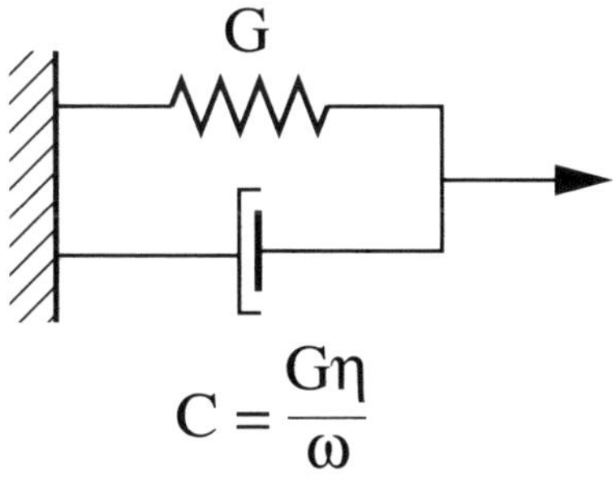

Figure 4.6. *Linear visco-elastic model*

As the visco-elastic models reveal hysteresis loops under harmonic stresses, it is tempting to represent soil behavior using such models (Figure 4.6) for those stresses. However, the linear visco-elastic model leads to an energy dissipation cycle depending on the stress frequency, which contradicts the experimental observations (section 4.1).

We must establish equivalence between the actual material and the model: the equivalence is based on the equal dissipative qualities of both the material and the model.

For one-dimensional stresses, the stress-strain relationship in the Kelvin-Voigt model is written as follows:

$$\tau = G\,\gamma + C\,\dot{\gamma} \qquad [4.3]$$

where G and C are the spring and dashpot constants, and γ and $\dot{\gamma}$ are respectively the strain and strain rate.

Under harmonic stress:

$$\gamma = \gamma_m\ e^{i\omega t} \qquad [4.4]$$

the previous relationship is written:

$$\tau_m = G\left[1 + i\,\frac{C\omega}{G}\right]\gamma_m = G^*\,\gamma_m \qquad [4.5]$$

where G is a complex modulus.

The dissipated energy during a loading cycle is equal to:

$$\Delta W = \pi \, C \, \omega \, \gamma_m^2 = \pi \, \text{Im}\,(G^*)\, \gamma_m^2 \tag{4.6}$$

where Im (G^*) refers to the imaginary part of G^*.

As C is constant, we note that the dissipated energy depends on the stress frequency $f = \omega/2\pi$.

If we refer to Figure 4.7, without going into too much detail, it is possible to define a damping term for a material that is frequency-independent, a standard expression of the energy ΔW dissipated during a cycle. The standardization is made owing to the elastic energy W accumulated during a cycle:

$$W = \frac{1}{2} G \gamma^2 \tag{4.7}$$

Therefore, we can define either a damping ratio β or a loss coefficient η, independent of the stress frequency:

$$\eta = 2\beta = \frac{1}{2\pi} \frac{\Delta W}{W} \tag{4.8}$$

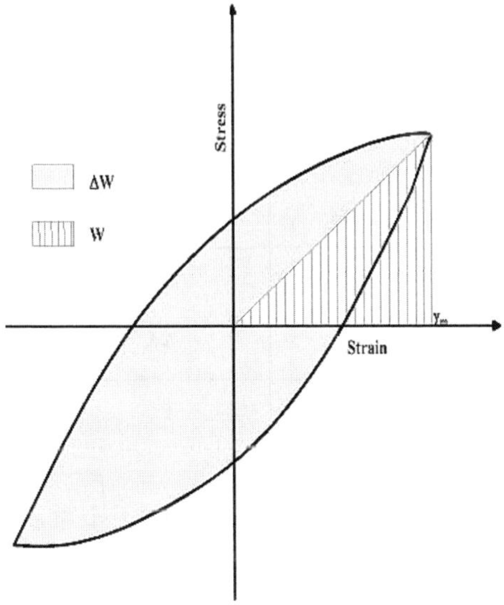

Figure 4.7. *Definition of standard damping*

The energy dissipated during a cycle then takes the form:

$$\Delta W = \pi \, G \, \eta \, \gamma_m^2 \quad [4.9]$$

The purpose of the equivalent linear visco-elastic models is to establish a relationship of the form:

$$\tau_m = G^* \gamma_m \quad [4.10]$$

where G^* is a complex modulus that has to be chosen so as to involve the same stiffness and energy dissipation properties as the real material. Several models have been proposed for this purpose. Their characteristics are summarized in Table 4.3.

The first two models are due to Berkeley's team [SEE 70]; the third was developed by [DOR 90]. We note from Table 4.3 that the first model respects the dissipated energy but overestimates stiffness; the second model respects stiffness but underestimates dissipated energy. Only the third model satisfies both parameters.

	Complex modulus $G^* = \tau/\gamma$	Energy dissipated during a AW cycle	$\lvert G^* \rvert$ modulus
Material		$\pi G \eta \gamma_m^2$	G
Model 1	$G[1 + i\eta]$	$\pi G \eta \gamma_m^2$	$G\sqrt{1+\eta^2}$
Model 2	$Ge^{i\theta}$ $\eta = 2 \sin \frac{\theta}{2}$	$\pi G \eta \gamma_m^2 \sqrt{1 - \frac{\eta^2}{4}}$	G
Model 3	$G\left[\sqrt{1-\eta^2} + i\eta\right]$	$\pi G \eta \gamma_m^2$	G

Table 4.3. *Characterization of equivalent linear visco-elastic models*

An alternate representation of the data in Figure 4.2 is obtained by plotting both the norm of the secant modulus of the hysteresis loop (G/G_{max}) and the equivalent critical damping ratio β, as a function of strain (Figure 4.8). This figure clearly shows the notion of strain threshold (γ_s), for which neither strain nor its dependence on a plastic index are constant.

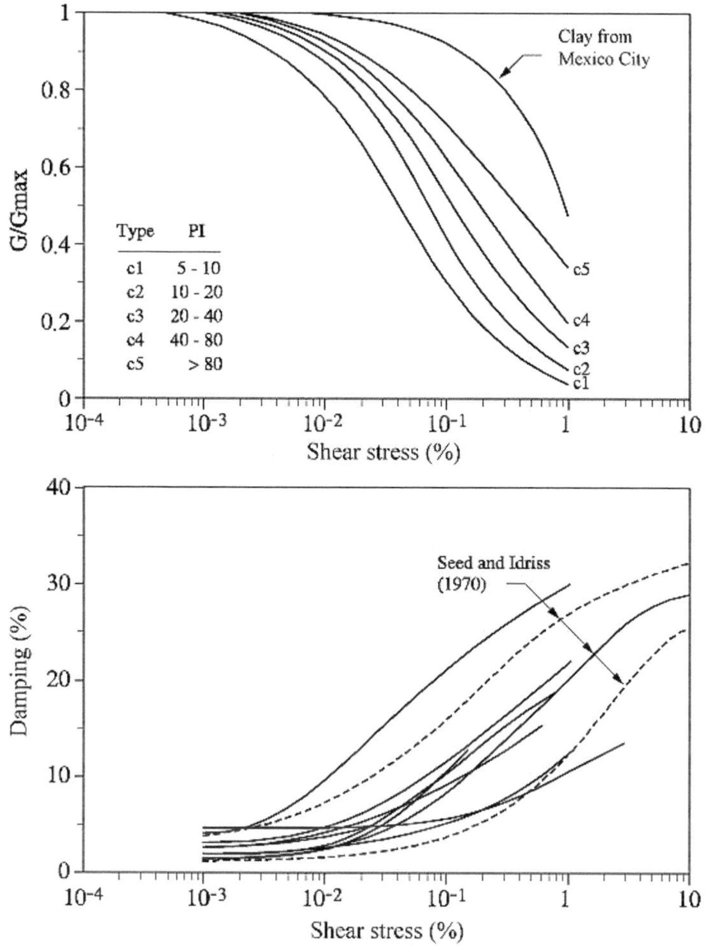

Figure 4.8. *Characteristic results for clay*

Correlatively, Figure 4.9 shows that for cyclic shear strains γ lower than $\gamma_v \approx 10^{-4}$, volume strain can be neglected. For saturated soils the result is an insignificant variation of the interstitial pressure, and therefore deterioration of properties is absent.

To conclude, within this strain range, soil behavior is definitely non-linear, though it stays markedly elastic because the permanent changes in its microstructure are negligible.

In the linear visco-elastic modeling described below, the properties of the material (secant modulus, damping) are adjusted to a “medium” strain level to approximately account for the non-linearity of the behavior.

The results developed above in the case of unidirectional stresses immediately become generalized in the case of harmonic stress with all the limitations for the experimental validity of such an extension that we have mentioned previously.

The behavior law is written in a form similar to a generalization of Hooke’s law:

$$\underset{\sim}{\varepsilon} = \underset{\sim}{\Lambda}^{*} : \underset{\sim}{\sigma} \qquad [4.11]$$

where tensor $\underset{\sim}{\Lambda}^{*}$ is formed thanks to the complex volume and shear moduli. The previous behavior law leads to the same shaped solutions as the linear elastic law, hence its unquestionable appeal.

In the case of an isotropic material, the behavior is entirely described using two moduli, G and B, and theoretically, using two damping ratios associated with the shear and volume strains, as well as their dependence on the amplitude of the shear strain. In practice, both critical damping ratios are chosen to be identical.

The variation of these characteristics with regard to strain takes the shape given in Figure 4.8. It is the conventional form for the results of the tests:

$$G = G(\gamma) \ , \ \eta = \eta(\gamma) \qquad [4.12]$$

As practice stands at the moment, measuring these values can only be performed under laboratory conditions with intact samples. The trial best suited to the range of strains required is the resonant column test.

Such models are extensively employed in common practice. When they are used together with an iterative process allowing values of modulus G and the loss coefficient η to be selected as those compatible with the average level of resultant strain, they yield strain and acceleration values that compare favorably with those obtained using more sophisticated models or observed during real earthquakes. Such models have the virtue of simplicity: they only require measurement of three parameters (one more than the elastic model): shear modulus, volume modulus and loss coefficient. As the soil has a non-linear behavior, these parameters depend on the state of stresses and strains (Figure 4.8). The main limitation of these models is the fact that they are unable to give irreversible strain values: those calculated by the model are necessarily equal to zero. Whenever a more accurate representation of the soil is necessary, use of non-linear models is inevitable.

4.2.3. *High strain domain non-linear models where $\gamma \geq \gamma_v$*

In this strain domain, important changes in micro-structure occur (rearrangement of grains), which generates significant volume and shear irreversible strains (Figures 4.3 and 4.9). These microstructure changes involve volume variations inside unsaturated materials, and an increase of interstitial pressure within saturated materials. This interstitial pressure increase can lead to canceling the actual stresses and therefore the soil strength: it is a liquefaction phenomenon. Apart from this loss of strength, the increase in interstitial pressure brings about a decrease in the initial stiffness, which is a function of the state of the real stresses withstood by the soil (Figure 4.10). Both factors combined – the stiffness loss and the loss of strength – make saturated soils subjected to amplitude strains in which γ is higher than γ_v highly degradable.

As this type of behavior generates significant irreversible strains, it can only be studied using non-linear models. Accumulated experience shows that strain hardening elasto-plastic models are the most appropriate for describing this behavior ([PRE 78], [PRE 87]); the models have to be written using real stresses as these govern soil behavior. In saturated soils, coupling of the fluid phase with the solid phase can be described within the frame of porous medium mechanics due to Biot's relationship.

Measuring the values that are used to formulate the behavior laws must be carried out in laboratory conditions, on intact samples, under strictly controlled environment conditions and over a range of stress paths that are all the more numerous because the behavior law includes a greater number of parameters. In any case, these tests are difficult to implement, because taking geo-material samples is always delicate and can induce sample modifications, especially deep sampling. The nature of the soil to be sampled (clay, marl) plays a part in the extent of the modification. The properties most susceptible to modifications are soil stiffness and the capacity of saturated sands to withstand liquefaction. At present, the most versatile tool for obtaining this data is the triaxial device, which makes it possible to reach important strain amplitudes with different stress paths.

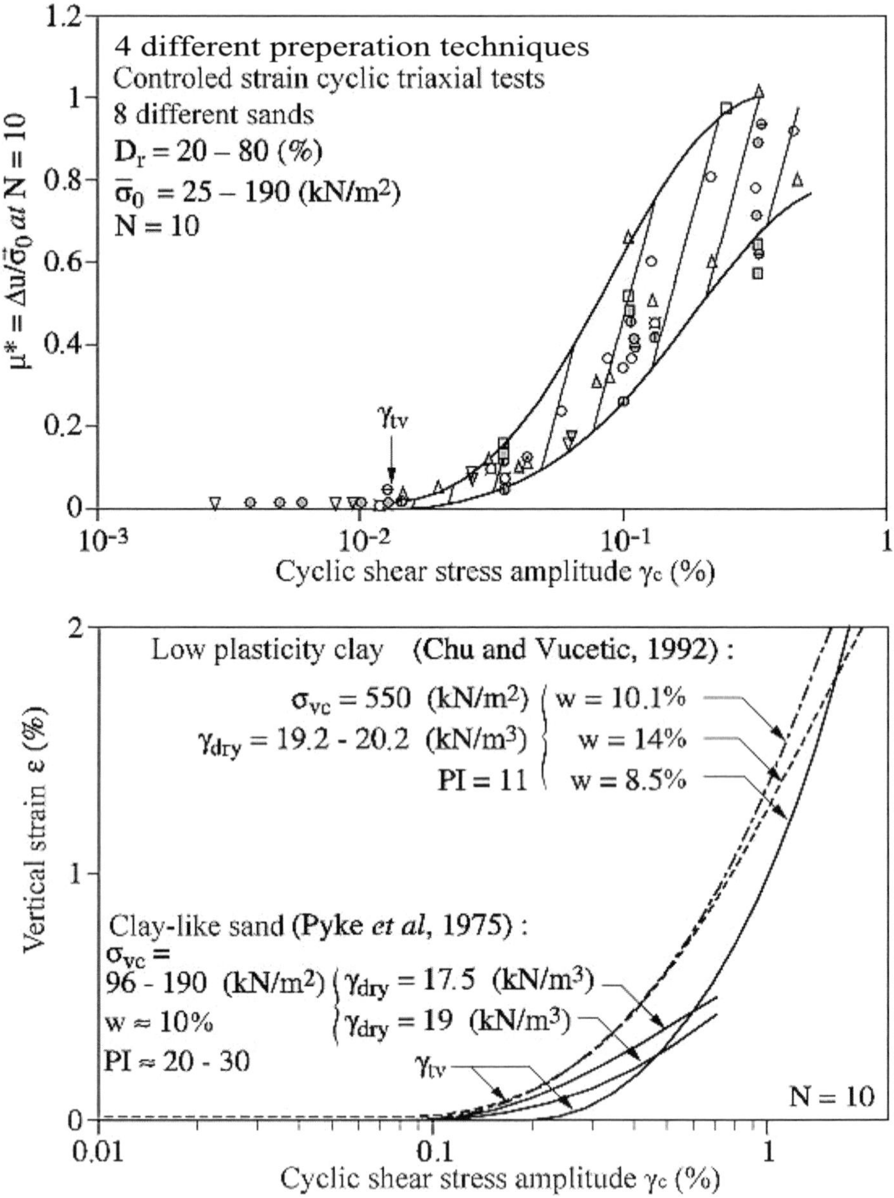

Figure 4.9. *Irreversible interstitial pressure and volumic strain*

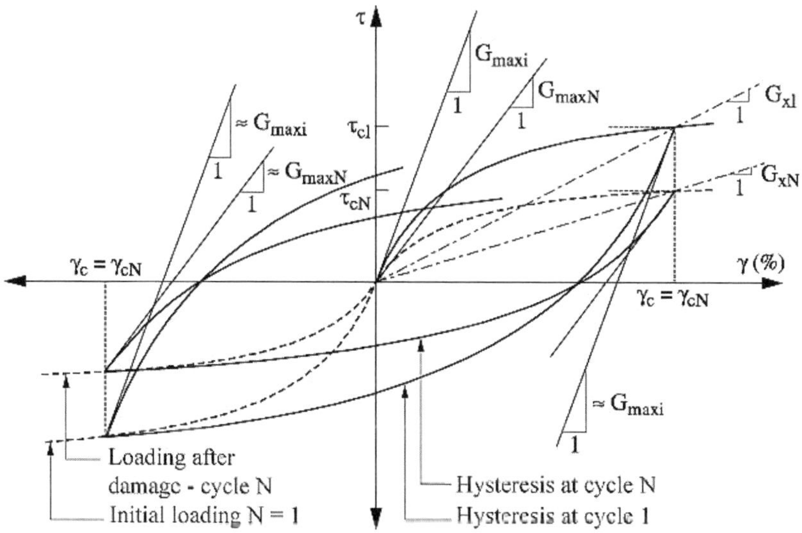

Figure 4.10. *Variation of hysteresis loops with the number of cycles*

4.3. Linear soil-structure interactions

For the study of the soil-structure interaction in linear systems, the models adopted to represent the soil are either the elastic model (section 4.2.1) or the equivalent linear visco-elastic model (section 4.2.2). The non-linearities in the behavior of the soil can be taken into account by selecting G shear modulus and β damping modulus values compatible with the average strain induced in open field by the seismic stresses. This approach involves neglecting additional non-linearities linked to the soil-structure interaction, such as soil plastification at the ends of superficial foundations, or along the pile shafts. Comparison of the results obtained with non-linear calculations shows this approach gives acceptable results when foundations show sufficient safety regarding their ultimate load.

4.3.1. *Illustration of the soil-structure interaction effect*

The influence of soil-structure interactions on the response of a building can be illustrated using the analog model shown in Figure 4.11 [WOL 85]. The structure is represented by a mass and a spring placed at a height h above the foundations. The bond between the structure and its foundations is represented by a stiff bar. The latter is lying on the ground, and its interaction is modeled via an impedance function for the foundations. In this model, what stands out is the simplified representation of impedance functions as springs and dashpots that are independent of frequencies; the dashpot theoretically explains both radiation damping and material damping as defined in section 4.2.2. With a view to simplifying the presentation, we assume that material damping can be neglected relative to the radiation damping (elastic behavior of the soil), which is a valid assumption for a homogenous medium and low to medium amplitude seismic stresses.

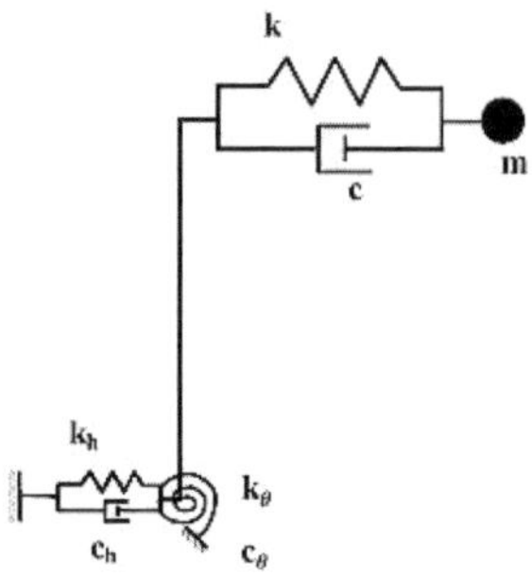

Figure 4.11. *Soil-structure interaction simplified model*

The system in Figure 4.11 has 3 degrees of freedom:

– the horizontal displacement u_1 of mass m;

– the displacement u_0 of the foundations;

– the rotation θ of the foundations,

and is subjected to a horizontal displacement of the supporting soil, with a pulsation harmonic ω and an amplitude u_g.

The dynamic balance equations of the system can be derived from Lagrange's equations [CLO 75] if we take q_i as generalized variables:

– $q_1 = u$, relative displacement of the mass in relation to A;

– $q_2 = u_0$, displacement of the foundations;

– $q_3 = \theta$, rotation of the foundations.

The relationship between the absolute displacement and u^t of the mass m and the previous variables is obvious:

$$u^t = u_g + u_0 + u + h\theta \qquad [4.13]$$

with T as the total kinetic energy:

$$T = \frac{1}{2} m \left(\dot{u}_g + \dot{u} + h\dot{\theta}\right)^2 \qquad [4.14]$$

and V is the potential energy:

$$V = \frac{1}{2}\left(ku^2 + k_h {u_0}^2 + k_\theta \theta^2\right) \qquad [4.15]$$

and δW is the work of non-conservative (damping) stresses:

$$\delta W = -\left(C\dot{u}\delta u + C_h \dot{u}_0 \delta u_0 + C_0 \dot{\theta}\delta\theta\right) \qquad [4.16]$$

Lagrange's equations are written as follows:

$$\frac{d}{dt}\left(\frac{\partial T}{\partial \dot{q}_i}\right) - \left(\frac{\partial T}{\partial q_i}\right) + \left(\frac{\partial V}{\partial q_i}\right) = \frac{\delta W}{\delta q_i} \qquad [4.17]$$

Using the previous notations and taking the acceleration, speed and displacement relationships into account, this means:

$$\dot{x} = i\omega x,\ \ddot{x} = -\omega^2 x \qquad [4.18]$$

When we introduce the critical damping ratios, we obtain:

$$\xi = \frac{i\omega C}{2k},\ \xi_h = \frac{i\omega C_h}{2k_h}\ \ \xi_\theta = \frac{i\omega C_\theta}{2k_0} \qquad [4.19]$$

$$\begin{cases} -m\omega^2 (u_0 + u + h\theta) + k(1 + 2i\xi)u = m\omega^2 u_g \\ -m\omega^2 (u_0 + u + h\theta) + k_h (1 + 2i\xi_h) u_0 = m\omega^2 u_g \\ -mh\omega^2 (u_0 + u + h\theta) + k_\theta (1 + 2i\xi_\theta)\theta = mh\omega^2 u_g \end{cases} \qquad [4.20]$$

When we introduce the following notations:

$$m\omega_s^2 = K\ ,\ m\omega_h^2 = k_h\ ,\ mh^2\omega_\theta^2 = k_\theta \qquad [4.21]$$

and then eliminate u_0 and θ between the three previous equations, we obtain:

$$\left[1 + 2i\xi - \frac{\omega^2}{\omega_s^2} - \frac{\omega^2}{\omega_h^2}\frac{1+2i\xi}{1+2i\xi_h} - \frac{\omega^2}{\omega_\theta^2}\frac{1+2i\xi}{1+2i\xi_\theta}\right]u = \frac{\omega^2}{\omega_s^2}u_g \qquad [4.22]$$

Taking into account the fact that ξ, ξ_h, $\xi_\theta << 1$, the previous equation becomes:

$$\left[1 + 2i\xi - \frac{\omega^2}{\omega_s^2} - \frac{\omega^2}{\omega_h^2}(1+2i\xi - 2i\xi_h) - \frac{\omega^2}{\omega_\theta^2}(1+2i\xi - 2i\xi_\theta)\right]u = \frac{\omega^2}{\omega_s^2}u_g \qquad [4.23]$$

Let us now consider a simple oscillator with 1 degree of freedom and the same mass m, with its characteristic pulsation $\tilde{\omega}$ and damping $\tilde{\xi}$ submitted to the harmonic displacement $\tilde{u}_g$ with a pulsation ω at its base (case of a structure embedded at its base). The harmonic response of the oscillator is given by:

$$\left(1 + 2i\tilde{\xi} - \frac{\omega^2}{\tilde{\omega}^2}\right)u = \frac{\omega^2}{\tilde{\omega}^2}\tilde{u}_g \qquad [4.24]$$

The equivalent oscillator will have the same response as the structure in Figure 4.11 if the following equations are verified:

$$\frac{1}{\tilde{\omega}^2} = \frac{1}{\omega_s^2} + \frac{1}{\omega_h^2} + \frac{1}{\omega_\theta^2} \qquad [4.25]$$

$$\tilde{\xi} = \frac{\tilde{\omega}^2}{\omega_s^2}\xi + \frac{\tilde{\omega}^2}{\omega_h^2}\xi_h + \frac{\tilde{\omega}^2}{\omega_\theta^2}\xi_\vartheta \qquad [4.26]$$

$$\tilde{u}_g = \frac{\tilde{\omega}^2}{\omega_s^2} u_g \qquad [4.27]$$

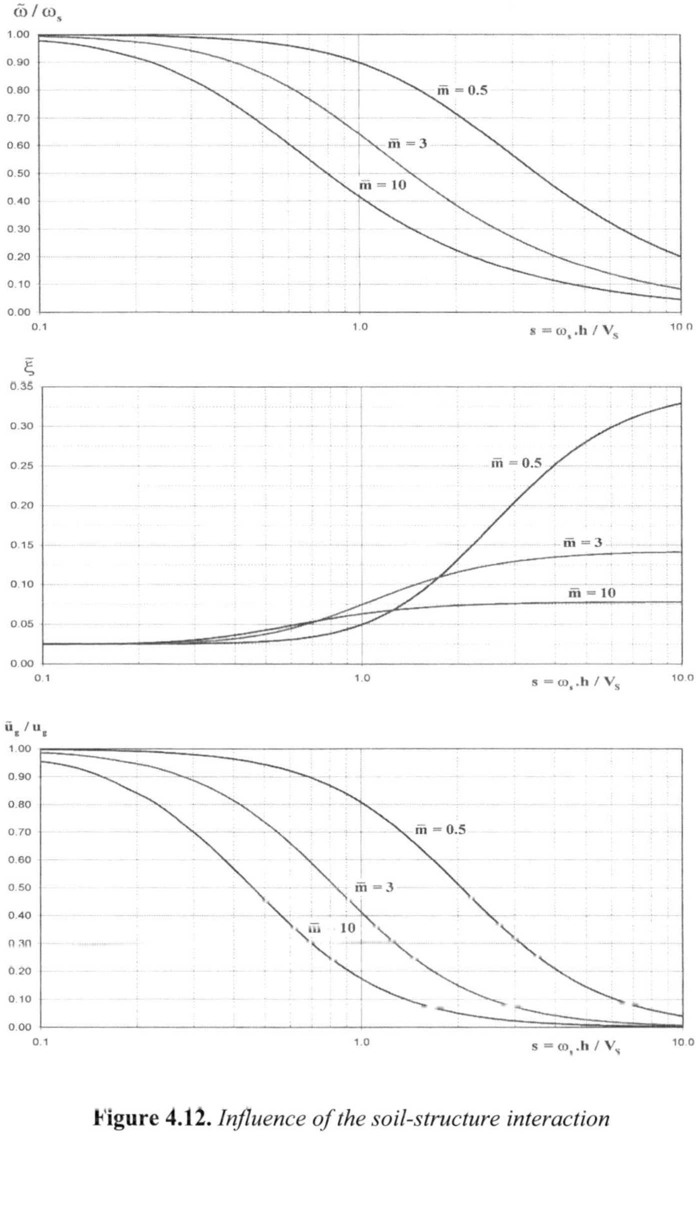

Figure 4.12. *Influence of the soil-structure interaction*

The previous equations are obtained by equating the real part with the imaginary part of equations [4.23] and [4.24] and for [4.27] by assuming a resonance situation ($\omega = \widetilde{\omega}$).

Equations [4.25] and [4.26] assume the soil-structure interaction results in:

– decreasing the characteristic pulsation ω_s of the embedded basis structure ($\omega < \omega_s$);

– increasing the damping of system ($\widetilde{\xi} > \xi$) with regard to the embedded basis structure;

– decreasing the effective incident stress at the basis of the structure ($\tilde{u}_g < u_g$).

The conclusions are shown in Figure 4.12, which represents a circular foundations lying on an homogenous elastic semi-space, and the relative variations $\tilde{\omega}/\omega_s$, $\tilde{\xi}$, $\tilde{u}_g/u_g$ are a function of the non-dimensional parameters:

$$\bar{h} = \frac{h}{r} = 1 \text{ , } s = \frac{\omega_s h}{V_s} \text{, } \bar{m} = \frac{m}{\rho r^3} \qquad [4.28]$$

in which r is the radius of the foundations, and ρ and V_s are respectively the volumic density and rate of the waves S within the soil (equation [4.2]).

Figure 4.12 clearly shows that the influence of the soil-structure interaction is all the more marked if the foundation soil is soft (increasing s) or if the structure is massive (increasing m).

4.3.2. *Expression of a soil-structure problem*

Before examining the different methods employed to take soil-structure interactions into account, the problem is worth formulating from a general point of view. This formulation aims at dealing with phenomena due to finite elements. In fact, the problem is so complex that resorting to digital methods cannot be avoided, but in the rest of this chapter, we will try to point out steps that can be dealt with analytically and those that are amenable to existing solutions.

Motion equations are obtained by referring to Figure 4.13, which schematizes a soil-structure unit.

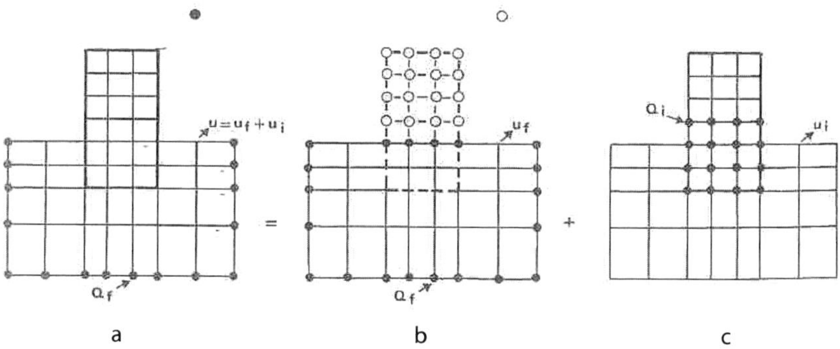

Figure 4.13. *Superposition theorem for the soil-structure interaction*

If we refer to the mass, damping and stiffness matrices as [M], [C] and [K], the equation of motion is written:

$$[M]\{\ddot{u}\}+[C]\{\dot{u}\}+[C]\{\dot{u}\}=\{Q_f\} \quad [4.29]$$

As the source of the motion (seismic focus) is generally not included in the model, the load vector $\{Q_f\}$ has values not equal to zero only on the outer border of the model.

When there is no structure, the equation of the open field motion is similar in shape to that described by [4.29]. With the f indexes designating the mass, damping and stiffness matrixes related to the open field alone, this equation will be:

$$[M_f]\{\ddot{u}_f\}+[C_f]\{\dot{u}_f\}+[K_f]\{u_f\}=\{Q_f\} \quad [4.30]$$

Equation [4.30] can only be solved by making certain assumptions about the nature and the direction of the incident field propagation.

Assuming that:

$$\{u\}=\{u_i\}+\{u_f\} \quad [4.31]$$

Equation [4.31] defines the interaction displacement $\{u_i\}$ which satisfies the equation:

$$[M]\{\ddot{u}_i\}+[C]\{\dot{u}_i\}+[K]\{u_i\}=-\{Q_i\} \quad [4.32]$$

with:

$$\{Q_i\}=[[M]-[M_f]]\{\ddot{u}_f\}+[[C]-[C_f]]\{\dot{u}_f\}+[[K]-[K_f]]\{u_f\} \quad [4.33]$$

The load vector $\{Q_i\}$ is determined from the open field displacements. For linear systems, the superposition theorem is illustrated in Figure 4.13 [LYS 78]: the interaction problem is divided into the sum of an open field soil response problem [4.30] with a source problem [4.32] where the applied forces $\{Q_i\}$ have components not equal to zero only at nodes common to the soil-structure. This last problem is analogous to a machine vibration problem. The total displacement for the integration problem is then given by equation [4.31].

Equation [4.33] clearly shows that there is interaction as soon as a mass or stiffness difference develops between the soil and the structure. To simplify matters, let us suppress the damping term in the equation and restrict the problem to one of a structure lying on the ground, being subjected to the vertical propagation of volume waves (shear or compression). Under such conditions, in the open field, all points of the ground surface will be driven by the same motion. If the foundations of the structure are infinitely stiff, the last term of the equation becomes equal to zero, and the load vector $\{Q_i\}$ is restricted to:

$$\{Q_i\}=[[M]-[M_f]]\{\ddot{u}_f\} \quad [4.34]$$

The stresses $\{Q_i\}$ applied at the basis of the structure generate a movement of the support that is equivalent to a force of inertia field within the structure. Consequently, interaction only results from the inertia developed inside that structure. This is referred to as the *inertial interaction*, and its effect is illustrated by the example in section 4.3.1.

On the other hand, let us consider a structure laid in the ground, the mass of which is equal to zero outside the ground, and equal (in value and repartition) to the mass of the soil for the part that is in the ground. The expression for loads $\{Q_i\}$ becomes:

$$\{Q_i\} = \left[[K] - [K_f]\right]\{u_f\} \tag{4.35}$$

These only result from the soil and structure stiffness difference for the part embedded in the ground. Even without any mass difference, there is interaction; it is referred to as *kinematic interaction*. It results from the stiffness of the foundations that prevents them from following motions imposed by the ground. We saw before that it was strictly equal to zero in some cases; it can be low in others (foundations on flexible piles) or highly significant (stiff structure strongly cross-braced and sunk deep in the ground).

In the most general case, interaction comprises both an inertial interaction and a kinematic interaction.

Figure 4.13 and the previous arguments illustrate the two best methods for solving the soil-structure interaction. Figure 4.13a corresponds to the overall methods, the solutions to which are obtained via direct resolution of equation [4.29]. They do not resort to any notion of superposition, and thus they are theoretically suited to non-linear problems. Alternatively, sub-structure methods rely either on the division of Figure 4.13b-4.13c, or on similar divisions to solve the problem step by step. These methods can indeed only be applied to linear problems suitable for superposition.

4.3.3. *Superposition theorem*

Dividing the soil-structure interaction into inertial and kinematic interactions (as discussed in section 4.3.2) not only has the advantage of allowing us to illustrate the essential phenomena, it also generates a resolution method based on the sub-structuring principle, the validity of which relies on superposition theorem [KAU 78], [ROE 73]. This theorem establishes that the response of the model in Figure 4.14 (on the left) subjected to a $\ddot{u}_g$ acceleration on its basis can be obtained the following ways:

– either in one step by solving the equation:

$$[M]\{\ddot{u}\} + [K]\{u\} = -[M]\{I\}\, u_g \tag{4.36}$$

where:

– $\{u\}$ represents the vector of the relative displacements of the system with regard to the foundation;

– $\{I\}$ represents a unit vector giving the direction of the stress $\ddot{u}_g$;

– $[M]$ and $[K]$ represent the mass and stiffness matrices of the system.

To simplify matters, the damping terms are omitted;

– or in two steps by expressing the vector of relative displacements $\{u\}$ as the sum of the "kinematic" displacements $\{u_{cin}\}$ and of the inertial displacements $\{u_{iner}\}$:

$$\{u\} = \{u_{cin}\} + \{u_{iner}\} \qquad [4.37]$$

and simultaneously solving the two equation systems [4.38] and [4.39] below:

$$[M_{sol}]\{\ddot{u}_{cin}\} + [K]\{u_{cin}\} = -[M_{sol}]\{I\}\ddot{u}_g \qquad [4.38]$$

$$[M]\{\ddot{u}_{inert}\} + [K]\{u_{inert}\} = -[M_{st}]\left[\{\ddot{u}_{cin}\} + \{I\}\ddot{u}_g\right] \qquad [4.39]$$

in which $[M_{sol}]$ and $[M_{st}]$ represent the mass matrices of the soil and structure parts:

$$[M] = [M_{sol}] + [M_{st}] \qquad [4.40]$$

The equivalence of [4.36], [4.38] and [4.39] is obtained by a mere addition of the two last values whilst taking [4.37] and [4.40] relationships into account.

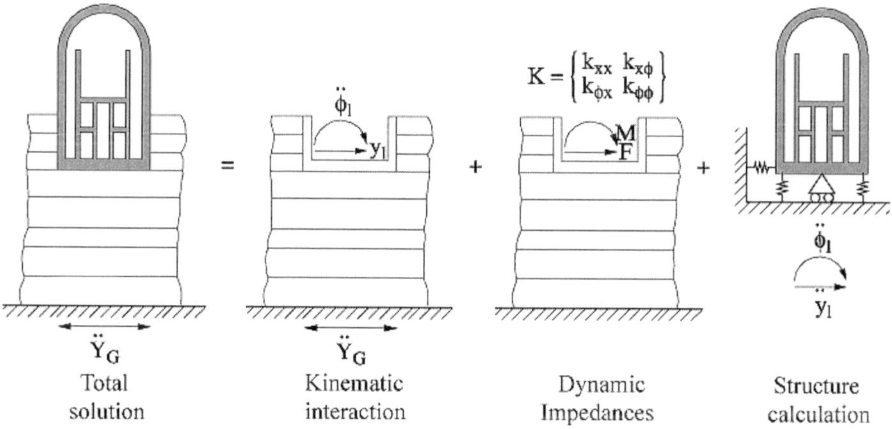

Figure 4.14. *Superposition theorem (from [KAU 78])*

Equation [4.38] gives the response of a structure without a mass to seismic stress $\ddot{u}_g$. The solution gives kinematic interaction displacements that are used as loads in [4.39], when imaginary inertial forces are applied to the structure.

When resolving equation [4.39], the modeling of the soil is indifferent: it can be represented either with finite elements or using a stiffness matrix representing foundations and the soil defined at the soil-structure interface. This stiffness matrix results from the condensation of all degrees of freedom of the soil at the interface [PEC 84]; the condensation is only possible for a resolution in the frequency field. Within such a framework, the stiffness matrix is formed from the complex moduli (section 4.2.2), which take damping into account. The stiffness matrix consists of a real part (representing the stiffness of the foundations) and a fictitious part that integrates all the damping phenomena (material and radiating). The terms of the matrix depend on the frequency.

In the case of a structure with rigid foundations, it is legitimate to replace the stiffness matrix (NxN), N being the number of interface nodes, with a (6x6) matrix that gives the rigid body movements of the foundations; this matrix is called an impedance matrix, and it can be conceptually represented by springs and dashpots depending on the frequency. The result is that the solution of the kinematic interaction problem is completely defined by rigid body movements of the weightless structure; the latter can then be replaced by weightless foundations subjected to the same seismic stress.

Examining the structure of equation [4.39] reveals that the solution $\{u_{inert}\}$ can be interpreted as the displacement vector that is related to a fictitious support submitted to the rigid body movements (translations and rotations) of the foundations.

Thus, with rigid foundations, the overall problem can be divided into three sub-problems:

– determining the motion of weightless rigid foundations subjected to the seismic stress, this step representing the solution to equation [4.38];

– determining the impedance matrix of the foundations, this matrix including a real part and a fictitious part which both depend on the frequency;

– calculating the dynamic response of the structure linked to the impedance matrix and its supports subjected to the kinematic interaction movement calculated in the first problem).

Insofar as the foundations are perfectly stiff, the procedure is strictly identical to that leading to the one-step resolution of the overall system (equation [4.36]). The interest of such division clearly appears whenever it is possible to simplify one of the three steps of the calculation.

The diffraction problem (step a) still exists except for structures with shallow foundations subjected to the vertical propagation of volume waves; in such cases, resolution of step a) is identical to that of the response of an open field soil profile, since the kinetic interaction is equal to zero. The second step solution can be avoided for certain configurations by using the results from impedance functions published in literature. The third step is essential; however, it is simpler and more familiar to engineers, as it originates in conventional dynamic analysis of the structures

4.3.4. *Practical modeling of the soil-structure interaction*

One of the main arguments proposed for using sub-structure methods rather than global methods is that they are both easy to use and less expensive. It is probably true for structures with shallow foundations submitted to vertical volume wave propagation; in this case, we have seen the kinematic interaction is equal to zero, which suppresses the diffraction problem resolution step, and that analytic or already published solutions to the impedance problem exist. Besides, if the reference movement is defined at the ground surface, the only possible step is dynamic analysis of the structure. Sub-structure methods have the advantage of allowing some modifications without making it necessary to start the whole analysis again: a modification of the characteristics of the structure does not require a new dynamic analysis of the structure; modifying the characteristics of the project earthquake enables the use of the impedance problem solution.

As soon as the study structure is either partially or completely buried, resolving the diffraction problem becomes complex; the rigorous solution becomes as delicate to obtain as the global solution to the problem. Nevertheless, simplified methods to take this kinematic interaction into account have been proposed in the case of stiff foundations ([HAL 75], [KAU 78]). It is nonetheless true that, in the most general case, sub-structure methods become less competitive than global methods. Besides, the latter can solve non-linear problems as well, yet they remain limited to two-dimension geometries.

In the rest of this chapter, we mainly endeavor to enhance particular aspects related to soil modeling. For structure specific problems, the reader can refer to [CLO 75]. However, we restrict ourselves to finite element calculations, which are the only ones possible when implementing either global or sub-structure methods requiring resolution of the diffraction problem.

4.3.4.1. *Model of soil behavior*

In practice, we use the equivalent linear visco-elastic model. This model has the advantage of allowing the non-linearities of the behavior to be taken into account owing to an iterative pattern. It is especially well suited to global methods; with each iteration, the properties of the soil (moduli and damping) are adjusted, within each element, to the average strain while the element is stressed.

When using sub-structure methods for linear systems, the non-linearities of the soil are taken into account by using those characteristics that are compatible with the average strain level of each open field element.

4.3.4.2. *Nature and propagation direction of the incident wave*

The commonly used model is the plane wave type, which assumes that the horizontal movement of the incident field is created by the vertical propagation of SH shear waves, and vertical movement is created by the vertical propagation of P compression waves. These assumptions give no indication of the resulting movement obtained after the waves reverberate on the structure. Besides, we can see that waves generated after interaction with the structure mainly consist of Rayleigh waves.

Other hypotheses related to the nature of the incident field are possible and can be taken into account. Breaking down the displacements into open field and interaction displacements shows that once the open field displacements have been calculated, the nature of the incident field no longer explicitly influences the formulation of the interaction problem ([4.32] and [4.33]). Sub-structure methods are quite well-adapted to taking diverse incident fields into account, providing the open field movement can be determined.

4.3.4.3. *Geometrical modeling of the medium*

The finite element formulation is standard; nevertheless, some conditions have to be respected. Transmitting high frequencies imposes a maximum dimension on the elements, at most equal to a fraction of the corresponding wavelength. Typically, we use a value between 1/8 and 1/5 of the wavelength:

$$h_{max} \leq \frac{1}{5} \text{ to } \frac{1}{8}\frac{V_s}{f_{max}} \qquad [4.41]$$

where f_{max} represents the highest frequency to be transmitted, and V_s the propagation rate of the shear waves. This criterion is generally applied to the vertical dimension of the mesh because, considering the generally admitted assumption of wave vertical propagation, the displacement field varies faster vertically than horizontally, especially some distance away from the structure.

The extension of the finite element mesh constitutes one of the most critical problems in the resolution of a dynamic problem involving propagation phenomena using the finite element method. As a matter of fact, without any special conditions, the side and lower limits of the model are open surfaces that completely reverberate the wave fronts that hit them. The energy carried by these waves is reverberated back to the structure instead of being carried ad infinitum inside the soil. As the only energy dissipation takes place through material damping, the model has to be extended so that the waves reverberated at the limits do not reach the structure while its response is being estimated. The procedure soon makes the calculation cost prohibitive.

To free us from such reverberations, some special devices called absorbing boundaries have been developed. Located at the ends of the model, these boundaries are supposed to represent the exact stress conditions existing at that limit, due to the presence of soil outside the model.

Generally speaking, the side boundaries of the model can be divided into local boundaries or consistent boundaries.

The local boundaries generally consist of localized dashpots, the characteristics of which depend on the mechanical properties of the medium around them. These boundaries do not couple the different degrees of freedom of the nodes along the boundary and perfectly absorb only the waves with a normal incidence [LYS 69]. They can advantageously be implanted into time or frequency calculations.

Unlike local boundaries, consistent boundaries couple all the degrees of freedom of the boundary nodes and perfectly absorb all kinds of waves. Formulating these boundaries ([KAU 74], [LYS 72] and [WAA 72]) involves frequency dependent terms; therefore, they can only be used for resolutions in the frequency field. There are no consistent boundaries to represent the effect of the semi-space underlying the model. This is why the lower boundary of the model is supposed to be rigid. If it is chosen to be deep enough (about a structure wide), the reverberation phenomena on this boundary become negligible. Actually, the field of the waves reverberated by the structure mainly consists of surface waves which die down fast with depth. Besides, we can take advantage of this property by having a moving lower boundary, with a variable meshing, which is deeper the lower the studied frequencies. In fact, low frequencies die down more slowly but they require coarser meshing to ensure a correct transmission of the wave [LYS 81].

4.3.4.4. *Digital integration pattern*

It is theoretically possible to choose a digital integration pattern that is either a time, a frequency or a mode pattern. These aspects will not be explained in detail below; for more information, see [PEC 84]. We will only mention that for linear problems, the frequency integration pattern is best suited to solving soil-structure interaction problems because of the formulation of the soil behavior law (section 4.2), the dependence on frequency of the impedance matrices used in sub-structure methods, and the absorbing boundaries used to estimate the kinematic interaction displacements or for global method resolutions.

However, if a time or mode integration pattern is preferred, the variation of impedances with frequencies has to be taken into account. This can be implemented either by successive iterations allowing us to adjust the impedance to the frequency of the soil-structure interaction mode, or by resorting to analog models using springs, dashpots and additional mass (Figure 4.15).

Such models, which originate in cone models [WOL 95], lead to very satisfying approximations of the impedance functions [PEC 94]. Nevertheless, due to the presence of an additional mass, they require the use of precautions related to the definition of the actual movement of the foundations. They can be equally used for either mode analysis or time integration.

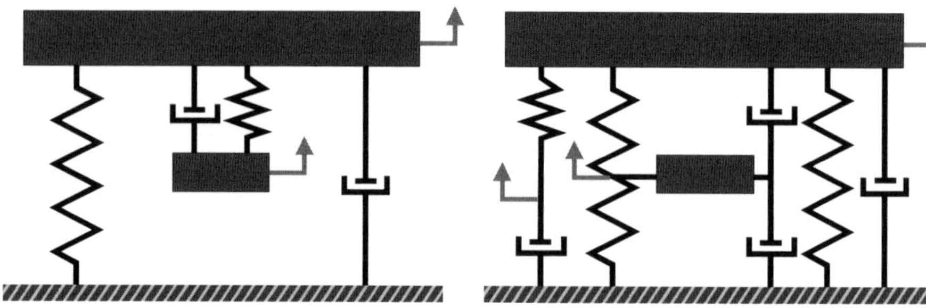

Figure 4.15. *Analog model for the impedance matrix*

4.4. Non-linear soil-structure interactions

In most cases, linear modeling of the soil-structure interaction represents a reasonable approximation of the phenomena. However, extreme situations do exist, in which taking into account the geometric or behavior non-linearities is the only possible approach representative of the phenomenon. Taking these phenomena into account currently represents a topic that is developing fast, and for which important progress has been achieved in recent years. Hereafter, we restrict ourselves to identifying the phenomena and superficially point to the way to deal with them.

4.4.1. *Geometric non-linearities and uplift of the foundations*

It has often been observed after major earthquakes that slender structures with shallow foundations show good behavior and seem to have been protected from seismic aggression by their propensity to move independently of the foundation soil; these additional "degrees of freedom" are provided by the possible uplift at the soil-foundation interface. This uplift is expressed by transformation of the kinetic energy into potential energy, due to the raising of the center of gravity, and through lengthening of the vibration periods. According to the frequency content of the stress, the longer vibration period can shift the vibration mode towards a lower amplification spectrum region.

4.4.2. *Non-linearities of behavior*

Such non-linearities come from the local plastification of soil near the foundation element. In the case of pile foundations, plastification takes place along the upper part of the pile shaft, at a height typically equal to 4 or 6 diameters; for superficial foundations, the plastification is likely to take place at the angles of the base, owing to the oscillation moment of the structure.

The extreme soil plastification case corresponds to mobilizing the ultimate capability of the foundations. If loading is applied permanently, such a situation would bring about failure through loss of the supporting capability of the foundations. Because under seismic stress, the applied loading varies with time, this temporary mobilizing of the strength capability does not lead to failure, but the appearance of irreversible displacements. These are not necessarily harmful to the good behavior of the structure. Their evaluation allows us to design foundations by relying on performance criteria (performance-based design) rather than on a safety criterion with regard to "failure" [PEC 00].

4.4.3. *Modeling the non-linear soil-structure interaction*

The most direct method for taking geometric or behavior non-linearities into account is still the finite element method. Thanks to a behavior law adapted to the materials and the soil-structure interface elements, this method is the most versatile. However, we should not lose sight of the fact that implementing this method is quite complex and requires modeling, digital analysis and soil and structural dynamics competences. Furthermore, in spite of considerable computer advances, the calculation times are still quite long. The result is that the method is best suited to verifications, but not to pre-dimensioning, as this requires many varying studies.

To get round such difficulties, sub-structuration methods similar in spirit to the methods developed for the study of linear phenomena have been developed recently. We merely present the outlines that give birth to the dynamic macro-element concept. For a more comprehensive presentation of the subject as well as an exhaustive bibliographic study of those methods, see [CRE 01a], [CRE 01b] and [CRE 02].

The general philosophy of the method involves defining two sub structures: the soil and the soil-foundation interface on the one hand, and the structure on the other. The division is made at the level of the foundations. The soil-interface sub-structure is conceptually sub-divided into the neighboring field and far field (Figure 4.16); the exact boundary between both sub-fields is unknown, but it does not explicitly intervene in the macro-element concept.

The far field corresponds to the region where the soil-structure interaction can be neglected; the behavior in this area is governed by the propagation of seismic waves and the energy dissipation there is mainly viscous radiative damping. Quite naturally, and by analogy with the impedance concept, the distant field is modeled using a spring and a dashpot independent of the frequency for simplification.

The neighboring field represents the medium part interacting with the foundations: all the behavior non-linearities (geometric, material) are concentrated with their potential couplings. This element is formulated in terms of global variables as the stress wrench acts on the foundations and its stiff body movements.

The whole displacement of the foundations equals the sum of an elastic displacement $\underline{u}^{el}$, a plastic displacement $\underline{u}^{pl}$ and a displacement linked to the uplift $\underline{u}^{up}$.

$$\underline{u}^{tot} = \underline{u}^{el} + \underline{u}^{pl} + \underline{u}^{up} \qquad [4.42]$$

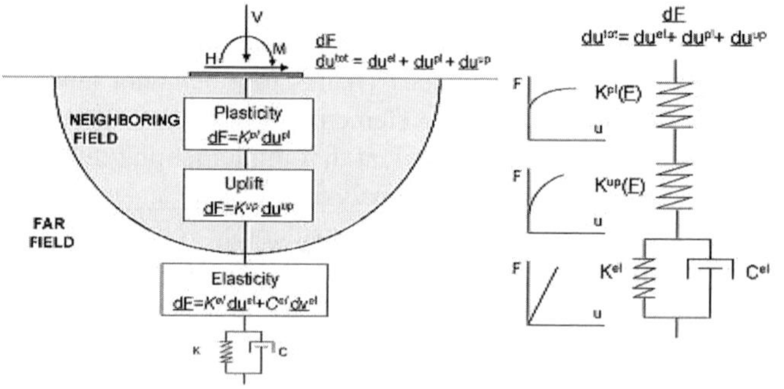

Figure 4.16. *Dynamic macro-element concept*

The plastic displacement $\underline{u}^{pl}$ is calculated on the basis of a conventional plastic model defined by the data of a failure criterion, a loading surface and a flow rule for strain hardening. The failure criterion is provided by the carrying capability of the foundations [PEC 97]; the strain hardening law combines both isotropic and kinematic strain hardenings and the flow rule is non-associated.

The displacement caused by the uplift is estimated from an uplift model for foundations lying on elastic soil [CRE 98].

The macro-element thus constituted is placed in series with the one representing the far field (Figure 4.16, on the left). Conceptually it can be represented by an assembly of springs and "brush springs" that couple all the degrees of freedom of the foundations.

Though for the moment it is limited to two-dimension geometries, this macro-element provides us with a simple and effective tool for the resolution of soil-structure interaction problems involving non-linear phenomena. Confronted with finite element digital simulations or with experiment results (CAMUS 4 model tested on the CEA's vibrating table), this model gives results that are quite satisfactory from a practical point of view, easy to use and less cumbersome than the finite element method.

4.5. Bibliography

[CLO 75] CLOUGH R.W., PENZIEN J., *Dynamic of Structures*, McGraw-Hill, 1975.

[COL 90] COLLECTIF, AFPS, Rapport de mission, Séisme de Loma Prieta, 1990.

[COL 02] COLLECTIF, EUROCODE 8, Design of Structures for Earthquake Resistance, Part 1: General Rules, Seismic Actions and Rules for Buildings, CEN/TC250/SC8/N317, May 2002.

[CRE 98] CREMER C., Elaboration of a Soil Structure Macro Element for Foundation Uplift, mémoire de stage DEA, ENS Cachan, 1998.

[CRE 01a] CREMER C., Modélisation du Comportement non Linéaire des Fondations Superficielles sous Séisme, Thesis, LMT Cachan, 2001.

[CRE 01b] CREMER C., PECKER., DAVENNE L., "Cyclic Macro Element of Soil Structure Interaction: Material and Geometrical Non-linearities", *International Journal for Numerical and Analytical Methods in Geomechanics*, 25, pp. 1257–1284, September 2001.

[CRE 02] CREMER C., PECKER A., DAVENNE L., "Modelling of Non Linear Dynamic Behavior of a Shallow Strip Foundation with Macro Element", *Journal of Earthquake Engineering*, Vol. 6, N°2, pp. 175–211, 2002.

[DOR 90] DORMIEUX L., CANOU J., "Determination of Dynamic Characteristics of a Soil based on Cyclic Pressumeter Test", *3rd Int. Symposium on Pressumeters*, Oxford, UK, pp. 159–168, 1990.

[HAL 75] HALL J.R., KISSENPFENNIG J.R., "Special Topics on Soil-Structure Interaction", *International Seminar on Extreme Load Conditions and Limit Analysis Procedures for Structural Safeguards and Containment Structures*, Paper U2.2, Berlin, 1975.

[HAR 78] HARDIN B.O., "The Nature of Stress-strain Behavior for Soils", *State of the Art, International Conference on Earthquake Engineering and Soil Dynamics,* Pasadena, USA, 1978.

[IDR 80] IDRISS J.M., KENNEDY R.P., "Analysis for Soil-Structure Interaction Effects for Nuclear Power Plants", Report by the Ad Hoc Group on Soil Structure Interaction of the Committee on Nuclear Structures and Materials of the Structural Division of ASCE, 1980.

[KAU 74] KAUSEL E., "Forced Vibrations of Circular Foundations on Layered Media", *Soils Publication* N° 336, Department of Civil Engineering, Massachusetts Institute of Technology, Cambridge, USA, 1974.

[KAU 75] KAUSEL E., ROESSET J.M., "Dynamic Stiffness of Circular Foundations". *Journal of Engineering Mechanics Division*, ASCE Vol. 101 N° EM6, 1975.

[KAU 78] KAUSEL E., WHITMAN A., MURRAY J., ELSABEE F., "The Spring Method for Embedded Foundations". *Nuclear Engineering and Design.* N° 48, 1978.

[LYS 69] LYSMER J., KUHLEMEYER R.L., "Finite Dynamic Model for Infinite Media", *Journal of Engineering Mechanics Division*, ASCE Vol. 95 N° EM4, 1969.

[LYS 72] LYSMER J., WAAS G., "Shear Waves in Plane Infinite Structures", *Journal of Engineering Mechanics Division*, ASCE Vol. 98 N° EM1, 1972.

[LYS 78] LYSMER J., "Analytical Procedures in Soil Dynamics", *State of the Art, ASCE Conference on Soil Dynamics and Earthquake Engineering*, Pasadena, California, USA, 1978.

[LYS 81] LYSMER J., RAISSI M., TAJIRIAN F., VAHDANI S., OSTADAN F., "Sassi – A System for Analysis of Soil-structure Interaction", Geotechnical report N° 81–02, University of California, Berkeley, USA, 1981.

[PEC 84] PECKER A., *Dynamique des sols*, Presses de l'Ecole Nationale des Ponts et Chaussées, 1984.

[PEC 94] PECKER A., "Seismic Design of Shallow Foundations", *State of the Art, Proceed. 10th European Conference on Earthq. Engin*, Ed. Balkema, Vienna, Austria. pp. 2905–2910, 1994.

[PEC 97] Pecker A. "Analytical formulae for the seismic bearing capacity of shallow strip foundations", *Seismic Behavior of Ground and Geotechnical Structures*, Seco and Pinto (Eds..), Balkema, 1997.

[PEC 00] PECKER A., PENDER M.J., "Earthquake Resistant Design of Foundations: New Construction", *Geo Eng 2000*, Vol. 1, pp. 313–332, Melbourne, Australia, 2000.

[PRE 78] PREVOST J.H. "Plasticity theory for soil stress-strain behavior", *Journal of Engineering Mechanics Division.* Vol. 104, N° EM5, 1978.

[PRE 87] PREVOST J.H., Modelling the behavior of geomaterials, Notes de cours", Non-linear soil mechanics and dynamic soil-structure interaction, Lausanne, Switzerland, 1987.

[ROE 73] ROESSET J.M., WHITMAN R.V., DOBRY R., "Model Analysis for Structures with Foundation Interaction", *Journal of Structural Engineering Division,* ASCE, Vol. 99 N° ST3, p. 89–416, 1973.

[SEE 70] SEED H.B., IDRISS I.M., "Soil Moduli and Damping Factors for Dynamic Response Analysis", *Report EERC 70-10*, Earthquake Engineering Research Center, 1970.

[WAA 72] WAAS G., "Earth Vibration Effects and Abatement for Military Facilities", Report 3, US Army Engineer WES, Vicksburg, Miss., USA, 1972.

[WOL 85] WOLF J.P., *Dynamic Soil Structure Interaction*, Prentice Hall, Englewood Cliffs, New Jersey, 1985.

Chapter 5

Experimental Methods in Earthquake Engineering

Introduction

Considering the vast size of most civil engineering structures, direct structure experimentation cannot generally be considered. Apart from a few particular cases of structure instrumentation on high-risk sites in which earthquakes are anticipated, model experimentation in the laboratory has been preferred. The most natural approach uses the shaking table, which involves reproducing the motion of the soil on which the model has been laid. By imposing acceleration on the table identical, for example, to the one measured on the ground surface during an earthquake owing to jacks, we can reproduce the behavior of actual structures. That behavior results from the properties of stiffness (evaluative in the case of concrete cracking and steel plasticity), of damping and of the distributed mass of the structure.

Nevertheless, directly transposing the results of model laboratory tests to actual structures encounters several difficulties, which include the need to use laws of similarity which are often ill-adapted to concrete material, the piloting of the table, and the accuracy of the accelerogram representation at the model level (considering the filtering carried out by the table itself).

Chapter written by Alain MILLARD, Pierre PEGON and Jean-Claude QUEVAL.

With the development of numerical methods, a new method has evolved: the pseudo-dynamic method, which is complementary to shaking table tests.

In principle, it involves numerically integrating, in the course of time, the equations of dynamics, written for the structure into a relative motion reference frame:

$$M\, a_r(t) + C\, v_r(t) + F_{int}\left[d_r(t)\right] = -M\, a_e(t)$$

where M and C represent the viscous damping and mass operators, F_{int} the inner forces (apart from viscous effects), $a_r(t)$, $v_r(t)$ and $d_r(t)$ the accelerations, speeds and displacements related to the structure with regard to its base considered as fixed, and $a_e(t)$ the driving acceleration of the same base.

At a given time, the equation is numerically integrated over a short period of time, giving an approximation of the displacement d(t). The displacement is then applied to the model, fixed at its base, through a few wisely located jacks that allow an estimation of the internal forces (F_{int}) which will be used for the numerical integration on the next interval.

The main features of the method are the fact that it is possible to work on models that are much larger than shaking tables, and that we can proceed slowly. Nevertheless, rate effects are only taken into account through the viscous damping operator, and the displacements are only imposed on the structure at a small number of points, which is not well adapted to real structures with a well-distributed mass.

As we may note, the theoretical concepts on which both experimental methods are based are rather simple. The actual difficulty lies in the interpretation of the results and their transposition to real structures, as has been demonstrated in numerous model tests; it also lies in the implementation aspects, which fundamentally condition the quality of the results. Thus, in the case of shaking tables for example, it is essential to control the electro-mechanical system that drives the table. In the same way, for the pseudo-dynamic method, the splitting into time periods as well as the integration algorithm of the dynamics equations have a considerable influence on the final quality of the results.

This is the reason why, in this chapter, the stress is deliberately laid on these aspects that may at first seem to be more "technological" (numerical as well as experimental), yet which remain linked to the dynamic behavior of the structures. The test examples given, obtained from the ELSA and TAMARIS European facilities, illustrate our remarks explicitly and concretely.

5.1. The pseudo-dynamic method

5.1.1. *Introduction*

Even if performed in a quasi-static way, the pseudo-dynamic test (PSD) on the one hand combines on-line calculations and control and on the other hand the continuous measurement of the properties of a given structure, allowing researchers to simulate the dynamic response of a structure in a realistic way.

The motion equations of a discrete system modeling the structure during the test are integrated by a step by step numerical algorithm. The viscous damping and inertia forces are modeled analytically, which is simple when compared to the task of modeling the non-linear forces that develop in the structure, and are measured experimentally. In most cases, it is impossible to model them accurately. This process allows both the hysteresis damping linked to irreversible strains and the damage of structural materials (typically the main dissipation source during an earthquake) to be taken into account.

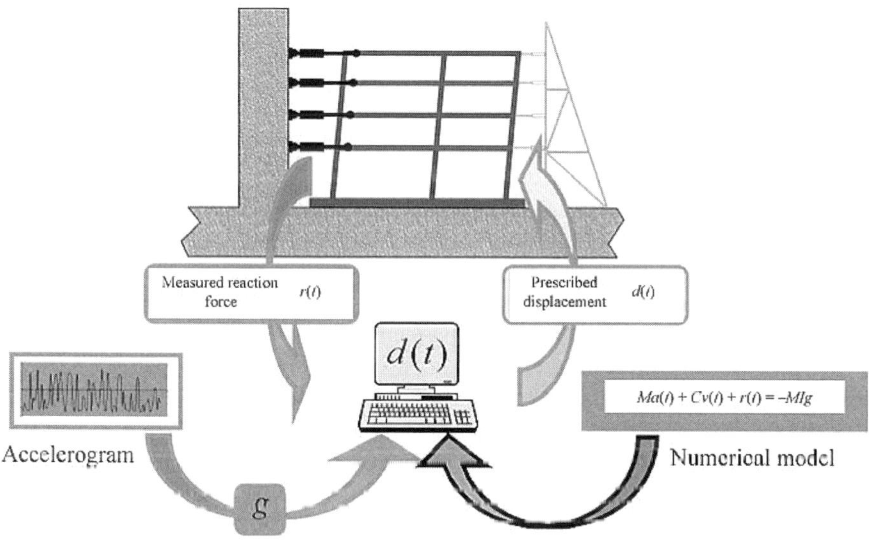

Figure 5.1. *Pseudo-dynamic method*

Let us consider the building in Figure 5.1. To simulate its response to an earthquake, a recording of the acceleration of an earthquake (either natural or artificial) is supplied as input to the computer in charge of the PSD method. The horizontal displacement of the building floors (where the mass is supposed to be concentrated) is calculated for a short time step using an appropriate time integration algorithm. These displacements are then imposed on the structure via servo-controlled hydraulic jacks leaning on a reaction wall. Load cells measure the load level needed to impose the required displacement on each floor, and these values are fed back to the computer to be used at the next time step.

Because the inertia is modeled, the experiments do not have to be carried out in real time, which allows us to test large size structures with relatively low hydraulic power. In this regard, PSD tests are complementary to those carried out with shaking tables which are conducted in real time on extremely reduced structure components or structure models.

5.1.2. *History of the PSD method*

The PSD method was created in Japan [TAK 75] over 30 years ago, and the basic implementations, as well as numerous improvements, were made. Initially, the integration methods used were the central finite difference method (explicit Newmark's method), then an implementation with operator splitting of the trapezoidal method (implicit Newmark's method) [NAK 90]. The latter allows the use of an unconditionally stable scheme for cases including stiff structures and a certain number of degrees of freedom.

The PSD method was further developed in the USA [SHI 84]. There, a host of improvements were introduced, including the use of iterative implicit schemes [SHI 91].

The experience accumulated in both Japan and the USA has shown that success is determined by the quality of the implementation of the PSD method (see [ELB 89], [NAK 90] and [SHI 87]). Many elements in this implementation have a strong tendency to bring about errors. The measurement or control errors tend to accumulate [SHI 87], and if they get too high, they can eventually overcome the response [NAK 87].

All of this takes place as if a differential equation system was integrated with a computer that would contain only a small number of significant digits.

5.1.3. *The ELSA laboratory*

The European Laboratory of Structural Assessment (ELSA) [DON 92] is a laboratory of the European Community Research Common Center (see Figure 5.2) and was inaugurated in 1992. It consists of a 16 m high and 21 m wide reaction wall and of 2 test stands that can withstand loads about a few thousand kN, i.e. the loads necessary to strain and seriously damage real size structures. Although the wall is by far the longest in Europe, there is one more or less similar in the USA, in San Diego, but both are surpassed in size by the Japanese Tsukuba 21 m high wall.

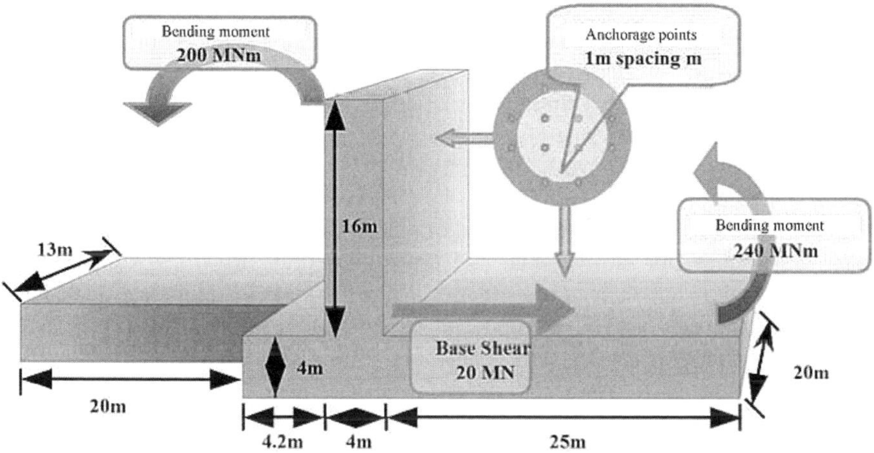

Figure 5.2. *Reaction wall at the ELSA*

The implementation carried out at the ELSA was the first to use fully numerical control algorithms for the motions of the jacks [MAG 91], [MAG 93]. These allow very accurate control and implementation of the different algorithms available for the numerical integration of motion equations.

It is not usually possible to test structures as big as bridges or oilrigs. Nevertheless, the seismic loading will often only damage some parts of them. For a lot of bridges, the damage will mostly concentrate at the base of the piers. In these cases, we can estimate that the remaining structure can be numerically modeled using the finite element method. It is then easy to combine the PSD test for one part of the tested structural portion with the time integration of the motion equations of the model of the remaining structure or numerical structure. For a bridge, the piers constitute the tested structure, whereas the bridge deck is the numerical structure. This kind of sub-structuration test forms the essence of the PSD method [DER 85], but had only been used in practice for small components in the early 1990s. At the ELSA large-scale tests were carried out for the first time [PIN 95a, PIN 96].

5.1.4. *Comparison with shaking tables*

At this stage, we can already compare the advantages and drawbacks of reaction wall PSD tests with shaking table dynamic tests.

Advantages: the PSD tests are carried out on real scale or only slightly reduced models, which are easy and cheap to build without any scale effect of the basic materials. The tests are carried out slowly, so they can be carefully observed. As they can be stopped at any time, the models can be loaded to failure and it is possible to potentially add instrumentation where it is necessary. It is also possible to use the sub-structuration method with them.

Disadvantages: PSD tests are carried out more slowly than they would be in real time, which does not suit all material types. We only drive few nodes in the structure, which is not an accurate situation when the mass is too widely distributed. Usually we can only control the model in the horizontal plane, which prevents the vertical components of earthquakes being taken into account.

5.2. The conventional pseudo-dynamic method

5.2.1. *Algorithms*

The basic hypothesis of the PSD method is that the dynamic behavior of a structure can be represented using a discrete model with a small number of degrees of freedom (dof). The motion equations of this idealized structure are expressed as a system of second order differential equations:

$$Ma(t) + Cv(t) + r(d(t)) = f(t) \qquad [5.1]$$

where M and C are respectively the mass and damping matrices, $a(t)$, $v(t)$ and $d(t)$ the acceleration, speed and displacement vectors, $r(d(t))$ the originally non-viscous internal forces and $f(t)$ the external forces applied to the structure.

An approximation of the solution to equation [5.1] can be obtained using a step by step direct integration method. The solution at each time step is obtained according to the values obtained at the previous time step, in order to have the displacement of the structure submitted to any external loading. From this point onwards, the time interval $[0.T]$ on which the integration has to be effected is divided into equal time steps of size Δt.

The integration schemes used at the ELSA belong to the so-called "β-Newmark" family. For these schemes, when the displacement, speed and acceleration vectors at time t are known, the values of the same vectors at time t + Δt are given by Newmark's formulae:

$$\begin{aligned} d^{n+1} &= \tilde{d}^{n+1} + \Delta t^2 \beta a^{n+1} & \tilde{d}^{n+1} &= d^n + \Delta t v^n + \frac{\Delta t^2}{2}(1-2\beta)a^n \\ v^{n+1} &= \tilde{v}^{n+1} + \Delta t \gamma a^{n+1} & \tilde{v}^{n+1} &= v^n + \Delta t(1-\gamma)a^n \end{aligned} \quad [5.2]$$

and verify the equilibrium system expressed in equation [5.1], shifted in time thanks to parameter α:

$$Ma^{n+1} + (1-\alpha)Cv^{n+1} - \alpha Cv^n + (1-\alpha)r^{n+1} - \alpha r^n = (1-\alpha)f^{n+1} - \alpha f^n \quad [5.3]$$

If we choose $\alpha = 0$, $\beta = 0$ and $\gamma = ½$, we obtain the well-known central finite difference method. The pattern is explicit, as it is possible to calculate d^{n+1} from known quantities then impose this displacement, measure the reaction load r^{n+1}, and finally calculate a^{n+1} and v^{n+1}.

This scheme is perfectly fitted to the PSD method. Unfortunately, it is conditionally stable. The solution may become unstable (i.e. increase indefinitely) when the value of the reduced frequency $\Omega_0 = 2\pi f_0 \Delta t$ is higher than 2. Here f_0 is the highest natural frequency of the dynamic system considered. If f_0 is high, the time step becomes very small. Then we have to impose many displacement increments on the structure, some of which are likely to be smaller than the accuracy of the control system, which may bring about very large errors.

The frequency f_0 may be high for stiff structures including a small number of dofs, or for most structures when the number of dofs becomes important. The excellent accuracy of the displacement checking system used at the ELSA (see section 5.2.2) allows test stiff structures owing to this explicit scheme without any real difficulty. The scheme is also used to the exclusion of all others by the continuous PSD method (see section 5.2). The number of dofs can become very high in sub-structuring tests (see section 5.2.3), and in these cases, other integration schemes have to be introduced.

If we choose $\beta = (1-\alpha)^2/4$ and $\gamma = (1-2\alpha)/2$, with $\alpha \in [-1/3, 0]$, we again find the α method [HIL 77]. This scheme is implicit, because the displacement d^{n+1}

depends on the acceleration a^{n+1} linked to the load r^{n+1}, which is a function of the displacement d^{n+1}. Generally it implies an iterative resolution process. Such an approach is used by the PSD method [SHI 91], even if it is not easy to implement in so far as the structure itself is involved in the iterations. The implementation can be considered as a modified Newton-Raphson method converging in a linear way at best. Actually a sub-relaxation coefficient is introduced into the successive iterations to force convergence in a monotonous way and without any parasitic discharge. The stability of this scheme will be discussed later.

Nevertheless, it is possible to introduce an implicit α-Newmark scheme without resorting to iterations. To do this we use an operator splitting (OS) method ([COM 97], [NAK 90]). This maintains the stability of the scheme, as it is implicit for the elastic part of the response, but requires no iteration whatsoever, as it remains explicit as far as the non-linear part is concerned. The OS method is based on the following approximation of the reaction force r^{n+1}:

$$r^{n+1}(d^{n+1}) \simeq \tilde{r}^{n+1}(\tilde{d}^{n+1}) + K^I(d^{n+1} - \tilde{d}^{n+1}) \qquad [5.4]$$

where K^I is a stiffness matrix which has to be higher than (or equal to) the tangent K^T stiffness matrix of the system. That condition ensures the unconditional stability of the scheme. In this case, the time step Δt can be any time step, and thus it is chosen according to the experiment carried out and not to verify any condition of stability. The α-OS scheme appears as the natural extension to the implicit case of the explicit scheme of the central finite differences.

As structures tend to soften, K^I can be chosen as the elastic stiffness matrix K^E or the initial stiffness matrix of the structure. The experimental installation obviously allows us to obtain a very good approximation of this matrix.

Eventually, we should note that the Newton-Raphson method of the iterative scheme also uses a K^I matrix with the same properties in terms of stability as those of the α-OS scheme. If the matrix is chosen properly, then the scheme will be unconditionally stable as well.

5.2.2. *Implementation at ELSA*

The checking system implemented at ELSA was quite original compared with previous analog implementations. As this system is entirely digital, it can easily be modified, improved and can take new types of sensors into account. Therefore, it is easy to adjust the system by introducing hybrid solutions to implement the PSD method, especially as far as the sub-structuration technique is concerned (see section 5.2.3).

The control system operates in the following way: each degree of freedom, corresponding to the displacement of a jack, is driven by a PC equipped with the interface needed to receive data from the experiment and send orders to the jack ([MAG 91], [MAG 93]). In the system, a digital displacement sensor measures the output signal. The real state of the system is then described by means of a variable y^k, which is updated at each step of the sampling. This variable is compared with the displacement orders d^k. In fact, the control algorithm operates from the $e^k = d^k - y^k$ error (and other information) to produce correction orders u^k. Finally, these numerical orders are transformed into a u*(t)* step-wise continuous signal, which is sent to the jack's servo-valve. From the displacement d^n at the beginning of the step to d^{n+1} at the end of the step, the control unit generates a function in the shape of a ramp followed by a plateau. This function represents the displacement path the jack will have to follow to gradually pass from d^n to d^{n+1}. The ramp is generated in the shape of a discrete sequence d^k that is used as orders. It is presented on the left in Figure 5.5.

In such cases, the control algorithm itself is a numerical implementation of a PID (Proportional Integral Derivative) analog method.

Note that the displacement rate for making the ramp is generally specified. The length of a loading step is therefore proportional to the displacement increment, thus it is unknown beforehand. On the other hand, the force sent back to the central processing unit is an average of values supplied by the acquisition of the last points of the displacements' plateau. The control units are geographically distributed according to the position of the jacks. Each control unit is connected to the central processing unit via a high bitrate optical fiber. Each unit works independently, in real time, for each of the checked dofs in the structure, and it is controlled by a central processing unit that integrates the motion equations.

The structural displacement is measured by an optical data line counter, engraved with a 2 µm resolution [MAG 91]. Units allowing a 0.5 or 1 m displacement are currently used. These digital sensors are directly connected to the control units, without resorting to any digital/analog conversion and without any analog pre-conditioning. The sensors therefore require no calibration and deliver a signal that gives a correct measure of the speed, by counting on a given time period (the sampling period, for example). The sharp measure of the sensors allows us to reach an accuracy of the displacement control around 50 μm.

It should be noted that the displacement sensors are mounted on a supporting structure independent of the movement of the jacks or those of the reaction wall on which they rest (see Figure 5.1). In fact, as the stiffness of the wall is not infinite, it bends with the motion of the tested structure. This means the lengthening of the jack is not characteristic of the motion of the tested structure alone, especially if we consider the accuracy of the displacement. Therefore, an independent reference system has to be introduced, and we should ensure its neutrality (for example, its mass should be such that no resonance system can be induced by the vibrations either of the hydraulic system or of the control system).

The hydraulic system which powers the jacks operates under a 210 bar pressure and delivers a 25-liter per second maximum flow. The jacks are two different types: maximum force: 0.5 and 1.0 MN; displacement: ± 250 mm and ± 500 mm.

5.2.3. *The sub-structuration method*

The PSD method is a hybrid method that combines the digital integration of the motion equations of a complex structure (condensed on a reduced number of dofs) with the measure of the reaction forces that results from the imposed motion. Despite its potential, direct experimentation on very large civil engineering structures such as bridges would be difficult: apart from the size problem, the simultaneous control of a large number of dofs would be quite laborious. Yet in order to deal with such cases, it is possible to expand the application field of the PSD method, at least whenever part of the structure can be modeled; this is called the sub-structuration procedure [DER 85].

This procedure takes advantage of the hybrid nature of the PSD method by combining the modeling of part of the structure (numerical structure) with the real testing of the remaining structure (tested structure). The procedure naturally applies to a bridge, as its biggest part, the deck can often be considered as elastic linear, and can therefore be modeled with finite elements. Only the piers subjected to damage will be tested in the laboratory. Another advantage, which is particularly welcome in the case of bridges, lies in the fact that we can impose asynchronous seismic loadings or loadings that have different amplitudes along the foundations.

The numerical part of the structure has to be modeled suitably, and then the resulting motion equation discrete system must be integrated in time. The numerical model can reveal a number of dofs that are far more important than those checked in the lab. Therefore, it is not straightforward to modify the software developed by the central processing unit implementing the PSD method and to expand it to the sub-structuration case.

An alternative solution, far more consistent with the decentralized character of the implementation at ELSA, has been chosen. It involves implementing the sub-structuration due to two processes using two different computers. The first process is carried out by the central processing unit responsible for the dofs tested in the laboratory, whereas the second process – the one in charge of the modeled part – operates with a working station implementing the finite element software. Both processes exchange minimum information via the network. The advantage of such an approach is obvious. The modifications in laboratory software are minor and the process dedicated to the numerical part does not specifically need to deal with the tested part. The communication between both processes has to be safe and has to potentially allow connections between two different operating systems, while being standard on the Internet. This is why Berkeley's communication system (*Berkeley sockets*) has been chosen. Practical details about this implementation can be found in [BUC 94].

We have already said that system described by equation [5.1] could be integrated in time by a step by step-type numerical scheme. The speed and displacement vectors at step (n + 1) are expressed as a function of the acceleration vector and vectors obtained at the previous time step (see equations [5.2]). If these expressions [5.2] are introduced into the α-off-center shape of the equilibrium described in [5.3], we obtain the following linear system for acceleration a^{n+1}:

$$\hat{M}a^{n+1} = \hat{f}^{n+1} \qquad [5.5]$$

where $\hat{M}$ is a pseudo-mass matrix and $\hat{f}^{n+1}$ a pseudo-force vector. If we take into account approximation [5.4] of the reaction forces of the OS scheme, the expression of both quantities is given by:

$$\begin{aligned} \hat{M} &= M + \gamma(1+\alpha)C + \beta\Delta t^2(1+\alpha)K^I \\ \hat{f}^{n+1} &= (1+\alpha)f^{n+1} - \alpha f^n - (1+\alpha)\tilde{r}^{n+1} + \\ &\quad \alpha\tilde{r}^n - (1+\alpha)\tilde{v}^{n+1} + \alpha\tilde{v}^n + \alpha(\gamma\Delta t C + \beta\Delta t^2 K^I)u^n \end{aligned} \qquad [5.6]$$

Now if we distinguish the data coming from the numerical sub-structure (S), from the data coming from the tested part (T), the [5.5] system can be rewritten as follows:

$$\begin{bmatrix} {}^{S}\hat{M}_{ij} & {}^{S}\hat{M}_{i\theta} & 0 \\ {}^{S}\hat{M}_{\delta j} & {}^{S}\hat{M}_{\delta\theta} + {}^{T}\hat{M}_{\delta\theta} & {}^{T}\hat{M}_{\delta J} \\ 0 & {}^{T}\hat{M}_{I\theta} & {}^{T}\hat{M}_{IJ} \end{bmatrix} \bullet \begin{bmatrix} a_j^{n+1} \\ a_\theta^{n+1} \\ a_J^{n+1} \end{bmatrix} = \begin{bmatrix} {}^{S}\hat{f}_i^{n+1} \\ {}^{S}\hat{f}_\delta^{n+1} + {}^{T}\hat{f}_\delta^{n+1} \\ {}^{T}\hat{f}_I^{n+1} \end{bmatrix} \quad [5.7]$$

where the indexes i, j refer to the dofs inside the numerical structure, δ, θ the dofs common to both parts and I, J the dofs inside the tested structure. Now if we condense the accelerations a_j, which are inside the numerical structure, we obtain the following control equation, for the whole dofs present in the lab:

$$\begin{bmatrix} ({}^{S}\hat{M}_{\delta\theta} - {}^{S}\hat{M}_{\delta j}\,{}^{S}\hat{M}_{ji}^{-1}\,{}^{S}\hat{M}_{i\theta}) + {}^{T}\hat{M}_{\delta\theta} & {}^{T}\hat{M}_{\delta J} \\ {}^{T}\hat{M}_{I\theta} & {}^{T}\hat{M}_{IJ} \end{bmatrix} \bullet \begin{bmatrix} a_\theta^{n+1} \\ a_J^{n+1} \end{bmatrix} = \begin{bmatrix} ({}^{S}\hat{f}_\delta^{n+1} - {}^{S}\hat{M}_{\delta j}\,{}^{S}\hat{M}_{ji}^{-1}\,{}^{S}\hat{f}_i^{n+1}) + {}^{T}\hat{f}_\delta^{n+1} \\ {}^{T}\hat{f}_I^{n+1} \end{bmatrix} \quad [5.8]$$

and the numerical structure is ruled by:

$${}^{S}\hat{M}_{ij}a_j^{n+1} = {}^{S}\hat{f}_i^{n+1} - {}^{S}\hat{M}_{i\theta}a_\theta^{n+1} \quad [5.9]$$

Equation [5.8] allows us to stress the differences between a simple PSD test and a test with sub-structuration. As far as sub-structuration is concerned, an additional mass matrix related to the connection nodes (term between brackets in the first member of [5.8]) should be added to the experimental pseudo-mass. In the same way, at each time step, the pseudo-force vector should include a contribution (term between brackets in the second member of [5.8]) that will become the numerical sub-structure. It is interesting to note that during the time-step loop, the data flow between the digital process and the experimental process is balanced. If the experimental part needs the contribution of forces on the connection nodes, the numerical process will require acceleration on the same connection nodes to be able to develop according to equation [5.9]. Let us also note that such an implementation is mainly asynchronous: each process has to wait for the other to be able to go ahead. That approach is natural within the frame of the conventional PSD method, where the implementation length of a loading step in displacement is not set *a priori*. As we will see, this is no longer the case with the continuous PSD method (see section 5.2).

The measurements supplied by networks of accelerometers installed in earthquake areas (see [OLI 91], for instance) clearly show that the ground motion is not uniform near its surface. Due to the reflections and refractions of seismic waves on soil layers that have different behaviors, measuring points not far away from each other can be subjected to important relative displacements.

Recent numerical studies (see [PIN 95b] for example) have shown that for reinforced concrete bridges, whether insulated or not, synchronous loadings lead to an upper bound of the response. Therefore, the traditional method based on that loading type would be particularly reliable. However, it should be noted that an asynchronous stress could substantially modify the distribution of damage inside a structure. It is thus important to be able to carry out tests for these asynchronous loadings. In the case of a synchronous-type seismic load, the equations of motion are written within a relative coordinates system linked to the moving base. The earthquake is taken into account through the system inertia forces that are introduced as external forces into the equation of equilibrium system [5.1]:

$$f = -MIb \qquad [5.10]$$

for which b is the intensity of the base acceleration and I the vector which takes into account the direction of the seismic loading at the level of each dof.

This approach is particularly well suited to PSD tests, as the base of the tested structure does not move. How can we process for an asynchronous test with the sub-structuration technique? The different parts of the tested structure, loaded differently at the base, will in no way be physically connected to one another. In the case of a bridge, only the piers of which are tested, the condition is verified: each pier will be tested within a different relative coordinates system, whereas the numerical structure, the motion of which will be described within an absolute coordinates system, will ensure the synchronization between the whole relative coordinates systems. In practice, as far as the tested structure is concerned, an asynchronous test will be carried out in a way similar to a synchronous test. On the other hand, the calculations related to the numerical structure are different [PEG 96a, PEG 00]. Furthermore, if a damping matrix is associated with the numerical structure, we will have to be careful about the introduction of a consistent formulation [PEG 96b], i.e. a formulation independent from the coordinates systems.

5.2.4. *Illustration*

An example of the experimental installation carried out for an irregular bridge test on a scale of 1:2.5 [PIN 95a, PIN 96] can be seen on the left in Figure 5.3.

Figure 5.3. *Typical structures tested at ELSA*

As the deck has been modeled, the piers can be put side to side, which allows a reduced space. In addition to horizontal jacks, vertical jacks can be seen at the front of the piers, where they apply the vertical loading due to the weight of the deck. On the right in Figure 5.3 is a real size 4-storey frame in a configuration including a filled in wall which was tested in the same period.

5.3. Continuous pseudo-dynamic method

5.3.1. *Continuous method principle*

The conventional PSD method requires considerable implementation time. The structure is stopped when it reaches the specified displacement, so the reaction force can be measured. The break, which can be about several seconds long, harms the regularity of the loading and causes interfering relaxation phenomena inside the structure. With the continuous PSD method [MAG 98], [MAG 00], the movement of the jacks is not deliberately stopped; therefore, it can follow the specified displacement quite accurately, because it no longer includes any discontinuity. The load can now be measured for each sampling period of the control system and the motion equations are immediately integrated, at the same frequency, to continuously provide a new displacement target. To achieve this goal, it has been necessary to insert the finite element method into the control program instead of the ramp generator.

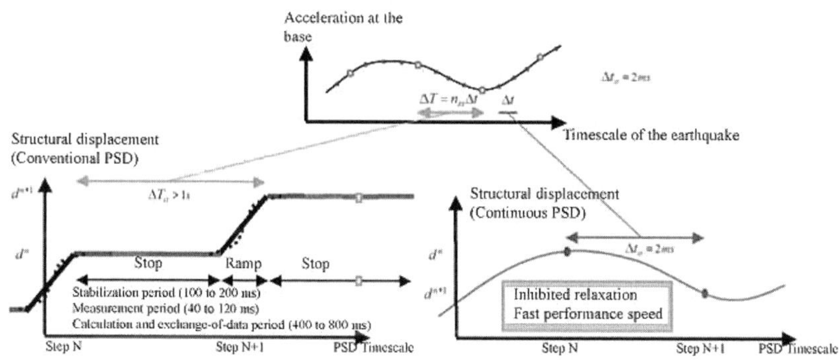

Figure 5.4. *Conventional and continuous pseudo-dynamic method*

We have seen that both PSD conventional and continuous methods use the same integration schemes. However, for each b_i acceleration value at the base (read in the recorded acceleration file with a sampling period ΔT), the continuous method will introduce a series of n_{ss} values with $b_{i+k/n_{ss}}$ ($0 < k < n_{ss}$) accelerations obtained by interpolation between b_i and b_{i+1}. For each ΔT large step, the step by step integration process will be carried out n_{ss} times, at the Δt_{st} operating period of the control system (typically 1 or 2 ms), as if λ sub-steps were introduced into the conventional method (see Figure 5.4).

The λ time dilatation factor between a PSD test and a real time test is given by: $\lambda = n_{ss} \Delta t_{st} / \Delta T$. For a test on a several-storey building, we will typically obtain: $n_{ss} = 500$, $\Delta t_{st} = 2\text{ms}$, $\Delta T = 5\text{ms}$, i.e. $\lambda = 200$. One second of the real earthquake will develop in 200 seconds in the test.

The results of a series of tests carried out on a 3 storey, 10.4 m high steel structure have shown that the PSD method could be implemented 10 to 20 times as fast as the conventional method, depending on the importance of the masses present in the experimental device. Let us actually note that besides the static stresses linked to the displacements, the force cells of the jacks measure inertia forces about λ^2 times lower than the real inertia forces, and therefore they can often be neglected. Nevertheless, when the mass of the tested structure is large and λ is smaller than ≈ 10, the inertial effects have to be taken into account when we try to resolve the equations of motion.

For a given test, Δt_{st} and ΔT are constant. Then the dilatation factor can be modified interactively just by modifying n_{ss} during the experiment. Actually, we always try to decrease λ as much as possible while keeping the control error within a satisfactory interval (typically < 50 μm), which indeed depends on the nature and the complexity of the test. In any case, as with the conventional PSD method, it is important to keep the errors as low as possible, and the motion regular. The absence of relaxation imposed on the structure by the PSD method definitely has a big part to play in it. Let us also note that as the reaction force is now taken into account with a high frequency (500 or 1,000 Hz), an excellent noise filter of this analog measurement is thus introduced, which is far better than in the case of the conventional PSD method where the number of measuring points of the force taken into account is much lower.

In practice, the values of λ can be close to the unit for a simple test (one dof in the lab with potential sub-structuration) with negligible forces of inertia in the experimental device. They can be much higher, up to 400 for a complex tested structure.

5.3.2. *Implementation at ELSA*

When tests with several jacks have to be carried out, it is necessary to achieve perfect synchronization of each jack. This requires some refinement in the architecture of the control system. The difficulty involves coordinating and carrying out for a specific time period of the control system (1 or 2 ms) all the tasks associated with a cycle of the PSD method: measuring, calculation of the movement and control of the displacement. Therefore, new control units, more powerful and more modular have been introduced, communicating with the computer in charge of the integration of the equations of motion, owing to a new broadband system.

As shown in Figure 4.5, each control unit consists of 3 modules integrated on the same card: signal processing and storing, acquisition and control, communication and exchange of data. These cards can then be plugged on the same single motherboard and work together owing to the data bus of the host computer, as well as with its processor that is in charge of the integration algorithm. All these processors (the slaves checking while the master is integrating) operate in real time with an appropriate operating system (TNT on 32 bits), synchronized by a clock. Moreover, as a major task via the network, the master processor also provides the interface with other computers working with more standard operative systems (Windows NT, 2000 or XP).

There lies another key point of the new architecture: easy data access. In using the possibilities of active object distribution via the network (DCOM technology by

Microsoft), it is possible to transfer the data acquisition, processing, storing and visualizing tasks via the master processor, and potentially the sub-structuration calculations to other computers in an integrated way. The data related to the configuration of the test, including the parameters useful for each control device, could be modified any time as well. Finally, it is possible to open an access to the test on the net, beyond the local network. This will give external users the opportunity to directly interact with the test. We can even contemplate sub-structuration tests where the different parts of a same structure are tested on different test-platforms.

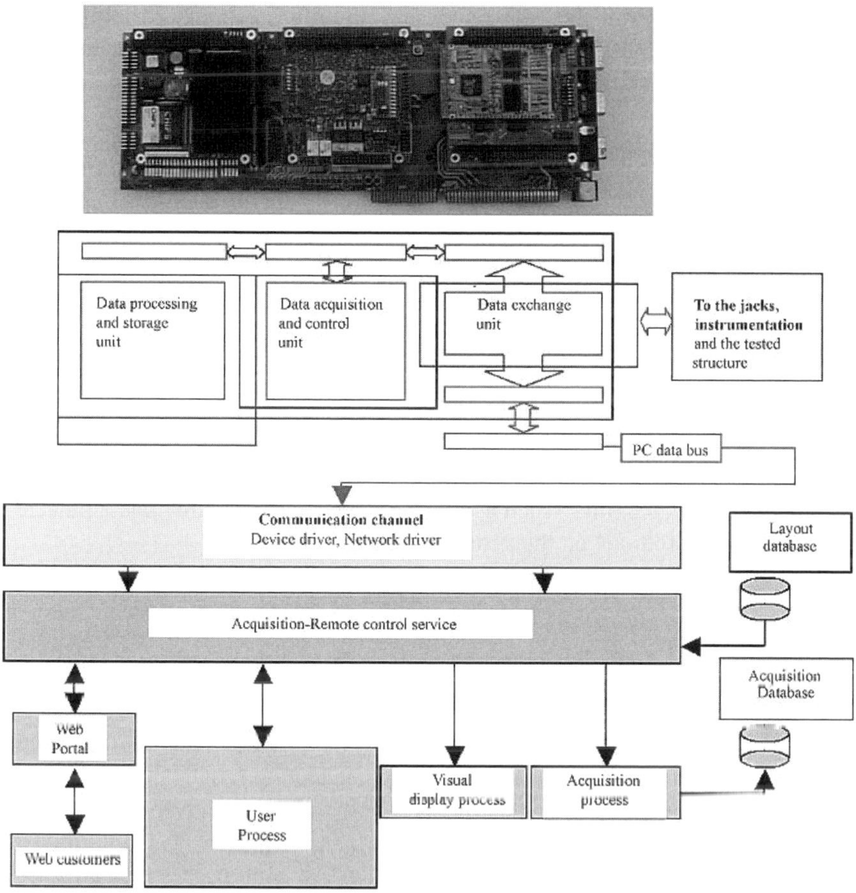

Figure 5.5. *Architecture of the new network tested at ELSA*

As we have already seen, in a PSD test, it is essential for the displacement control to be carried out in an accurate way. As a matter of fact, during the step by step process, errors can accumulate until they distort the response of the structure. Moreover, if the test procedure is applied to a structure, the response of which strongly depends on the stress rate, a sharp measure of the speed will also be important in order to correctly assess the reaction forces linked to the phenomenon.

We should also note that the parameters characteristic of the tested structures are generally not well known. As a consequence, the control algorithms have to work within an extremely variable environment including sudden and generally completely unforeseen disruptions. In such situations, whenever we try to impose a position tracking control, the PID fixed gain algorithm cannot operate efficiently enough on the whole possible behavior range of the tested structure. It then becomes necessary to use alternative control algorithms.

Thus, to ensure sufficient accuracy of the displacement and speed of the jacks, a recent adaptive algorithm developed at the University of Bristol [STO 90a, STO 90b, STO 92] has been inserted into the control systems. The principles of the MCS (*Minimal Control System*) algorithm which ensures a minimum synthesis of the control are the following:

– no assessment of the parameters of the system to control is required (not even on-line). The control algorithm requires neither any adaptive observer nor any control by recovery;

– the parameters (unknown) of the system to control can vary with time, even with a relatively wide bandwidth;

– the unknown quantities (such as the speed), which are highly contaminated by noise, can anyway retroact on the system to control;

– the implementation of the algorithm is very simple.

5.3.3. *Sub-structuration for the continuous method*

With the conventional PSD method, running the test was mainly asynchronous. A test step in the laboratory could last between 1 and 4 seconds depending on the size of the imposed displacement. This context was extremely advantageous for sub-structuration, since the loading process of the numerical sub-structure could also operate in a completely asynchronous way. At last, as both processes wait for each other, the calculation time of the digital part could even become longer than the implementing time of a loading step in the lab. It was then quite easy to carry out iterative non-linear calculations at the level of the numerical process.

The context completely changes when we consider the continuous PSD method. The experimental process is synchronous and the physical time between two integration time steps is extremely short: 1 or 2 ms. As it is difficult to carry out a non-linear calculation step in so short a period of time, a delayed approach like the one used previously (see section 5.2.3) is no longer possible, as the experimental process cannot wait for the numerical process any longer. Therefore, the coupling method should be synchronous and the data exchange should take place without any interruption.

The integration algorithm for the whole structure (both numerical and experimental) should have the following properties: be second order, be stable within the limits of stability of the explicit scheme implemented for the experimental part, introduce minimum dissipation and operate with non-stop data exchange.

To do so, we implement a partitioned approach that features the following characteristics ([MAG 98], [MAG 00]):

– an implicit scheme for the numerical part, coupled with the explicit scheme of the central finite differences used for the experimental part;

– two different time steps for each part of the structure (Δt for the experimental part and ΔT for the numerical part);

– sub-cycling of the explicit part with a number of sub-steps which can be very large ($\Delta T / \Delta t \approx 1{,}000$);

– assessment in advance of the state of the numerical structure, and interpolation of the forces to be imposed on the experimental structure by the numerical part to limit the digital dissipation of the scheme.

The master computer carries out the integration of the equations of motion of the numerical structure as its priority task. On each 1 or 2 ms cycle of the control system, the remaining time is used for the basic acquisition and data transfer tasks. These basic tasks can be given priorities and the main one can be dedicated to the calculation of the motion of the numerical structure. This strategy has been used to obtain preliminary results for a simple structure [MAG 98]. An alternative consists of having the calculations of the numerical structure carried out by another computer: this strategy has successfully been used in a recent test campaign on bridges.

5.4. Final comments

As we have seen, modeling plays a crucial part for the PSD method: choice of the system of the equations to integrate and model (whether linear or not) of the numerical part in the sub-structuration method. This part does not stop here [PEG 98]. Modeling can be used to carry out the design of the experiment itself. As it allows seismologists to carry out a dynamic simulation (by keeping all the dofs of the model) as well as a PSD simulation (by condensing the dofs of the mass matrix), it is used to optimize the number and positions of the jacks, in order to choose the best-suited integration scheme and the best integration step. It allows us to ensure that the jacks will not saturate as far as either forces or displacements are concerned. At last, experiment is often used to identify models which will later be used either for parametric studies or to select a reinforcement technique.

Implementing the PSD method has often led to the propagation of excessive errors. The extreme care taken for its implementation at the ELSA, both in its conventional and its continuous version, showed that it could be used with confidence for quite varied situations: concrete, metallic, masonry or mixed constructions, bridge piers, variedly shaped walls, structures including various passive insulation systems, models of historical monuments, reinforcement and repair technical studies.

The PSD method itself remains a research topic, as is shown by the recent publication of new integration schemes [CHA 02], [WAN 01] or by the involvement of ELSA in the continuous method where aspects of the regularity of loading and sub-structuration keep on being improved.

5.5. Shaking table tests

5.5.1. *Introduction*

Several techniques can be used to test reinforced concrete structures in earthquakes and especially:

– centrifuge tests;

– pseudo-dynamic tests;

– shaking table tests.

The shaking table tests consist of placing the structure on a table operated in order to reproduce a given seismic load.

A shaking table generally consists of:

– a table;

– a guiding system along 1, 2 or 3 directions;

– a loading system, with 1 or several degrees of freedom;

– a system generating pressurized hydraulic oil;

– a servo-control system;

– a driving system;

– a reaction basemat;

– acquisition and processing means.

The table plate is generally square. It is made of welded assemblies with reinforcements in order to have free-free frequencies as high as possible (over 50 Hz).

The table can be loaded in a horizontal mono-axial mode, in a vertical-horizontal bi-axial mode or in a three-axial mode. It is guided by sliding rail systems or directly by transverse jacks. In the case of a sliding rail system, the maximum loading frequencies are lower than 100 Hz because of the operating gaps.

As for multi-axial tables, these are driven directly by jacks in all directions. There are geometrical compensations between the various axes in order to have perfectly rectilinear displacements in the three directions. All the jacks are equipped with hinges. The hinges limit the stiffness of the table as well as its rotations.

The hydraulic power system allows us to supply the jacks with pressurized oil (210 bars) with outflows varying according to the number of pumps in service.

In order to control the table, electronics bays are needed to drive the jacks. The jacks are controlled with closed loops. The basic closed-loop control system consists of:

– an amplifier for the control of the servo-valve and the servo-controller;

– feedback of a LVDT displacement sensor to measure the position of the piston.

A servo-controller compares the feedback signal of the LVDT displacement transducer with a reference signal (control or program). For any difference between the feedback signal and the reference, the servo-controller injects an error signal into the servo-valve in order to bring the difference back to zero. The performance of the basic control is limited because of the frequency response of the different components and the maximum outflows of the servo-valves

Whenever the tested specimens are very big and heavy and the oil outflows are quite important, it is necessary to use multi-stage servo-valves using two closed-loop control circuits: an outer loop and an inner loop.

In big servo-valves, there is usually a pilot stage (outflow from 1 to 60 liters/min), which is used only to drive an enslaved stage allowing an important outflow (up to 700 liters/min capacity). A displacement sensor on the enslaved stage provides a signal proportional to the position of the enslaved stage and then roughly proportional to the output flow of the servo-valve. There also exists an inner loop identical to the previous one but which uses the LVDT feedback of the pilot stage.

In spite of the settings of the various loop gains, this kind of control does not allow us to obtain a flat transfer function within a wide frequency range. This is the reason why three variable controls are used on some systems.

The system combines the information from each of the following signals:

– displacement of the low frequency;

– speed in the middle frequency range;

– acceleration for high frequencies.

The servo-controller subtracts the feedback for each of the three variables of the corresponding reference signal, which thus form three error signals. The three errors are then amplified and added in order to make up a global error signal making it possible to drive the system. In order to stabilize the control system, pressure-measuring feedback is injected at the level of the error signal. The technique is generally used on the hydraulic circuit to improve the stability of the system. The function stabilizes the high frequencies produced during the seismic tests to provide more stable acceleration and above all it allows us to reduce the influence of the resonance frequency of the oil column of the jacks.

For multi-axial tables where several jacks enable the table to be driven in a given direction, the control systems are split into degrees of freedom, that is, that the reference signals do not correspond to the signals sent to each jack but to the control in one degree of freedom. From the reference signal and the geometrical positions of the jack, the control system generates a control signal per jack.

In spite of the control loops, it is difficult to obtain a flat transfer function for the whole frequency range. Therefore, a computer is used. From a pre-test (impulse or random-loading type) it measures the transfer functions for each degree of freedom. Then, from these transfer functions, the reference signal is corrected and a control signal is determined, which makes it possible to obtain the desired accelerogram on the table. After each test, the actual response is measured on the table and compared with the reference signal. An iterative calculation system allows the control signal to be corrected again, in order to converge to the desired response.

As the jacks have to move important masses (a table and model), they are fixed on a reaction basemat, the mass of which is generally at least 20 times bigger than the mass of the models to be tested. Furthermore, to avoid generating any vibrations inside the test hall, the reaction basemat is laid on suspension set at a very low frequency (< 1 Hz). This suspension consists of springs or anti-vibration masses and dampers.

Finally, digital acquisition methods enable us to digitize and store all the responses of the various sensors implanted on the models.

5.5.2. *Characteristics and performance of shaking tables*

Generally, due to the large dimensions and masses of tested structures, the loading system consists of one or several electro-hydraulic jacks. In fact, they are the only equipment that allows the application of high-amplitude motions to important-mass models at low cost and in a small space. In general, the jack strokes are about 250 mm, compared with the electro-dynamic strokes of vibrating pots, which are smaller at 50 mm. The other advantage of jacks is that they allow very low frequency motions and even continuous strokes. One of their negative aspects is seen if we consider the frequency response of the servo-valves: the maximum responses are lower than 100 Hz (500 Hz at the most for some jacks with limited maximum strokes).

The performance of shaking tables is determined by three limiting factors:

– displacement limit corresponding to the maximum stroke of the jack or the jacks;

– a speed limit corresponding to the maximum outflow supplied to the jacks. The outflow is restricted by the maximum flow provided by the hydraulic pumps or by the maximum flow possible through the servo-valves;

– an acceleration limit that is a function of the mass loaded on the table of the mass of the table itself and of the maximum force of the jacks.

The maximum force of the jacks depends on the section of the piston, the oil pressure supplied by the pumps and the head losses in the servo-valves. The pressure commonly used in hydraulic groups is 210 bars, and at high frequency, head loss in the servo-valves is about 40 bars.

Thus, the maximum acceleration on the table is given by:

$$\gamma = F/\left(M_t + M_s\right) \qquad [5.11]$$

with:

– M_t = mass of the table;

– M_s = mass of the tested structure;

– F = maximum force of the jacks.

The maximum speeds reached by hydraulic systems are about 1 m/s. For large installations, the maximum outflows cannot be supplied exclusively by the hydraulic pumps, which deliver a constant outflow. During tests with sinusoidal loads, the average necessary outflow Q_m is given by $Q_m = 2\ Q_k/\pi$. (Q_k being the peak outflow). For *time history* tests, i.e. when we try to reproduce a seismic motion versus time (acceleration or displacement) for a limited period, it is possible to increase the speed by using auxiliary accumulators, which are often placed as reserve powers on the hydraulic network. Thus, once pressurized, they allow us to supply the system for short moments with the amount of oil necessary to pass the speed peaks.

Shaking tables have another limit, namely a frequency limit. The frequency range can be limited by:

– the first frequency specific to the table;

– the dynamic response of the servo-valves;

– the resonance frequency of the oil column.

Figure 5.6. *General view of the 4 shaking tables of the TAMARIS facility in Saclay*

As far as the eigenfrequencies of the table is concerned, it is a direct function of the dimensions of the table and its design features. The most common tables are between 1 m x 1 m to 3 m x 3 m in size. The largest tables are 6 m x 6 m large (CEA/Saclay in France, Berkley in the USA). A few exceptional tables can reach 15 m x 15 m (Tadotsu in Japan). Plans are underway to allow coupling between several large-size tables. The tables themselves are generally made of steel, though some are made of pre-stressed concrete. As they are quite heavy, they are problematic as far as the mass or the maximum acceleration are concerned. In order to improve performance, tables can be made of aluminum or composite materials.

With small shaking tables that use small servo-valves, the maximum frequencies due to the responses of servo-valves are about 100 to 150 Hz. With other installations requiring high-outflow servo-valves, the maximum frequencies are about 50 to 100 Hz. To improve the frequency efficiency with some big installations, it is possible to use several smaller servo-valves mounted in parallel, themselves used as drivers to operate the racks allowing high outflows.

The resonance frequency of the oil column is due to the compressibility of the compressed oil inside the jack chambers. Although oil has a very high bulk modulus, the stiffness of the oil column is low enough for the frequency of the oil column to be within the relevant frequency range of the table.

The stiffness of the oil column is given by the formula $K = 4\ S^2\ B/V$ where:

– S is the section of the jack piston;

– B is the oil bulk modulus (= 1.4×10^9 N/m²); and

– V is the whole volume of oil enclosed in the jack.

The oil column frequency is then given by $F = 1/2\pi.\ \left(K/\left(M_t + M_s\right)\right)^{1/2}$.

For large shaking tables, the frequency ranges from 8 to 20 Hz depending on the load (for example, the frequency of the oil column of the AZALEE table in the vertical direction with a 50 ton model is about 20 Hz).

Another limitation is due to the stiffness of the table. Though it is reinforced and therefore quite stiff, the table is supported either by vertical hinged jacks or by truss rods or sliding rails. Such equipment has a finite stiffness which influences the behavior of the model and which should be taken into account for the calculation of the eigenfrequencies of the model. In many cases, a 6 Hz frequency model will have, once laid on the table, a bending frequency of 5 to 5.5 Hz.

Figure 5.7. *Test with a 1/3-scale load-bearing wall model on the AZALEE table*

The advantage of using a shaking table lies in the fact that it is the only test means that allows simulation of structure inertia with a distributed mass. In fact, in the case of pseudo-dynamic tests, the jacks only apply forces on nodes where all masses are supposed to be lumped. On the other hand, because the jacks have to set both the table and the model in motion, it is quite easy to see that the limit is linked to the maximum load supported by the table. In Europe, several laboratories carry out shaking table tests: Table 5.1 gives the characteristics of the tables they use for testing civil engineering structures.

The largest European table is located at the CEA in Saclay. It offers 6 degrees of freedom, is 6 m x 6 m in dimension, and can support a 100-ton structure. The biggest table is Japanese and is located in Tadotsu. It is 15 m x 15 m and can support 1,000 tons. A few projects with large tables are being studied in the USA (a multi-axial, 4 m x 4 m, long stroke (1 m) table at the University of Buffalo) and in Japan (a multi-axial long stroke table that can support 1,200 tons).

Characteristics	BRISTOL England	LEE/NTUA Athens Greece	MASTER ISMES Italy	LNEC Portugal	AZALEE CEA/Saclay France
Dimensions (m)	3 x 3	4 x 4	4 x 4	5.6 x 4.6	6 x 6
Mass of the table (tons)	3	10	11	40	26
Mass of the specimen (tons)	15	10	30	40	100
Maximum height (m)	4	11			12
Number of degrees of freedom	5	6	6	3	6
Translations	X,Y,Z	X,Y,Z	X,Y,Z	X,Y,Z	X,Y,Z
Rotations	θx, θy, θz	θx, θy, θz	θx, θy, θz		θx, θy, θz
Horizontal displacement (mm)	± 150	± 100	± 100	± 175	± 125
Horizontal speed(mm/s)	± 700	± 1,000	± 550	± 200	± 1,000
Horizontal acceleration (g)	± 4.5	± 2	± 3	± 1.8 and 1.1	
Vertical displacement (mm)	± 150	± 100	± 100	± 175	±125
Vertical speed (mm/s)	± 700	± 1,000	± 440	± 200	±1,000
Vertical acceleration (g)	± 7	± 4	± 2	± 0.6	

Table 5.1. *Characteristics of the main European tables*

Other smaller tables exist in several private European laboratories (including SOPEMEA in France, ANSALDO and ENEL in Italy and others in Spain, Germany, etc.).

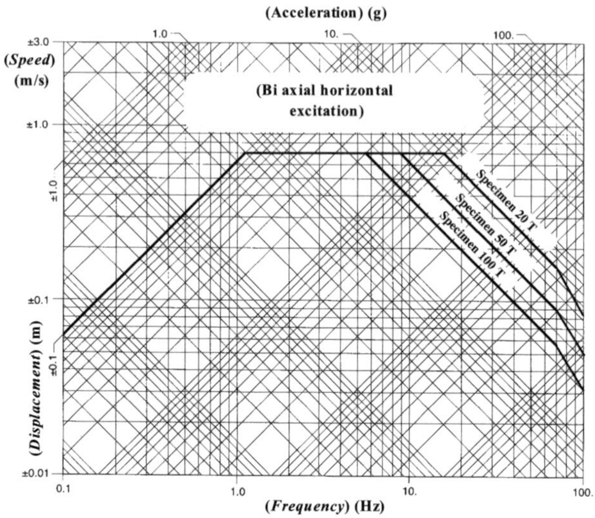

Figure 5.8. *Maximum performance of the AZALEE table under horizontal-horizontal biaxial load*

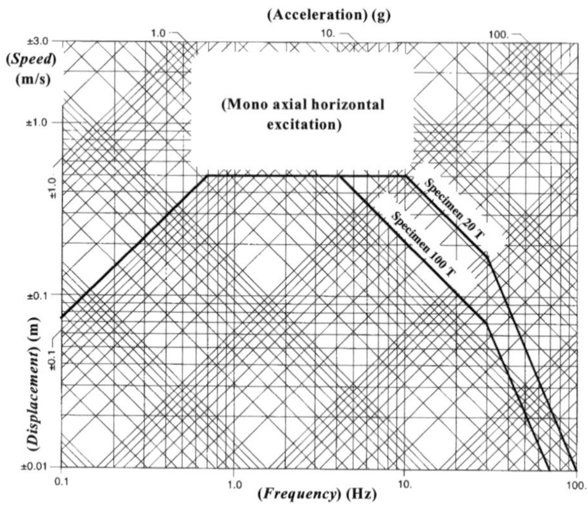

Figure 5.9. *Maximum performance of the AZALEE table under horizontal mono-axial load*

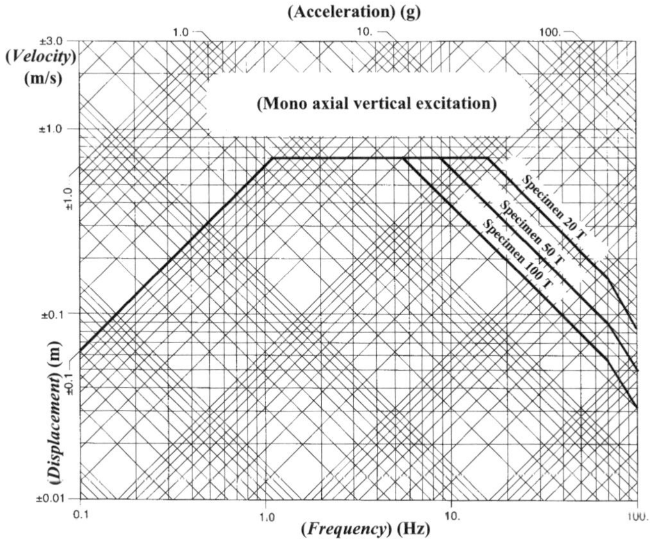

Figure 5.10. *Maximum performance of the AZALEE table under vertical mono-axial load*

5.6. Laws of similarity

As discussed above, the main limitation of shaking tables is the mass they can support. In many civil engineering cases, because of the masses and dimensions, it is impossible to carry out tests on real-size models.

For reinforced concrete, the minimum 6-cm thickness seems to be a limit value that allows the preservation of the mechanical characteristics of concrete. For conventional civil engineering models, this gives a reduced 1/3 scale. With a reduced scale, it is also necessary to represent the metallic frames, which raises the question of the accuracy with which it represents the mechanical characteristics of the steels. As small-diameter steels are not manufactured in the same way as large-diameter steels, they have different characteristics, particularly for the elongation limit as well as for the steel-concrete bond.

The tables tests carried out on reduced-scale models involve using laws of similarity to retrieve the stresses and the strains.

In most cases, the materials are kept identical, as it is impossible to find other materials with the same multiple characteristics of concrete.

Depending on the tests and models, two laws of similarity are used for shaking table seismic tests.

The first law of similarity is a speed law. It can be used when gravity does not influence the model's behavior or when permanent stresses do not prevail. For the same speed, the same stresses will also be found in the material. For a geometrical reduction scale n, this involves applying a horizontal acceleration that is n times as high as for scale-1 and is contracted in time by an n coefficient. As gravity remains constant, the vertical acceleration (or gravity) is then n times as low as it should be.

For civil engineering tests, generally, the gravity effect is important. Therefore, to get back to the normal loads and the stresses, it is necessary to apply a vertical acceleration that will have the right horizontal acceleration ratio. As gravity cannot be modified, another law of similarity called the acceleration law is used. In this case, both horizontal and vertical acceleration levels applied are similar to scale-1 acceleration, and the signals are contracted in time by a factor $\sqrt{n}$. In order to find the right stresses, because the masses have to be in $1/n^2$ (instead of $1/n^3$, if we keep the same material), it is necessary to load the floors or the reduced-scale model with additional masses.

5.7. Instrumentation

During seismic tests on civil engineering models, the following conventional instrumentation is used:

– accelerometers (mainly controlled or piezo-resistive types to have the lowest possible pass-band or a pass-band starting from zero);

– wire displacement transducers to allow measurement, even when the model has transverse displacements, and to allow measurement of important displacements;

– LVDT-type displacement transducers to measure crack openings;

– strain gages stuck on the metallic frames during manufacturing or stuck directly onto the concrete.

This instrumentation enables us to determine:

– the overall behavior of the model, by measuring the eigenfrequencies, the eigenmodes, the overall displacements and the accelerations at different points of the model; and

– the local behavior, by punctually measuring the strains and crack openings;

– the displacement transducers allow us to measure either the relative displacements (if they are fitted to a supporting frame loaded on the table or directly fitted on the model) or absolute displacements (when they are fitted on a supporting frame outside the table).

Accelerometers stuck at different points allow acceleration to be measured at a given point for determining torsion accelerations, or calculation of the average acceleration at floor level by summation, differentiation or averaging. Thanks to minute software that takes the dimensions of the model into account, it also allows us to obtain normal and shear force values at each floor and each level.

5.8. Loading

The models placed on a shaking table can be submitted to three different types of loading:

– sinusoidal scanning loading for measuring the eigenfrequencies of the model with the associated damping;

– random noise loading for measuring the eigenfrequencies of the model together with the associated damping;

– time history loading where time accelerations are applied in a mono-axial or multi-axial way.

With sinusoidal loading, scanning is continuous. The scanning rate is adjusted according to the damping of the structure to avoid being too fast to measure the resonance peak correctly (rates of about 1 or 2 octaves/min are commonly used for damping ranging from 2 to 5%). During the tests, a control system allows the table to be driven with constant acceleration over the whole frequency range. The acceleration control can take place on the peak value of the basic signal, i.e. on the signal measured on the table and band-pass filtered with a filter focused on the loading frequency, or on the value of the return signal (in which case, the return is disrupted by all the distortions measured on the table).

Damping is calculated either:

– from the width of the resonance peak at – 3 dB using the formula:

$\beta = \Delta F / 2F_0$, where ΔF is the peak width and F_0 is the resonance frequency; or

– from the resonance over-voltage using the formula:

$\beta = 1/2Q$, where Q is the resonance over-voltage ratio.

For random noise loading, we start from a flat spectrum in g^2/Hz. The software generates a random signal whose frequency band energy corresponds to that of the spectrum defined for the tests. During the test, driving software checks that the energy level remains constant within each frequency range throughout. Acquisition is carried out using a certain number of samples (a minimum of 64 to obtain a 5% accuracy). The duration of the samples depends on the analysis frequency band: it is longer when we want to analyze low frequencies. For each sample, a PSD is calculated and an average is obtained for all samples. Comparing the PSDs of the sensors with the PSD of the table sensor enables the identification of the resonance frequencies. This kind of test is generally less disadvantageous for reinforced concrete structures: unlike a sinus test, it stays at the resonance frequency for a shorter time and the maximum amplitudes reached are less important.

The *time history* tests involve applying a given seismic load to the table and in reproducing it as well as possible. The load can correspond to any of the following:

– a real earthquake recorded on site (real accelerogram);

– acceleration derived from pre-dimensioning calculation;

– acceleration generated from the spectrum of oscillators (synthetic accelerogram) for a given damping.

5.9. Conclusion

In conclusion, the advantages of shaking table tests are as follows:

– they allow simulation of inertia loads on a structure with a distributed mass;

– they allow tests to be carried out with vertical loads;

– they reveal some dynamic behaviors (vertical normal loads due to the opening and closing up of cracks);

– it is not necessary to use calculations to take the structure damping into account since they give the real response of the structure.

However, their drawbacks include the following:

– limitation of the loaded mass, which dictates carrying out tests on reduced scale models, which may not be truly representative;

– the stiffness of the table influences the eigenfrequencies of the model;

– the soil-structure interaction is difficult to take into account (generally, the table is considered as hard soil (rock-type));

– driving and reproducing the accelerogram on the table when the model is very heavy is difficult, deteriorates noticeably during the test for the current driving systems, and cannot take into account sudden variations in the mechanical characteristics of the models.

5.10. Bibliography

[BUC 94] BUCHET P., PEGON P., "PSD Testing with Substructuring – Implementation and Use", *JRC Special Publication*, No. I.94.25, JRC, Ispra, Italy, 1994.

[CHA 02] CHANG S.-Y., "An improved on-line dynamic testing method", *Engineering Structures*, Vol. 24, pp. 587–596. 2002.

[COM 97] COMBESCURE D., PEGON P., "α-operator splitting time integration technique for pseudodynamic testing: error propagation analysis", *Soil Dynamics and Earthquake Engineering*, Vol. 16, pp. 427–443, 1997.

[DER 85] DERMITZAKIS S.N., MAHIN S.A., "Development of Substructuring Techniques for On-Line Computer Controlled Seismic Performance Testing", *Report N° UCB/EERC-85/04*, Berkeley, USA, 1985.

[DON 92] DONEA J., JONES P.M., MAGONETTE G., VERZELETTI G., "The pseudo-dynamic test method", in *Recent Advances in Earthquake Engineering and Structural Dynamics*, V.E. Davidovici (Ed.), Ouest Editions, Nantes, France, pp. 769–780, 1992.

[ELB 89] EBERG E., The Pseudodynamic Test Method – Numerical Algorithms and Implementation, Doctoral Dissertation, Division of Structural Engineering, The Norwegian Institute of Technology, Trondheim, Norway, 1989.

[HIL 77] HILBER H.M., HUGHES T.J.R., TAYLOR R.L., "Improved numerical dissipation for time integration algorithms in structural dynamics", *Earthquake Eng. Struct. Dyn.*, Vol. 5, pp. 283–292, 1977.

[MAG 91] MAGONETTE G., "Digital control of pseudo-dynamic tests", in *Experimental and Numerical Methods in Earthquake Engineering*, J. Donea and P.M. Jones (Eds), Kluwer Academic Publishers, Dordrecht, The Netherlands, pp. 63–99, 1991.

[MAG 93] MAGONETTE G., "The digital control of pseudodynamic tests", *Experimental Methods in Earthquake Engineering and Structural Dynamics*, J. Mazars and E. Booth (Eds.), AFPS-SECED, Saint-Rémy-lès-Chevreuse, France, pp. 50–62, 1993.

[MAG 98] MAGONETTE G., PEGON P., MOLINA F., BUCHET P., "Development of fast continuous pseudo-dynamic sub-structuring tests", *Proc. of the Second World Conference on Structural Control*, 28 June and July 1998, Kyoto, Japan, 1998.

[MAG 00] MAGONETTE G., PEGON P., BUCHET P., BONELLI A., "Continuous pseudo-dynamic testing with non-linear sub-structuring", *Proc. of the Second European Conference on Structural Control*, 2–6 June 2000, Paris, France, 2000.

[NAK 87] NAKASHIMA M., KATO, H., "Experimental error growth behavior and error growth control in on-line computer test control method", BRI-*Report* No. 123, Building Research Institute, Ministry of Construction, Tsukuba, Japan, 1987.

[NAK 90] NAKASHIMA M., KAMINOSONO T., ISHIDA M., ANDO K., "Integration Technique for substructure pseudodynamic test", *Proc. 4th U.S. National Conference on Earthquake Engineering*, Vol. 2, Palm Springs, California, USA, pp. 515–524, 1990.

[NAK 93] NAKASHIMA M., AKAZAWA T., SAKAGUCHI O., "Integration method capable of controlling experimental error growth in substructure pseudo dynamic test" (in Japanese), *J. of Struct. Constr. Engng. AIJ*, Vol. 454, pp. 61–71, 1993.

[OLI 91] OLIVEIRA C.S., HAO H., PENZIEN J., "Ground motion modelling for multiple-input structural analysis", *Structural Safety*, Vol. 10, pp. 79–93, 1991.

[PEG 96a] PEGON P., "PSD testing with sub-structuring: the case of asynchronous motion", *JRC-Special Publication*, No. I.96.42, Ispra, Italy, 17 pages, 1996.

[PEG 96b] PEGON P., "Derivation of consistent proportional viscous damping matrices", *JRC-Special Publication*, No. I.96.49, Ispra, Italy, 26 pages, 1996.

[PEG 98] PEGON P., "The numerical simulation of buildings subjected to seismic loading: application of an integrated modelling strategy", invited lecture, *Proc. 11th European Conf. on Earthquake Engng.*, Paris, France, 1998.

[PEG 00] PEGON P., PINTO A.V., "Pseudo-dynamic testing with sub-structuring at the ELSA laboratory", *Earthquake Engineering & Structural Dynamics*, Vol. 29, pp. 905–925, 2000.

[PIN 95a] PINTO A.V., VERZELETTI G., PEGON P., MAGONETTE G., NEGRO P., GUEDES J., "Pseudo-dynamic testing of large-scale R/C bridges", *EUR Report* n° 16378 EN, JRC, Ispra, Italy, 1995.

[PIN 95b] PINTO P.E., "Research and practice on seismic design of bridges in Europe", *Proc. of the U.S. National Seismic Conference on Bridges and Highways*, San Diego, California, USA, 1995.

[PIN 96] PINTO A.V. (Ed.), "Pseudo-dynamic and shaking table tests on RC bridges", *ECOEST/PREC8 Concerted Action Report No. 5 (EC-HCM)*, LNEC, Lisbon, Portugal, 1996.

[PIN 02] PINTO A.V., PEGON P., MAGONETTE G., MOLINA J., BUCHET P., TSIONIS G., "Pseudodynamic tests on a large-scale model of an existing RC Bridge using non-linear substructuring and asynchronous motion", *EUR report* No. 20525 EN, Ispra, Italy, 2002.

[SHI 84] SHING B., MAHIN S.A., "Pseudo-dynamic test method for seismic performance evaluation: theory and implementation", *Report No.UCB/EERC-84/01,* Berkeley, California, USA, 1984.

[SHI 87] SHING B., MAHIN S.A., "Cumulative experimental errors in pseudo-dynamic tests", *Earthquake Eng. Struct. Dyn.*, Vol. 15, pp. 409–424, 1987.

[SHI 91] SHING B., VANNAN M.T., CATER E., "Implicit time integration for pseudo-dynamic tests", *Earthquake Eng. Struct. Dyn.*, Vol. 20, pp. 551–576, 1991.

[STO 90a] STOTEN D.P., BENCHOUBANE H., "Empirical studies of an MRCA algorithm with minimal controller synthesis", *Int. J. Control*, Vol. 52, No. 4, pp. 823–849, 1990.

[STO 90b] STOTEN D.P., BENCHOUBANE H, "Robustness of a minimal controller synthesis algorithm", *Int. J. Control*, Vol. 51, N° 4, p. 967–983, 1990.

[STO 92] STOTEN D.P., "Implementation of MCS on a servo-hydraulic testing machine", *Proc. I Mech E – Part 1, Journal Sys. Cont. Eng*, Vol. 206, No. 13, pp. 189–194, 1992.

[TAK 75] TAKANASHI K., UDAGAWA K., SEKI M., OKADA T., TANAKA H., "Nonlinear earthquake response analysis of structures by a computer-actuator on-line system", *Bull. of Earthquake Resistant Structure Research Centre* No. 8, Institute of Industrial Science, University of Tokyo, Japan, 1975.

[WAN 01] WANG Y.-P., LEE C.-L., YO T.-H., "Modified state-space procedures for pseudo-dynamic testing", *Earthquake Engng Struct. Dyn.*, Vol. 30, pp. 59–80, 2001.

Chapter 6

Experiments on Large Structures

Introduction

Realistic modeling of the behavior of civil engineering structures is the first and most difficult step in dynamic or seismic analysis. The parameters such analysis can define include mass distribution, damping characteristics, the stiffness of the strength system as regards inertia, the influence of secondary elements and various interaction phenomena. Forced or ambient vibration testing of large civil engineering structures are generally used to validate these parameters at the design stage.

The ability of particular numerical methods or mathematical models to represent the different interaction phenomena that determine the dynamic behavior of civil engineering structures can only be judged by comparing the forecasts they give with well-documented experimental results. Thus, dynamic tests allow the creation of experimental databases that can be used to develop and validate specific digital models, taking particular conditions or special interactions into account.

From a practical perspective, the dynamic testing of real structures is very important when assessing the safety of large works such as dams and bridges, because they characterize the fundamental properties of a structure, like the mass and stiffness at certain vibration frequencies, and also provide information regarding damping. Variations in these properties give a good indication of the changes the

---

Chapter written by Patrick PAULTRE and Jean PROULX.

physical properties of the works will undergo. It is thought that soon, dynamic testing will be used to not only to detect, locate and quantify damage, but also to assess the likely productive life of all civil engineering works. This chapter briefly describes the instrumentation used to perform dynamic testing. Both forced vibration and ambient vibration tests, widely used for assessing civil engineering structures, are presented. Some examples of real-life applications on large structures – a cable-stayed suspension bridge, a pedestrian footbridge, a large gravity dam and a large arch dam – are used to illustrate different test techniques. Tests used to detect damage in a civil engineering structure are also explained, and the most recent research in this area is described.

6.1. Instrumentation

The main consideration in any experimental investigation is the accuracy and reliability of the instrumentation being used. In a dynamic test, all the instruments can modify the measured signals. Such modifications must be taken into account during data processing. To do this, it is necessary to consider all the components of the instrument, from the measuring device to the analog-to-digital converter of the acquisition device. Appropriate corrections can then be made at the data processing stage. A review of the different instruments used for tests on large civil engineering structures is presented and includes accelerometers, hydrophones, strain and displacement gauges, and connection wires and data acquisition devices.

Force-balanced or piezoelectric accelerometers are commonly used for structure testing. At the low frequencies typically associated with civil engineering structures, the most commonly used devices are of the former type, generally seismic-mass accelerometers whose frequencies range from 0 to 300 Hz, the 0 to 50 Hz range being the most common. The advantage of using such accelerometers is that load amplification is not required.

An accelerometer behaves as a low-pass filter that reduces the amplitude and modifies the high-frequency phase of the measured signal. Figure 6.1 shows the frequency response curve of an accelerometer with a cut-off frequency of 50 Hz (3 dB reduction, or about 30% of the amplitude of the measured signal).

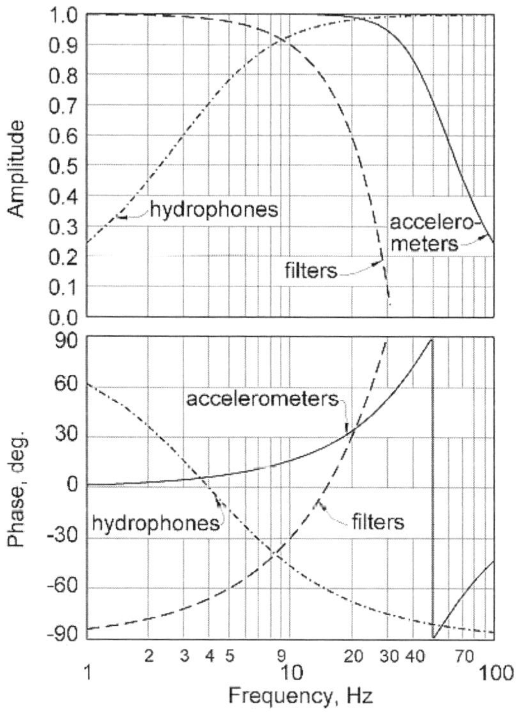

Figure 6.1. *Frequency response of some measuring devices*

Several types of dynamic acquisition devices are available. We recommend use of devices that can carry out several acquisitions simultaneously (*track and hold*), thereby eliminating some signal aligning corrections. Acquisition devices must be fitted with *anti-aliasing* "hardware" filters before the analog-to-digital conversion, to prevent frequencies near the acquisition devices getting into the range of interest, because of aliasing during the conversion into digital format. As an example, Figure 6.2 illustrates a 0.4 second long 95 Hz signal. Suppose that the sampling frequency of the data acquisition device were 100 Hz, the black spots representing the digitized values. The system would then record a 0.2 s-period signal instead of the real one, and the latter would have aliased to 5 Hz. In this case, use of a 50 Hz "hardware" filter would ensure the original 95 Hz signal almost completely died down before being digitized.

Controling the cut-off frequencies of these filters is essential, as they may change from test to test. The frequency response curve of the hardware filter should also be determined in order to make the necessary corrections when processing the signal. A frequency response curve for a filter with a 20 Hz cut-off frequency is shown in Figure 6.1.

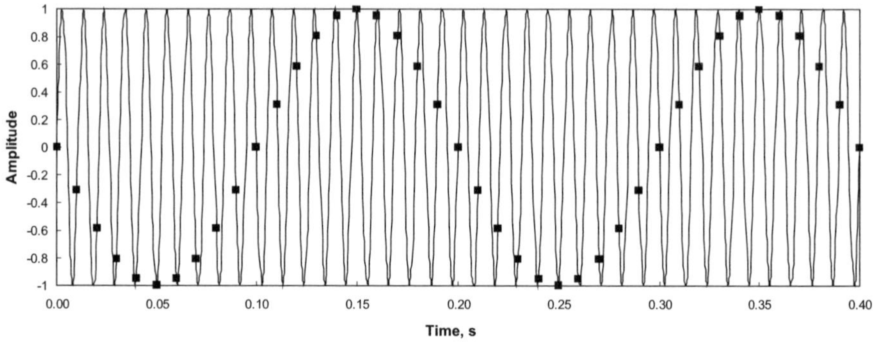

Figure 6.2. *Frequency aliasing*

Hydrophones are high-accuracy pressure sensors used to measure hydrodynamic pressures in dam reservoirs. These sensors behave as high-pass filters, as Figure 6.1 shows. The strain and displacement sensors do not require any measuring corrections, but the signal from such instruments has to be amplified before acquisition.

The different instruments are connected to the data acquisition device via several-conductor electric wires, which can reach several hundred meters in length due to the large dimensions of typical structures. These wires often cause noise with or without any load amplifier. Noise originates mainly from three sources: electromagnetic noise, ground loops and tribo-electric effects.

Electrical wires conveying large amounts of alternating current cause electromagnetic effects. Such cables are surrounded by powerful electromagnetic fields that vary over time, and these induce static voltages in parallel cables. We recommend the use of sheathed wires with a conducting surface, which protects the inner part from external electrostatic and electromagnetic fields. The wires carrying signals should be placed as far as possible from electrical wires, and the use of wound cables liable to turn a transformer should be avoided.

A ground loop arises when a common connection inside a system is grounded to more than one point. This happens, for example, when the shielding of the wire carrying the measured signal is grounded at both extremities. Both groundings are unlikely to have the same potential, and the potential difference causes currents to flow within the loop. The signal will therefore be modulated by the potential difference, which will induce noise in the measuring system. We can avoid forming such a loop by grounding only one end of the signal carrying wires.

Tribo-electricity effects are due to the static electricity produced by friction between two different surfaces. Thus, the bending and folding of a cable, a shock or pressure can create an electrical charge between conducting wires and the outer shield, or between the conducting wires themselves, which generates a voltage between both ends of the wire. For this reason, we recommend use of low-noise wires. These are generally tested along their whole length by manufacturers, to guarantee low noise levels. When adding connectors at the ends of these wires, we should take care not to create leaking spots. Using carbon tetrachloride as a solvent and xylene as a cleaning agent is also helpful.

6.2. Dynamic loads

The development of reliable testing methods and high-frequency analog-to-digital data recording devices has facilitated the dynamic testing of large civil engineering structures. Two main dynamic loading methods are used.

The first test involves applying a harmonic load, via an unbalanced mass, as an exciter (the principle of which is illustrated in Figure 6.3). The harmonic forced response is measured and the process is repeated over a given frequency range. The amplitude of the harmonic force applied to the structure is proportional to the squared excitation frequency. The main drawback of this method is that it is difficult to generate an appreciable harmonic force at low frequencies, and impossible to obtain the static response of the tested structure. The testing method proceeds with rather low increments of the excitation frequency to obtain a sufficient frequency resolution, which involves protracted experiment times.

Despite this, the advantages of the method are appreciable, which explains why it is still used so often. To begin with, the whole power available for excitation concentrates at only one frequency. It also enables structure linearity to be tested by the successive introduction of different amplitude harmonic forces for the same excitation frequency: this is easily achieved by varying the mass or its eccentricity.

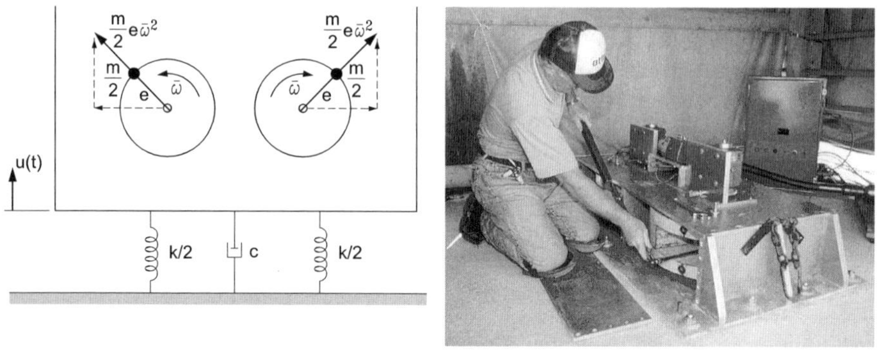

Figure 6.3. *Schema of an unbalanced rotating mass exciter connected to a structure and installation of an exciter on the crown of a concrete dam*

The second method involves applying a stationary random excitation and measuring the stationary response. This modal test technique is used quite widely in mechanics in aeronautics. It is seldom used on civil engineering structures because applying the random load and measuring it accurately is far from easy.

An alternative to this method is being used increasingly in civil engineering and involves using ambient environmental loads as exciters, choosing an accelerometer as the *input* reference signal and treating all the others as *outputs*. These approaches are called *output signal-only methods* or *under ambient load test methods*.

6.3. Data processing

Dynamic tests aim to determine the main dynamic properties of a structure, including its resonance frequencies, modal damping and vibration modes, by processing signals from it in both time and frequency fields. How is the data obtained actually processed?

Forced vibrations: in the case of forced vibration tests, data processing is relatively easy; it involves establishing the frequency response curves. These are generally better defined with ambient vibration tests and allow a better assessment of damping owing to the half-power bandwidth method. Frequency response curves are obtained from the raw data gathered under forced vibrations (see Figure 6.19 section 6.7.2). The main steps involved in data processing under forced vibrations are given below:

– For each excitation frequency, we must:

- determine the exact excitation frequency and the force generated by the exciter,

- calculate of the amplitude and phase delay for this frequency for each measuring unit. This can be done using the least squares method. The amplitude is standardized with regard to the force, for the latter is not constant as it varies with the frequency square number. From this step, the signal from a measuring unit at a given frequency is replaced by only two values, namely the amplitude and the phase,

- correct the amplitude and the phase delay according to the frequency responses of the measuring and recording instruments (Figure 6.1).

– Likewise, for each measuring unit, we must:

- establish the frequency response curves (amplitude and phase),

- identify the resonance frequencies (peaks) and of the corresponding phases,

- determine of the modal damping associated with each frequency identified owing to the *half-power bandwidth method*,

- determine the forms of resonance from the amplitudes and phases of each measuring unit. Continuous duty accelerations are transformed into displacements and drawn, taking into account the phase delays in relation to a reference point (generally the excitation point).

Examples of frequency responses obtained under forced vibrations for tests on buildings and large dams are presented in the following sections.

Ambient vibrations: tests under ambient vibrations involve far more complex data processing, as calculating frequency responses is conducted in the frequency domain. The method involves calculating the frequency content owing to Fourier transform and obtaining power spectral density (PSD) curves. A maximum recording period should be used to optimize resolution within the frequency domain, and to identify neighboring modes. Rectangular, tapering extended cosine or exponential windows can be applied to the recorded signals. Exponential windows favor the first part of the recorded signal, whereas cosine windows favor the central portion.

To soften the effects of local vibrations, a time and frequency average is used. We therefore obtain average spectral density curves for several positions on the structure. In the case of real test modal techniques, once the excitation source is measured, it is possible to calculate the *cross-spectrum* (the ratio of Fourier transforms of the input and output signals). However, measuring the excitation source is often impossible and it is necessary to use a technique similar to modal techniques. This involves selecting a reference measuring position that will stay fixed during all the tests, and using this to calculate the cross-spectra of the obtained signals with regard to the reference signal. From this, we can estimate a coherence function from the cross-spectra [BEN 00]. This indicator varies from 0 to 1. A high value, associated with a peak on the power spectral density curves, reveals a mode specific to the structure, whereas local vibration modes lead to low values. Some examples of power spectral densities and coherence factors are shown in Figure 6.14.

Resonance can be directly identified from PSD curves using the coherence function as an additional indicator. It is possible to draw the eigenmodes for the amplitudes and phases associated with each resonance, as is the case with the forced vibration tests. Damping is often difficult to estimate because of the amount of noise in PSD curves, as the stress level of the structure under ambient vibrations is generally very low. To overcome this, a new method for identifying dynamic properties in the time domain has been developed, and it has been shown to provide an improved assessment of damping and the vibration modes from ambient vibration signals [AND 99].

6.4. Application to buildings

The finite element method is used to model very complex large civil-engineering structures all over the world. The ability of any mathematical model to represent the real behaviors of the structure can only be validated by comparing analytical results with experiment results. Obviously, such an approach cannot be achieved at the project definition stage or during designing, but it does make it possible to verify designing hypotheses. This method was used during the rebuilding of Montreal's Olympic Stadium roof, as the dynamic characteristics of the Stadium's tower had to be determined to extremely high precision. Besides determining the dynamic characteristics, repeated vibration tests can also be used as an instrument to measure damage in a structure. This technique has been successfully applied to a two-storey building subjected to higher and higher seismic stress levels causing appreciable damage in the structure.

6.4.1. *The slanting tower at the Montreal Olympic Stadium*

The main objective of the forced vibration tests carried out on the tower of the Montreal Olympic Stadium was to obtain vibration frequencies and modes to update a three-dimensional finite element model of the structure.

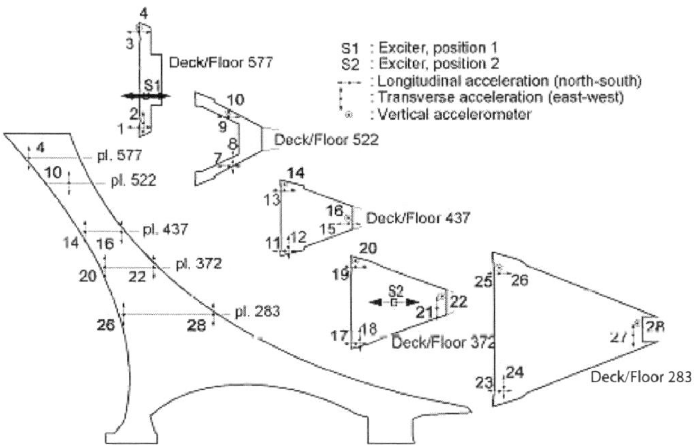

Figure 6.4. *Instrumentation of Montreal's Olympic Stadium slanting tower*

Researchers were especially interested in assessing the modal damping, since the lower part of the tower had been built in concrete (up to 133 m high), whereas the higher part was later completed in steel. Figure 6.4 shows a front view of the tower, and plane views of the instrumented levels.

Frequency sweeping from 0.5 to 9 Hz was carried out with an unbalanced gyrating mass exciter located (i) at the top of the tower (level 577 or 176 m high) and (ii) at the top of the concrete section (level 372 or 133 m high). With a 4 Hz frequency, the eccentricity of the masses was varied from 100 to 40% to reduce the excitation level. Accelerometers were placed on those levels and three others to measure the longitudinal, transverse and vertical responses.

Figure 6.5 shows the frequency responses obtained with standard amplitudes and phases at station 3 (longitudinal and vertical). The arrangement of the instruments was crucial for mode assessment and to allow distinction between torsion modes and bending modes.

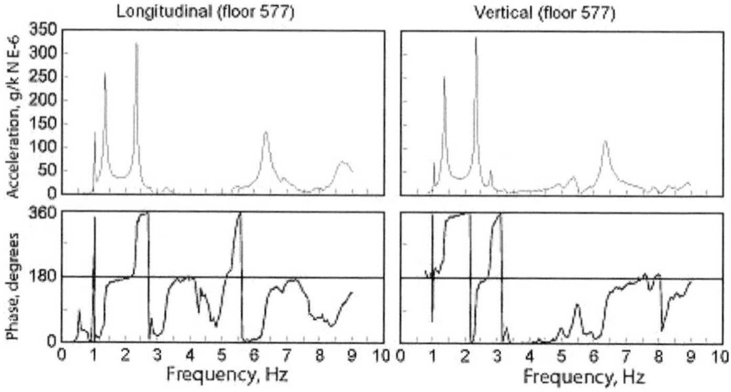

Figure 6.5. *Frequency response of the higher floor in the slanting tower*

Four torsion modes including the fundamental mode, and four bending modes were identified over the frequency range used. Three other coupled modes could also be observed. Ambient responses were also measured under the effect of wind, and the results were correlated to the dynamic properties obtained under harmonic vibrations. Vibrations due to the motion of the outer panoramic lift were recorded, both to estimate the amplitude of the resulting accelerations and to identify the specific modes excited when it is operating.

In total, 11 experimental modes were identified, among them the first torsion mode at 1.02 Hz and the second bending mode at 1.35 Hz. This enabled a three-dimensional finite element model, including the interaction with the flexible roof, to be constructed. It predicted a fundamental bending mode and a second torsion mode. Modal damping was identified for all modes, and this enabled calibration of Rayleigh's damping on the first mode (1.29%) and the eleventh mode (2.55%) to be calculated.

6.4.2. *Reinforced concrete building*

A real-size reinforced concrete building was constructed at the Sherbrooke University structure laboratory, as part of a project to assess the performance of very high efficiency concrete structures under seismic loads. The building was subjected to increasing seismic load levels using the pseudo-dynamic test method presented in this book. Here we only discuss the second objective of that study, which was to measure the damage caused by dynamic testing. The signature of the undamaged building was obtained and used as a reference to determine the damage.

Figure 6.6. *Real size very high efficiency concrete building after pseudo-dynamic tests*

The building was 5 m long in the E-W (East-West) direction, and 4 m long in the N-S (North-South) direction. Each floor was 3 m high. The building was a stiff nodal element structure supporting bidirectional slabs made from 70 MPa high strength concrete (see Figure 6.6).

The building was equipped with strain and displacement gages, and accelerometers set on both floors to measure acceleration in the three directions. The harmonic forced vibration tests were carried out with an unbalanced gyrating mass exciter placed on the roof, slightly off-axis, in order to load the torsion modes. Figure 6.7 illustrates the instruments on both floors of the building, the location of the exciter and those of the loads added to represent the weight of the service loads. Vertical accelerations allowed the identification of the vibration modes associated with the floors.

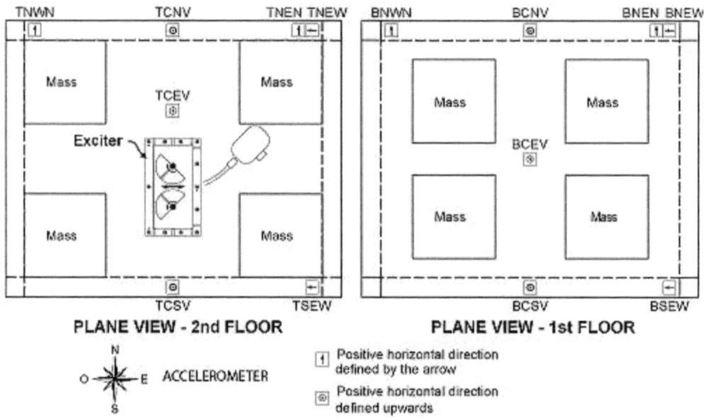

Figure 6.7. *Instrumentation of the building*

Figure 6.8 presents the frequency response curves measured on the first floor and the roof before damage, and after applying the standardized El Centro earthquake signal with a 0.129 g maximum acceleration. The effects of the damage caused by the earthquake are obvious in the figure and mostly involve an amplitude reduction, a notable increase of the modal damping (by a factor of 2) and a downward shift of the vibration frequencies. It is possible to quantify the damage and to locate it from this data, using appropriate system identification techniques [DOE 96]. Ultimately, we obtain damage "quantities" according to the stiffness reduction of the structural components (frames either parallel or perpendicular to loading directions, for example).

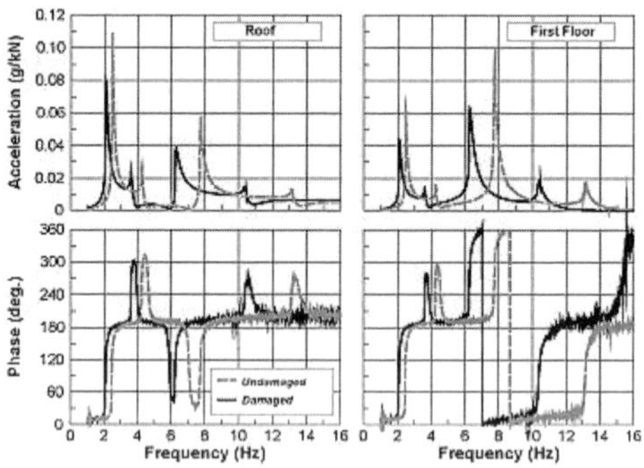

Figure 6.8. *Frequency responses of the undamaged and damaged building*

6.5. Bridge application

Bridges are probably the best characterized civil engineering structures as far as behavior is concerned, and specific dynamic test techniques have been developed for this type of structure. The tests are used for long-term damage and behavior monitoring as well as damage assessment following an earthquake. The following section examines two dynamic test applications on bridges (a pedestrian footbridge and a mixed cable-stayed/suspension bridge).

6.5.1. *Pedestrian footbridge*

As a rule, pedestrian footbridges are lightweight, slender structures that sometimes have vibration problems. In order to select the right vibration reduction techniques (passive or active damping devices), it is essential to know the dynamic properties of the structure and to estimate the vibration level with service loads as well.

The 60 m long pedestrian footbridge in Sherbrooke (Figure 6.9) is one of the earliest structure applications built using reactive powder concrete. It consists of a post-tensioned truss, the diagonal RPC beams of which are confined in thin stainless steel tubes, and a (30 mm) ultra thin slab. The innovative features of the structure have given birth to a novel behavior monitoring and instrumentation program.

Figure 6.9. *RPC pedestrian footbridge (Sherbrooke, Quebec, Canada)*

Figure 6.10 illustrates the instrumentation of the footbridge, and the locations of the vertical acceleration gages under a load due to pedestrians or pedestrian groups crossing.

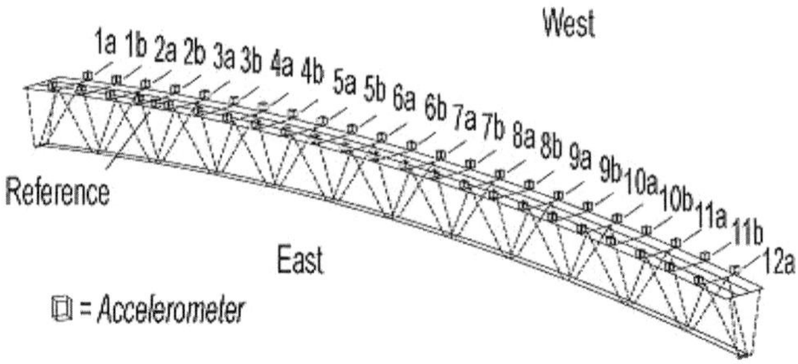

Figure 6.10. *Instrumentation of the footbridge*

Two different kinds of loading were used for the tests. In the first series of tests, several pedestrians walked randomly across the footbridge, walking in an unsynchronized way, in order that no vibration mode was favored. This was done in order to extract the dynamic properties (frequencies and damping and vibration modes) of the bridge under ambient vibrations. 120 s samples were recorded at 100 Hz for this part of the test.

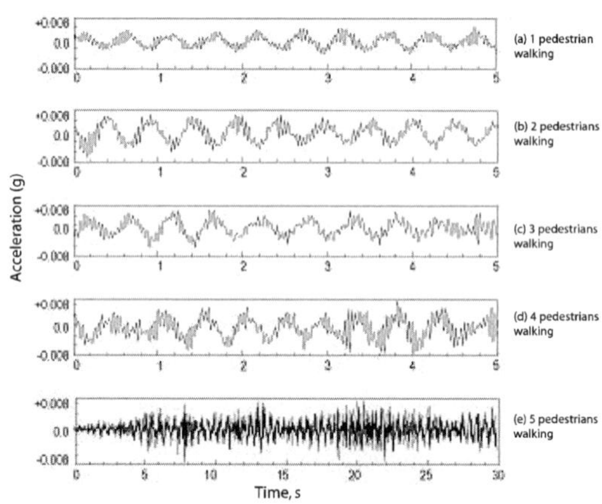

Figure 6.11. *Signals measured on the footbridge with different loading cases*

The purpose of the second series of tests was to estimate the vibration level under service loads. Pedestrian groups performed several "controlled" crossings, at different speeds and with different numbers of pedestrians on the footbridge. Figure 6.11 illustrates the signals measured while pedestrians were crossing, and Table 6.1 gives the vertical acceleration maximum values measured on the bridge deck.

Values respectively ranging from 0.022 g (0.22 m/s^2) to 0.14 g (1.40 m/s^2) could be observed with walking and running pedestrian groups. Design codes often specify the comfort criteria from critical accelerations under service loads, and frequently depend on the resonance frequencies.

Dynamic tests allow these criteria and their relationships with the dynamic properties of footbridges to be checked. For this footbridge, the critical acceleration specified by British standards (BS 5400) and by the Eurocode (EC2.2) is 0.76 m/s^2, whereas the critical acceleration specified by the Ontario code (ONT 83) is 0.48 m/s^2. These values can be exceeded in certain loading cases specified in Table 6.1.

Loading cases	Maximum accelerations (g)	Maximum accelerations (m/s^2)
1 pedestrian, walking	0.0135	0.132
2 pedestrians, walking	0.0133	0.130
3 pedestrians, walking	0.0159	0.156
4 pedestrians, walking	0.0219	0.215
Pedestrian group, walking	0.0222	0.218
1 pedestrian, running	0.0384	0.377
2 pedestrians, running	0.0604	0.593
3 pedestrians, running	0.1053	1.033
4 pedestrians, running	0.1352	1.326
Pedestrian group, running	0.1425	1.398

Table 6.1. *Maximum accelerations on the footbridge*

6.5.2. *A mixed cable-stayed/suspension bridge*

As far as road bridges are concerned, dynamic tests can take place with controlled traffic, using test vehicles (lorries and trailers) with a specified axle load, or under ambient vibrations generated by the wind or steady traffic. Bridge responses are measured by accelerometers, strain gauges and displacement transducers. Generally, the strains and displacements are used to obtain the dynamic amplification factor, whilst accelerations are used to obtain the dynamic properties, even if the accelerations can also be integrated to provide displacement responses.

Built in 1947, the suspended Beauharnois bridge is 177m long, and crosses one of the supply canals of an hydroelectric power station south-east of Montreal. The bridge underwent important structure modifications after a steel orthotropic deck slab and a mixed cable-stayed suspension system were installed. Dynamic tests were conducted on the new structure to calibrate a finite element model capable of taking the non-linear behavior of the cables and suspension into account.

The eastern part of the symmetric bridge deck was instrumented (Figure 6.12). Low-frequency accelerometers were placed at each suspender cable to measure the vertical, transverse and longitudinal responses of the bridge deck and of the east tower, under the influence of ambient and controlled traffic loads. Displacement transducers were placed at position 14, close to the east tower, and strain gauges placed on the main trusses.

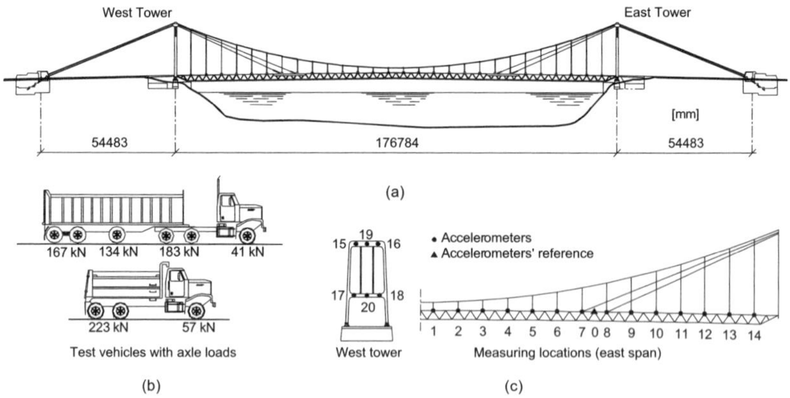

Figure 6.12. *Instrumentation of the bridge and test vehicles*

Two lorries and a trailer were used to cross the bridge at 75 different speeds, according to various positions on the bridge. The data recorded during the tests as well as the data obtained with usual traffic were used to calculate the dynamic amplification factor, a parameter which had rarely been calculated for stay-cable and suspension bridges.

Figure 6.13 shows the displacement versus the time, obtained thanks to LVDT n°2 (Position 14) with one lorry crossing at a speed of 76 km/h (48 mph). The maximum dynamic response was then derived from the measured signal, as shown in the figure. The maximum static response was obtained by filtering the dynamic response with a low-pass filter. The cut-off frequency and the bandwidth of the filter were calibrated from responses measured during low-speed crossings (quasi-static tests). These tests allow the measurement of the static deformation of the bridge and an equivalent to the line of influence to be measured at a selected point.

The dynamic amplification factor can be determined, as indicated in Figure 6.13, and an average calculated for each series of vehicles crossing, by working out the relationship between both response parameters. The same procedure is applied for strain measures. As is often the case with bridges, the positions of the two displacement transducers on the bridge were limited by the depth of the river and its currents. To overcome this problem, accelerations were integrated to give a rough estimate of the displacements to assess values of the dynamic amplification factor in different locations.

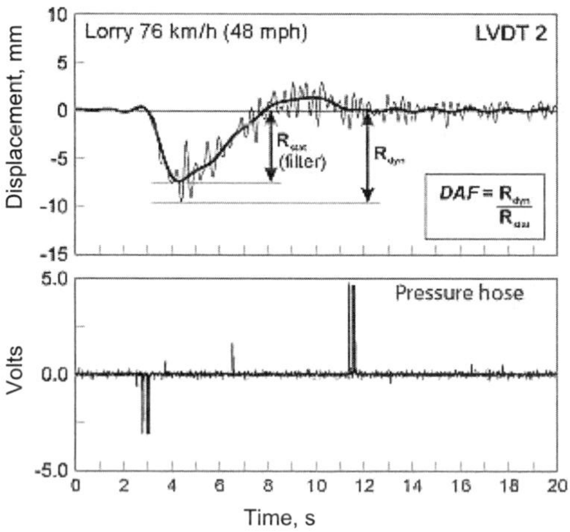

Figure 6.13. *Assessment of the dynamic amplification factor (DAF)*

Vehicle	Strain		Displacement*		Displacement**	
	Min.	Max	Min.	Max	Min.	Max.
10-tire lorry	1.19	1.20	1.10	1.11	1.06	1.07
Lorry and trailer	1.08	1.09	1.03	1.04	1.01	1.02
2 lorries in line	1.09	1.10	1.05	1.06	1.03	1.03
2 lorries side by side	1.10	1.10	1.06	1.07	1.04	1.05
All tests	1.12	1.12	1.07	1.07	1.03	1.04

*Measured displacements (LVDT) **Average of the integrated accelerations

Table 6.2. *Average values of the DAF*

Table 6.2 presents the average dynamic amplitude factor values corresponding to strains measured on the main trusses, for the displacements measured (LVDT at position 14), and the integration of the average accelerations recorded for all positions. The highest values were obtained for a single lorry, whereas strain dynamic amplification factor values were higher than the values obtained for displacements. This can be explained by the fact that strains are affected more by the contribution of higher modes than displacements, and also because the lowest vibration frequencies are below the 2 to 5 Hz range generally associated with commercial vehicles.

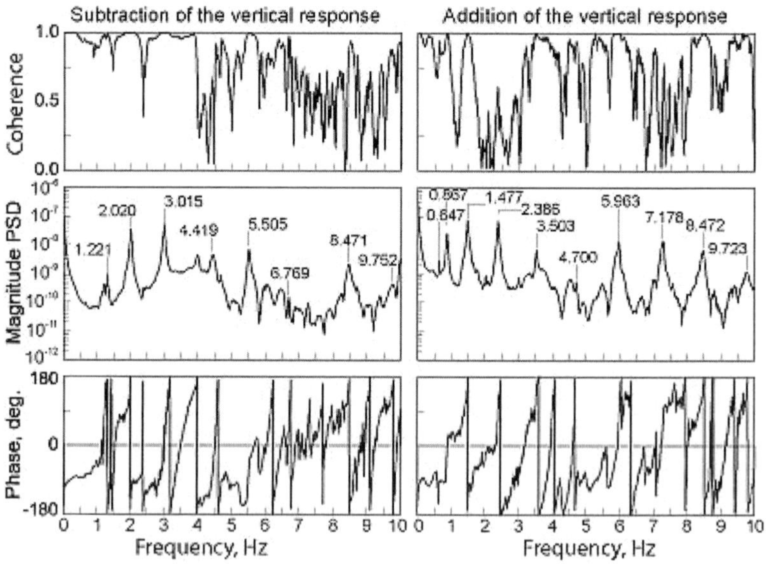

Figure 6.14. *Separation between torsion and bending modes*

Ambient vibration tests (steady traffic) were also carried out. As many as 70 ambient vibration response measurements were recorded for 120 seconds with a 100 Hz sampling frequency. The frequency content and coherence functions for vertical acceleration are shown in Figure 6.14. The responses of both sides of the bridge deck were added and subtracted to separate bending and torsion modes, which are often quite close to each other. Graphs on the right illustrate the frequency contents of the added data leading to the identification of bending modes, whereas those on the left illustrate the results obtained with subtracted answers where torsion modes are isolated. A total of 23 bridge deck modes (10 in bending, 8 in torsion, 4 lateral, 1 longitudinal) and 8 tower modes could be identified.

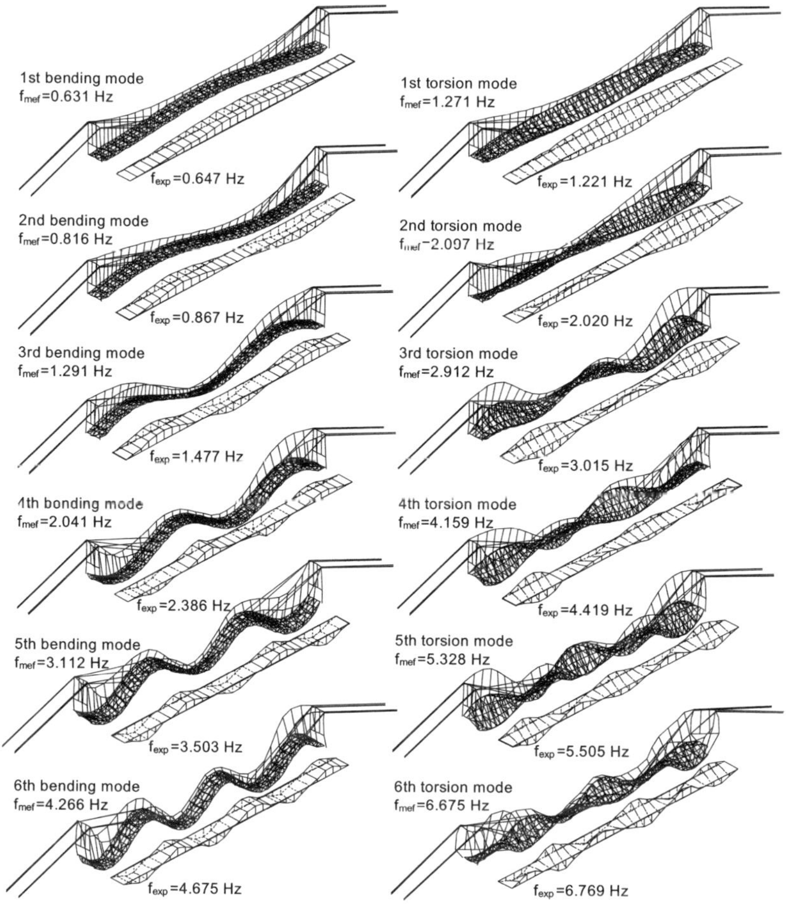

Figure 6.15. *Comparisons between the bridge vibration modes measured and calculated*

To calibrate the numerical model, initial tensions within the main suspenders and stays were calculated using a program aimed at the non-linear analysis of suspension bridges, which was used to rehabilitate the structure. A three-dimensional model was used to carry out an analysis of eigenvalues; finally, the results from the numerical model and its accuracy were checked against the experimental results (modes and frequencies).

Thanks to this method, the experimental modes of the bridge deck and the tower were predicted with an average accuracy of 5%. The first four torsion and bending frequencies, the experimental and numerical modes, together with the corresponding frequencies are illustrated in Figure 6.15. The determining factors for the calculation of eigenvalues and vectors in this case were found to be the cable tensions and the density of the highway

6.6. Application to large dams

The owners of large dams are increasingly concerned about the seismic safety of their works, and the best methods to assess them formed one of the main themes of the ICOLD Congresses (Large Dam International Commission).

In situ dynamic tests are acknowledged as the most reliable ways to assess the dynamic properties of dams [HAL 88, COL 90]. With recent advances in finite element modeling techniques that take complex interaction phenomena between different sub-structures into account (dam, foundation, reservoir, ice covering), compiling a reliable database on the dynamic behavior of dams is now greatly simplified. Ambient and forced vibration tests are also quite useful, but are carried out at significantly lower loading levels.

Recording accelerometer networks are already in place on some large dams ([DAR 95], [DAR 01]) and allow the behavior of these works to be studied in real time, especially during reservoir filling and emptying cycles. It is widely known that hydrodynamic pressures and wave absorption at the bottom of the reservoirs play an important part in the dynamic response of concrete dams, using hydrodynamic pressure measurements made during forced vibration tests on gravity and arch dams ([DUR 88], [PRO 97]).

The following sections describe three forced vibration test campaigns undertaken to assess the seismic behavior of large concrete dams, and highlight the use and efficiency of the non-destructive dynamic techniques used.

6.6.1. *Assessment of a response spectrum on the crown*

The Beauharnois gravity dam is located near Montreal (Quebec, Canada) and runs across the Saint-Laurent river. This 20m high concrete structure includes 36 alternators, and has a crown that is almost 1 km long. It was constructed in three phases in 1930, 1950 and 1960. The gravity dam is a channel-head with the power station juxtaposed downstream. The latter consists of concrete foundation, including the turbines and alternators, and a metallic superstructure.

A dynamic test campaign was carried out in collaboration with the owners, Hydro-Quebec, in order to determine the response spectrum on the dam's crown. Important works had been planned on the site and a building was to be erected on the crown to shelter the numerous transformers and other electrical equipment. It was therefore expedient to estimate the response spectrum on the crown, and this was assessed using data regarding the dynamic properties of the works.

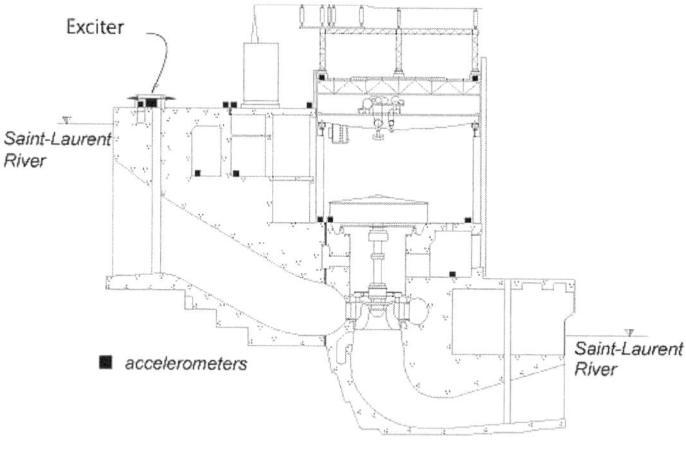

Figure 6.16. *Sectional drawing of the dam with test instrumentation*

As the Beauharnois dam produces a large amount of the total energy supply to the City of Montreal, stopping a group during the test period was out of the question. However, as repair work was underway on one of the alternators, the necessary measurements were made on an inoperative system, without contamination from noise due to the operation of the alternator and the water flowing through turbines. Nonetheless, some noise from the neighboring groups was measured.

An unbalanced mass exciter used to generate a horizontal harmonic force on the crown was placed inside the water intake valve shelters. Figure 6.16 is a sectional drawing showing the location of the accelerometers. As the test results were used to calibrate a two-dimensional finite element model, it was necessary to place instruments on the water intakes (upstream), the concrete power station and its metallic superstructure.

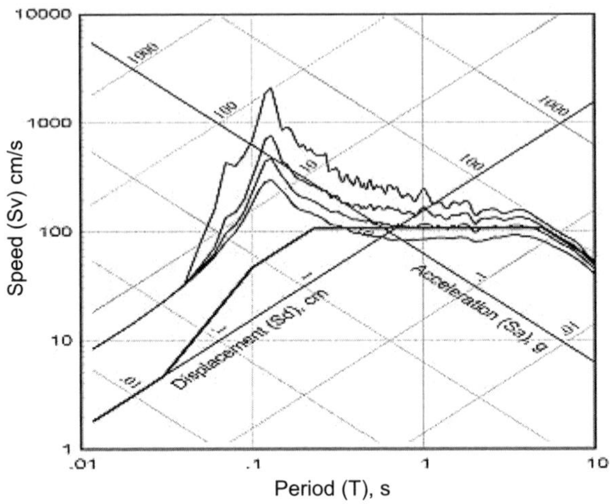

Figure 6.17. *Response spectrum on the crown of Beauharnois dam*

The frequency responses from each station were used to extract the resonance and damping of the system (the process will be illustrated for another dam in the next section). As the numerical model developed for the structure was two-dimensional, the identification was carried out from the first mode of the dam observed at a frequency of 8 Hz. The model was then used to estimate the influence of the different sub-structures in the dynamic response, by calculating the frequency response for a unit harmonic load on the crown, and comparing it to the accelerations measured during the tests.

Once the model had been calibrated with a reasonable confidence coefficient, it was used to calculate the response spectrum for a structure located on the crown. Several earthquakes, based on the dimensioning spectrum for the Montreal region, were generated and applied to the model. The crown's responses were then used to assess several spectra for different earthquakes: the average values for ten artificial earthquakes are shown in Figure 6.17. The resultant response spectrum was then used to design buildings located on the crown of the dam.

6.6.2. *Study of foundation-ice-reservoir-dam interactions*

Forced vibration dynamic tests provide well-defined frequency response curves; thus, the identification of dynamic properties is easier than it is with ambient vibration tests. Indeed, such tests require significant logistics, namely using an exciter and a generator set capable of supplying the necessary power (about 8 kW). Nevertheless, the accuracy of the results enables the validation of the numerical models, taking interactions between the different sub-structures into account, and thus facilitates validation of the other modeling techniques used to estimate the seismic response of hydraulic works.

From this perspective an ambitious project was undertaken, in collaboration with Hydro-Quebec, aimed at studying the dynamic interaction phenomena taking place between the dam, the foundation and the reservoir ([PRO 97], [PAU 02]). These sub-structures interact during an earthquake and play quite an important role in the dissipation of energy. Moreover, in Nordic countries, the ice cover represents another sub-structure liable to influence the dynamic response of the system.

The 84 m high Outardes 3 density-dam was chosen for the test campaign. It is located in the northern part of Quebec and its reservoir is covered with ice almost 6 months a year. Very little experimental data on the dynamic behavior of structures with such ice coverings exists, so the test campaign was also geared towards assessing the effect of ice on the dam's response. Forced vibration measurements were carried out in July and February, with daily temperatures ranging from –10 to –15°C in the latter case. The ice cover was measured and found to reach a thickness of up to 1.5 m along the upper bank.

The unbalanced mass exciter used for the tests can generate a force up to 90 KN within a 0 to 20 Hz range. Figure 6.18 illustrates the instruments used for the forced vibration tests. As the aim is to identify as many resonance values as possible within a given frequency range, it is sometimes necessary to change the exciter's location. When placed at the center of the structure, the latter will tend to favor symmetric modes. Although modifying the orientation of the stress in the horizontal plane is possible, it is sometimes necessary to move the excitement point to excite certain anti-symmetric modes as well.

The exciter positions used for the Outardes-3 tests are indicated by double arrows in Figure 6.18, and the accelerometers locations are identified by cubes. The latter were placed on the crown, in various locations in the inspection galleries and on the lower bank in order to model a three-dimensional representation of the resonance shapes.

To assess the reservoir-dam interaction, the hydrodynamic pressures generated by the motion of the dam and wave propagation inside the reservoir were measured. These pressures were collected by hydrophones mounted at regular intervals on a cable. The upper part of Figure 6.18 (summer tests) shows the cable network from which the hydrophones were suspended. The main cable was used to connect both banks of the reservoir, and secondary cables were placed perpendicular to the crown, linking the dam with the main cable. This floating assembly allowed a series of hydrophones to be suspended in several places inside the reservoir, and the hydrodynamic pressure response at 0, 30, 60 and 90 m from the upper bank was obtained. The distance between each transducer was 15 m, and ranged from 15 m to 75 m underwater. For the winter tests, with the ice cover as shown at the bottom of Figure 6.18, the cable network was unnecessary and the ice was merely drilled in several places to obtain measurements similar to those conducted in summer.

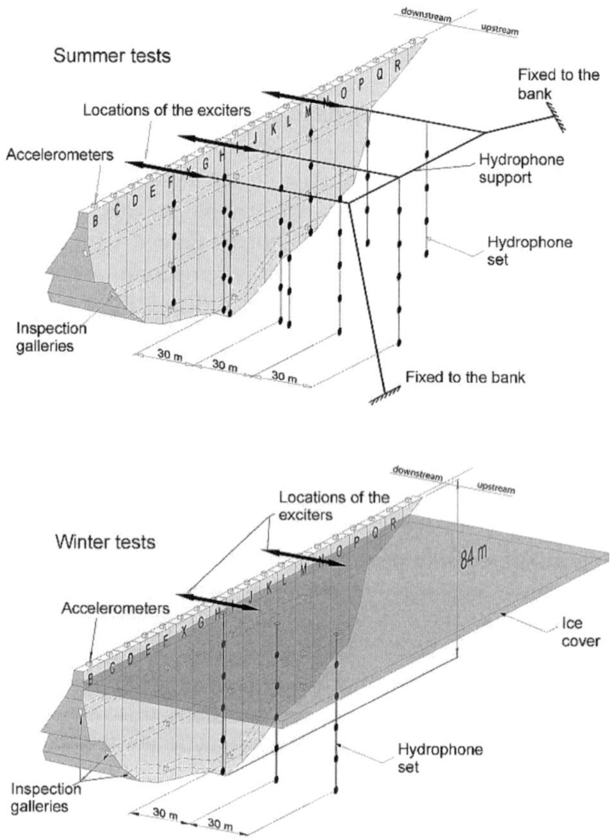

Figure 6.18. *Instrumentation of the Outardes 3 dam both in summer and in winter*

Figure 6.19 illustrates the typical forced vibration signals obtained. The hydrodynamic pressure is shown in the first two graphs ((a) and (b)) for two different depths along the upper bank, whereas the acceleration response obtained on the lower bank is illustrated in (c). A 5 V pulse is emitted for each mass rotation (graph (d)), which allows the exact rotation frequency of the exciter to be assessed (a little over 8 Hz in this case).

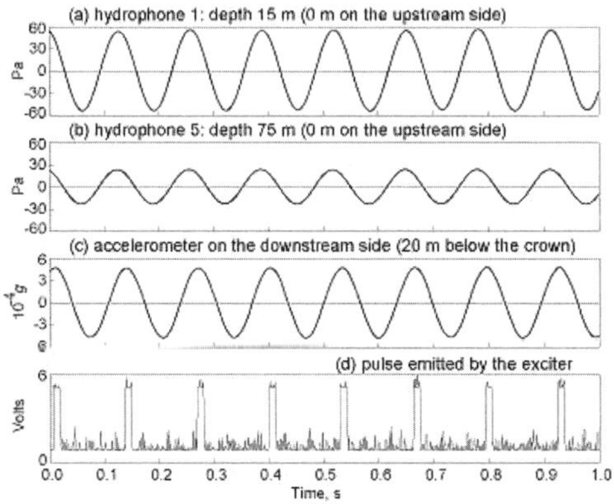

Figure 6.19. *Examples of signals measured under forced vibrations*

The signals were used to calculate the frequency responses for each measuring position and to extract the dynamic properties of the dam. Both graphs at the top of Figure 6.20 show the responses obtained on the crown for two exciter positions. By comparing the amplitudes and phases for the summer (dotted lines) and winter tests (solid lines), it is clear that amplitudes are far smaller when there is ice, and that resonance frequencies are generally decreased. Therefore, the ice cover imparts an additional stiffness to the reservoir-dam set and also increases modal damping indicated by smoothing of the peaks.

Figure 6.20 also illustrates the displacement effect of the excitation point. For results obtained in summer, the first resonance (which corresponds to a symmetric mode) is more loaded in a central position (block H) than in the quarter point position (block M). Conversely, the second resonance (associated with an anti-symmetric mode) is almost invisible at the central stimulation position and highly stressed at the other.

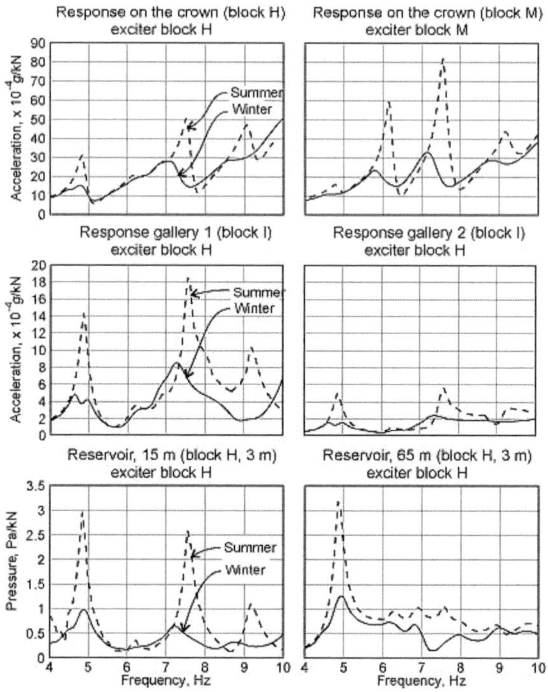

Figure 6.20. *Frequency responses on the crown*

Comparing both series of tests reveals a new mode near the first resonance. This is mostly visible with responses obtained inside inspection galleries, as illustrated by the diagrams in the middle of Figure 6.20. Amplitudes are drawn at the same scale to highlight the response reduction in passing from one gallery to another one. The double peak is clearly identified in the case of the first gallery, 20 m below the crown.

Ice cover also causes modifications similar to hydrodynamic pressures. The diagrams at the bottom of Figure 6.20 show responses along the upper bank of the reservoir (actually, 3 m away from the bank in the winter tests) at respective depths of 15 m and 65 m. The damping increase and resonance shift are obvious, as in the case of acceleration responses: the new mode can be observed as well. It should be remembered that the water level difference between both series of tests can be neglected and cannot account for the resonance frequency variation.

Resonance modes are obtained from amplitudes and phase delays for the different measuring positions on the crown. They make it possible to detect the appearance of a new mode due to interaction between the dam and the ice covering. This has a symmetric shape, as is the case with the first mode observed in summer.

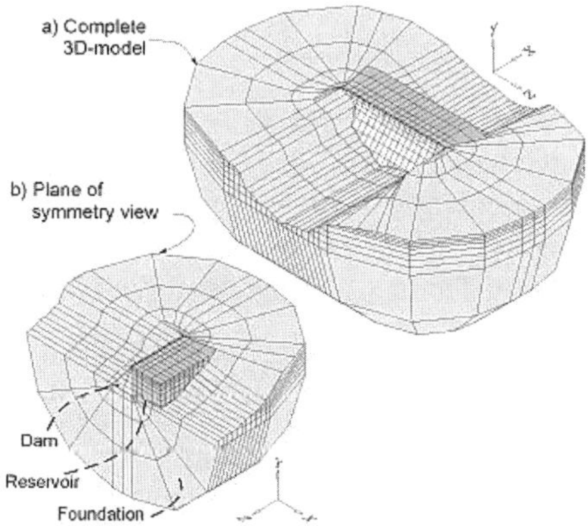

Figure 6.21. *Finite element model for the dam, the reservoir and the foundation*

These results have been used to calibrate a finite element model developed for the dam-reservoir-foundation set. This model is illustrated in Figure 6.21, which shows the mesh for the three structures. In this study, EACD-3D software [FOK 86], designed for three-dimensional seismic analysis of concrete dams, was used to carry out a parametric analysis. Effects due to dam and foundation stiffness, water compressibility (energy dissipation through upstream wave propagation) and dam-reservoir interactions were studied. The frequency responses on the crown, inside galleries and in the reservoir were compared and proved each sub-structure made its own contribution to the dynamic response of the construction works [PRO 97].

The results of the study were also used to develop a two-dimensional model of the ice covering [BOU 02], which was added to EAGD software used for the seismic analysis of gravity dams [FEN 84]. A parametric study confirmed the appearance of a new mode, together with the stiffness and damping modifications previously observed during the tests.

6.6.3. *Study of the effect of the water level inside the reservoirs*

Dynamic test techniques can be used to study the effect water level inside reservoirs has on the dynamic response of concrete dams. A multiple test campaign was completed on a large dam in Switzerland, as part of a collaboration between the Swiss Federal Water and Geology Office, Electricité de France, Hydro-Quebec and Emosson S.A.

Forced vibration measurements were carried out on the Emosson arch-dam, located on the French-Swiss border near Martigny. This 180 m high arch (represented by a finite element model in Figure 6.22) was submitted to forced vibrations by an unbalanced mass exciter, and the hydrodynamic pressure and acceleration responses for different water levels in the reservoir were derived [PRO 01].

The yearly variations of water levels inside the reservoir are shown in Figure 6.23. The test periods – there were four of them – were chosen with "empty" and "full" lake conditions, plus two intermediate levels. Access to the dam, which is located in an Alpine region, is limited by a snow covering which generally still prevails in spring during the empty reservoir conditions. The reservoir fills up in summer, and usually reaches its maximum operating level (roughly 225 x 10^3 m^3) in September.

Dynamic tests were conducted in a way similar to those described in the previous section. The frequency responses enabled several resonance values within frequencies ranging from 0 to 10 Hz to be identified, and derivation of the corresponding modal damping.

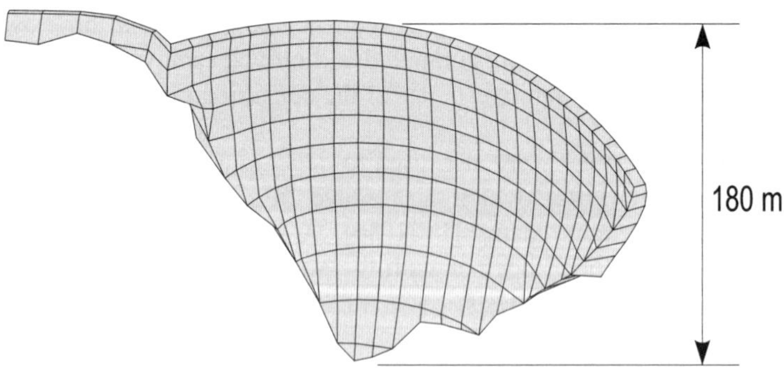

Figure 6.22. *Model of the Emosson arch-dam*

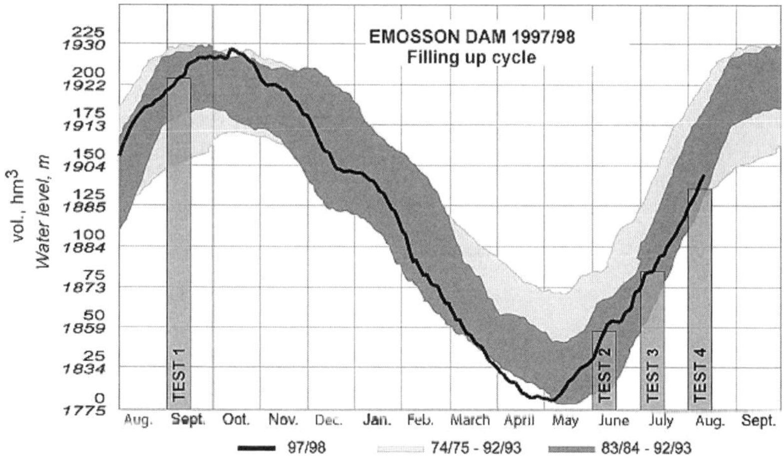

Figure 6.23. *Selection of the test dates during the filling-up cycle (source: Emosson S.A.)*

Figure 6.24 shows the variation of frequencies with water levels for the first seven modes. As the water level rises, resonance frequencies should decrease with the added mass from the reservoir. However, in the lower part of the curves in Figure 6.24, it is evident that the frequencies rise with the water level, and the trend gets more marked for higher modes. The same phenomenon was also observed during the ambient vibration tests at the Mauvoisin dam ([DAR 01], [DAR 00]).

This phenomenon has been attributed to an increase in the stiffness of the dam due to construction joints being tightened by the increased hydrostatic pressure. When water reaches a certain level, the phenomenon is counterbalanced by the added mass of the reservoir and the resonance values decrease. Figure 6.24 shows the level was reached between July and August.

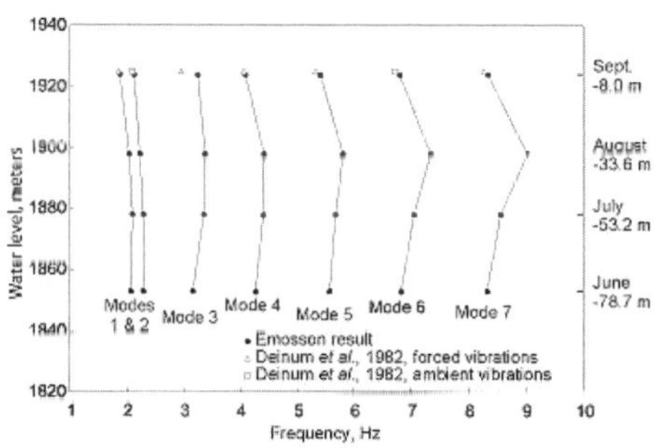

Figure 6.24. *Effect of the water level on the reservoir*

As for the project previously described, the experimental results have been used to develop and calibrate a finite element model of the same type as that shown in Figure 6.21. The predictions of this model, in terms of the frequency response of the acceleration on the crown, have been compared with test results for each measuring position. The main objective of this project was to model the influence of water level variation inside the reservoir on the dynamic response. For this structure, a 10% reduction of the dam stiffness simulated the opening of the joints in the dam due to a decreasing hydrostatic pressure.

6.7. Conclusion

A numerical model is only valid if it can accurately represent the behavior of a structure. Representing the dynamic behavior of a civil engineering structure is far more difficult than representing its static behavior, and an engineer's experience is required to determine the important parameters governing the response of the structure. As far as dimensioning or verifying the safety of a large structure is concerned, all the techniques presented in this chapter have been shown to constitute reliable and economical methods to confirm the hypotheses used for the building up of physical models. Furthermore, these techniques are at the root of new damage prediction methods in which multiple tests are used to follow the behavior of a structure over time.

6.8. Acknowledgements

The authors would like to acknowledge the support of the Civil Engineering and Natural Science Research Committee in Canada, the Quebec Technology and Nature Research Fund, as well as that of industrial partners who have contributed to the different research projects on the dynamic behavior of large civil engineering structures, notably the Ministry of Transport in Quebec, Hydro-Quebec, the Swiss Water and Geology Federal Office, Electricité de France and Emosson S.A.

6.9. Bibliography

[AND 99] ANDERSEN P., BRINCKER R., PEETERS B., ROECK G. (DE), HERMANS L., KRÄMER C., "Comparison of System Identification Methods Using Ambient Bridge Test Data", *17th International Modal Analysis Conference (IMAC)*, Kissimee, Florida, USA, pp. 1035–1041, 1999.

[BEN 00] BENDAT J.S., PIERSOL A.G., *Random Data: Analysis and Measurement Procedures*, third edition, Wiley Interscience, New York, USA, 2000.

[BOU 02] BOUAANANI N., PAULTRE P., PROULX, J., "Two-Dimensional Modeling of Ice Cover Effects for the Dynamic Analysis of Concrete Gravity Dams", *Engineering and Structural Dynamics*, 31, pp. 2083–2102, 2002.

[COL 90] COLLECTIF, US National Research Council Panel on Earthquake Engineering for Concrete Dams, *Earthquake Engineering for Concrete Dams: Design, Performance, and Research Needs*, National Academy Press: Washington, DC, USA, 1990.

[DAL 93] DALY J.W., RILEY K.G., MCCONNELL K.G., *Instrumentation for Engineering Measurements*, second edition, John Wiley & Sons, New York, USA, 1993.

[DAR 95] DARBRE G.R., "Strong-motion Instrumentation of Dams", *Earthquake Engineering and Structural Dynamics*, 24, 1101-1111, 1995.

[DAR 00] DARBRE G.R., SMET C.A.M. (ED), KRAEMER C., "Natural Frequencies Measured from Ambient Vibration Response of the Arch Dam of Mauvoisin", *Earthquake Engineering and Structural Dynamics*, 29, pp. 577–586, 2000.

[DAR 01] DARBRE, G.R., PROULX J., "Continuous Ambient-vibration Monitoring of the Arch Dam of Mauvoisin", *Earthquake Engineering and Structural Dynamics*. 31, pp. 475–480, 2001.

[DEI 82] DEINUM P.J., DUNGAR R., ELLIS B.R., JEARY A.P., REED G.A.L., SEVERN R.T., "Vibration Tests on Emosson Arch Dam, Switzerland", *Earthquake Engineering and Structural Dynamics*, 10, pp 47–470, 1982.

[DOE 96] DOEBLING S.W., FARRAR C.R., PRIME M.B., SHEVITZ D.W., *Damage Identification and Health Monitoring of Structural and Mechanical Systems from Changes in their Vibration Characteristics*, Los Alamos National Laboratory, USA, LA-13070-MS, 127 pages, 1996.

[DUR 88] DURON Z.H., HALL J.F., "Experimental and Finite Element Studies of the Forced Vibration Response of Morrow Point Dam", *Earthquake Engineering and Structural Dynamics*, 16, pp. 1021–1039, 1988.

[FEN 84] FENVES G., CHOPRA A.K., EAGD-84, "A computer program for earthquake analysis of concrete gravity dams", *Report No. UCB/EERC-84/11*, University of California, Berkeley, USA, 1984.

[FOK 86] FOK K.L., HALL J.F., CHOPRA A.K., EACD-3D, "A computer program for three-dimensional earthquake analysis of concrete dams" *Report No. UCB/EERC-86/09*, University of California, Berkeley, USA, 1986.

[HAL 88] HALL J.F., "The Dynamic and Earthquake Behavior of Concrete Dams: Review of Experimental Behavior and Observational Evidence", *Soil Dynamics and Earthquake Engineering*, 7(2), pp. 58–121, 1988.

[HAR 88] HARRIS C.M. (Ed.), *Shock and Vibration Handbook*, third edition, McGraw-Hill, New York, USA, 1988.

[MCC 95] MCCONNELL K.G., *Vibration Testing: Theory and Practice*, John Wiley & Sons, New York, USA, 606 pages, 1995.

[PAU 95] PAULTRE P., PROULX J., TALBOT M., "Dynamic Testing Procedures for Highway Bridges Using Traffic Loads", *ASCE J. Struc. Eng.*, 121(2), pp. 362–376, 1995.

[PAU 97] PAULTRE P., PROULX J., "Dynamic Testing of Large Scale Structures", *International Association for Bridges and Structural Engineering (IABSE)*, 7(1), p. 29–34, 1997.

[PAU 02] PAULTRE P., PROULX J., CARBONNEAU C., "An experimental evaluation of ice cover effects on the dynamic behaviour of a concrete gravity dam", *Earthquake Engineering and Structural Dynamics*, 31, pp. 2067–2082, 2002.

[PRO 97] PROULX J., PAULTRE P., "Experimental and Numerical Investigation of Dam-Reservoir-Foundation Interaction for a Large Gravity Dam", *Canadian Journal of Civil Engineering*, 24(1), pp. 90–105, 1997.

[PRO 01] PROULX J., PAULTRE P., RHEAULT J., ROBERT Y., "An Experimental Investigation of Water Level Effects on the Dynamic Behavior of a Large Arch Dam", *Earthquake Engineering and Structural Dynamics*, 30(8), pp. 1147–1166, 2001.

Chapter 7

Models for Simulating the Seismic Response of Concrete Structures

7.1. Introduction

This chapter will present the principles of the mode most frequently used for the simulation of the collapse behavior of reinforced concrete structures subjected to seismic loading. Non-linear calculation methods have proved their utility for the preparation and interpretation of dynamic, pseudo-dynamic and static tests on shaking tables and reaction walls, but now they are starting to be used to validate certain design rules, as well as for the seismic reassessment of existing structures which are more vulnerable to earthquakes.

The quality of this type of modeling, which is not currently regulated, depends on numerous factors. Thus, these models should take into account:

– the material effects, which may give rise to local non-linearities;

– the structural effects (mass distribution and behavior of the bonds);

– the environmental effects (support-structure interaction).

Within this context, two aspects affect the modeling, i.e. the structure discretization and the local behaviors of the materials and "interfaces", either

---

Chapter written by Didier COMBESCURE, Nicolas ILE, Jacky MAZARS and Jean-Marie REYNOUARD.

between the materials themselves or between the structure and its support. We should endeavor to deal with both aspects consistently to achieve relevant results.

We present below a few examples of the opportunities afforded by the literature to deal with analysis of the seismic operation of concrete structures. Generally, non-linear models are divided into two large families, depending on the scale represented: on the one hand, global and semi-local simplified models (fiber and multilayer models), based on bar or beam elements, and on the other hand, more comprehensive models that rely on volume or surface finite element meshing.

Each has advantages and drawbacks shown revealed by applications, implying different types of structures tested in the laboratory on a shaking table or with a reaction wall owing to a pseudo-dynamic method.

7.2. Different discretization families

7.2.1. *Global modeling*

Such models reproduce the behavior of a whole structure element, such as a beam, column, concrete wall or masonry wall, a whole storey or even a whole building. The laws of behavior are generally uniaxial and link a global strain-curvature, shear, strain, extension, relative displacement with its associated global stress: moment, shear load or normal load.

Multilinear curves are often used to represent the behavior of reinforced concrete elements: concrete cracking follows the elastic behavior; it brings about lower stiffness then steel plastification and sometimes concrete crushing or steel failure. Simplified laws allow us to reproduce the main phenomena observed with cyclic or seismic loading (decrease of the overall stiffness, strength degradation under cyclic loading, recovered stiffness when cracks close).

Figure 7.1 shows a shear law for the study of not very slender reinforced concrete walls, as well as the modeling of a masonry in-filled wall owing to two diagonal connecting rods. These global laws generally apply to beams or diagonal beams.

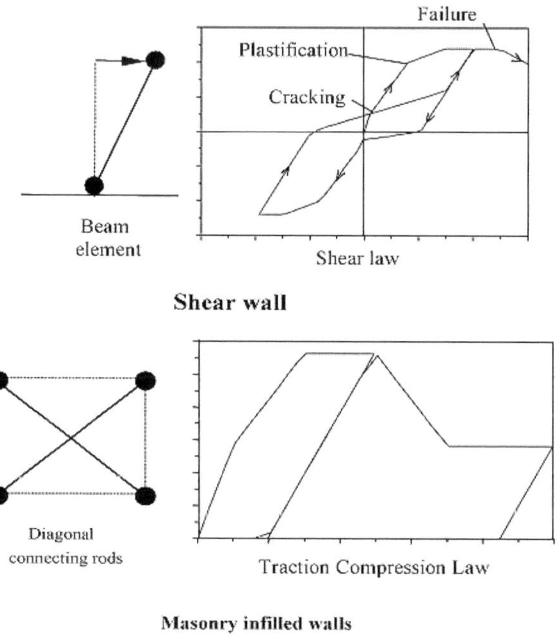

Figure 7.1. *Overall models of shear and masonry in filled walls*

7.2.2. *Semi-global modeling*

Multi-fiber and multi-layer models are examples of semi-global models: they rely on beam elements and cover the conventional laws of beam theory – the hypothesis of plane sections that remain plane for instance. Here, the inner stresses are calculated from uniaxial laws that reproduce the behavior of each material (concrete, steel, etc.) instead of empirical laws. Such an approach by of this "simplified method" frames results in generating problems of moderate size and with a quality standard that is respectable when compared to those obtained using conventional 2D or 3D finite element calculations, particularly if the discretization, the behavior models and the boundary conditions are well-defined.

Multi-layer discretization is well suited to dealing with plane problems involving slender structures. Later, among the applications, we will present an example illustrating "multi-layer" elements and including a damage model that will be described further in section 7.4.2. These elements include two nodes and consist of superimposed layers with non-linear behavior, specialized according to whether they include reinforcements or not (Figure 7.2). The Navier-Bernouilli hypothesis on cross-sections attributes a uniaxial behavior to the layers [MAZ 98]. As far as layers are concerned, introducing reinforcements relies on a partial pressure law that implies non-sliding between steel and concrete [LAB 91]. Thus, inside each layer:

$$\varepsilon = \varepsilon_{steel} = \varepsilon_{concrete}$$

E_b concrete modulus

$$E = \left(1 - C_{a/b}\right) E_b + C_{a/b} E_a$$

E_a steel modulus

$$\varepsilon_{inel.} = \left(1 - C_{a/b}\right) \varepsilon_{inel.\ concrete} + C_{a/b} \varepsilon_{inel.\ steel}$$

E homogenized (a + b) modulus

$$C_{a/b} = \frac{steel\ area}{concrete\ area}$$

$\varepsilon_{inel.\ concrete}$ concrete inelastic strain

$\varepsilon_{inel.\ steel}$ steel plastic strain

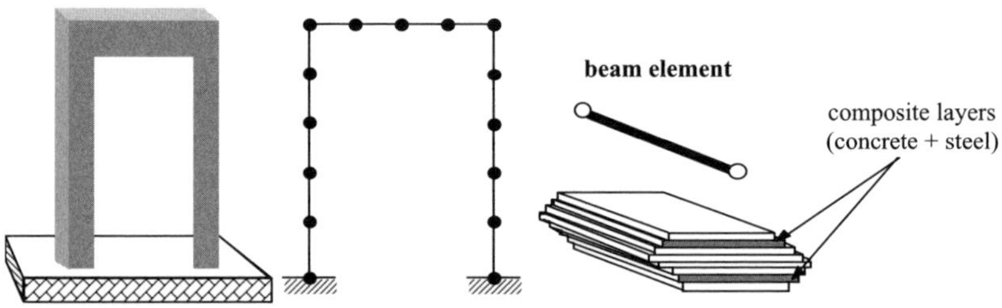

Figure 7.2. *Principle of discretization by multi-layer beam elements*

Following the same principle, *multi-fiber discretization* is more efficient, because it makes it possible to study the behavior of structures, for slender elements at least. We will present the principle behind writing this type of element more thoroughly, based on a Timoshenko beam element with shear strains ([GUE 97], [PEG 93]).

The shear and axial strains inside fiber i are written respectively as functions of the general strains of the beam element and of the section geometry:

In 2D (multi-layer beams):

$$\left(\varepsilon_x\right)_i = \varepsilon_x - y_i \cdot \varphi_z$$

and

$$\left(\gamma_y\right)_i = \gamma_y$$

In 3D (multi-fiber beams):

$$\left(\varepsilon_x\right)_i = \varepsilon_x - y_i\ \varphi_z + z_i\ \varphi_y$$

$$\left(y_y\right)_i = \gamma_y - z_i\ \phi_x$$

$$\left(\gamma_z\right)_i = \gamma_z + y_i\ \phi_x$$

where ε_x is the average axial strain, y_i is the distance between fiber i and the average fiber, ϕ_y and ϕ_z are the bending curvatures ($\phi_y = d\theta_y/dx$ and $\phi_z = d\theta_z/dx$), $\phi_x = d\theta_x/dx$ is the unit angle for torsion, and γ_y and γ_z are average shear strains.

The bending moments M_y, M_z and the normal force N_x are obtained by integrating the axial strains $\left(\sigma_x\right)_i$ corresponding to the strains $\left(\varepsilon_x\right)_i$:

$$N_x = \int_S \sigma_x dS,\ M_y = \int_S z\ \sigma_x dS \text{ and } M_z = -\int_S y\ \sigma_x dS \qquad [7.1]$$

In the case of a Timoshenko beam element, shear loads T_y, T_z and the torsion moment M_x are calculated by integrating the shear stresses $\left(\tau_y\right)_i$ and $\left(\tau_z\right)_i$ derived from $\left(\gamma_y\right)_i$ and $\left(\gamma_z\right)_i$:

$$T_y = \int_S \tau_y dS,\ T_z = \int_S \tau_z dS \qquad [7.2]$$

$$M_x = \int_S \left(y\ \tau_z - z\ \tau_y\right) dS \qquad [7.3]$$

For a shear elastic material, the behavior law is written as follows:

$$\left(\tau_y\right)_i = \alpha_y\ G\left(\gamma_y\right)_i \text{ and } \left(\tau_z\right)_i = \alpha_z G\left(\gamma_z\right)_i \qquad [7.4]$$

where G is the elastic shear modulus: $G = E/2(1+\nu)$, and parameters α_y and α_z are correction factors that take non-uniform shear stress distribution into account.

We observe that all kinematic hypotheses presuppose the absence of any warping. Torsion behavior does not respect Saint-Venant's theory, and the shear strain distribution is only valid for circular sections.

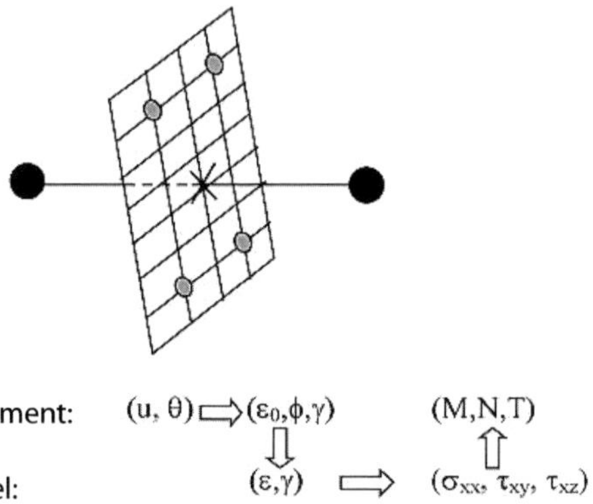

Figure 7.3. *Principle of multi-fiber models*

When beam elements without any shear strain are used (Bernouilli's assumption), the shear forces are generally chosen to be equal to the bending moment derivative.

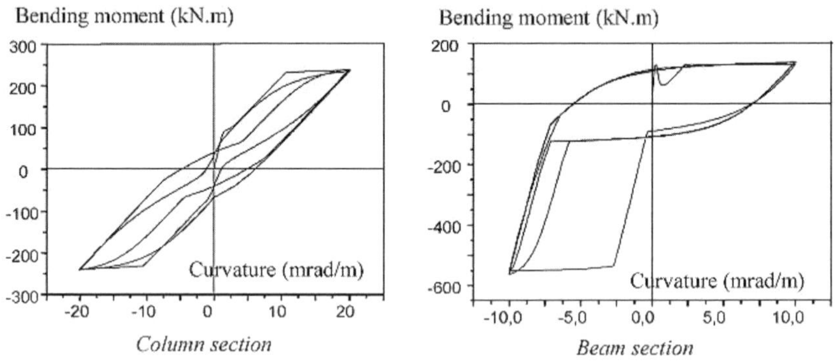

Figure 7.4. *Example of a curvature-moment relationship*

Multi-fiber or multi-layer models have several advantages in comparison to global models. First, the parameters of the model are the characteristics of the material and the section geometry instead of the curvature-moment or displacement-stress global curve. Thus, a section analysis with a fiber model allows us to identify the first-load curve of a curvature-moment global model. Figure 7.4 shows the results of the section analysis of a column and a beam.

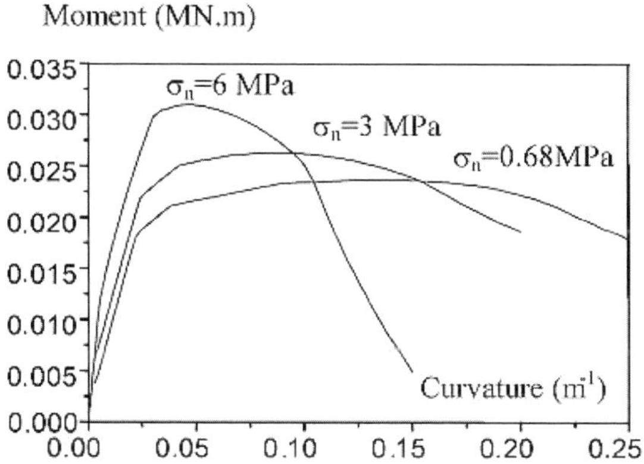

Figure 7.5. *Influence of the axial stress on the resistance and ductility of a column*

Secondly, modeling takes into account the effect of the normal stress on both the bending moment and the ductility of the section. The curvature-moment relationships in Figure 7.5 derive from the analysis of a column section submitted to different axial stresses, and clearly show not only the section modulus increase, but also the decrease of ductility caused by the normal stress rise.

7.2.3. *2D and 3D fine models*

The subject here involves the most exhaustive finite element description possible. Among others, mass elements will be used in 3D, as well as shell elements, interface or joint elements or even plane or asymmetric elements for associated 2D problems (see Figures 7.19 and 7.30). For further details, refer to the wide literature on the subject.

At this level of modeling, each material – masonry, concrete, reinforcement, mortar joints or dry joints – can be represented. Associating these laws can then allow us to predict the global behavior of a structure as well as its local response, but heavy calculation often restricts their use to the analysis of a structure element, geometrically simple specific structures or structures with simplified loading (static loading equivalent to seismic forces for instance). Hence, the use of such models should be reserved for the detailed analysis of the operation of a studied object. As we shall see later, this can be the case for better understanding the behavior of a model during an experiment, or for deepening the study of an existing structure (when re-assessing nuclear or industrial facilities or a dam for example). This necessarily implies both accurate discretization work and choice of constitutive models and boundary conditions. Such modeling can also prove useful in identifying the parameters of the global or semi-global models described above.

7.3. Behavior laws for concrete

The quality of modeling mainly relies on the ability of constitutive laws to reproduce the physical phenomena characterizing complex composite materials like reinforced concrete. For 2D and 3D modeling, these constitutive laws lean on such theories as the theory of damage and plasticity. Uniaxial laws are usually enough for global and semi-global models. Hereafter we describe and explain the principles of a few laws used for concrete and steel.

7.3.1. *Semi-empirical mixed models*

Semi-empirical laws are generally uniaxial and are directly based on experimental observations. They allow reproduction of post-cracking softening, the unilateral behavior of concrete (renewal of stiffness when the cracks close up), as well as softening after compression strength has been reached, taking this softening confinement and the compression strength into account. Below, we present a law inspired by Hognestad's model [HOG 51].

During compression, or with monotonous loading, the law is the parabolic type, becoming linear with softening (Figure 7.6a). A plateau with residual stress not equal to zero can be defined after the softening. When using civil engineering conventions (compression positive strain), the following values are obtained:

For $0<\varepsilon<\varepsilon_{c0}$,

$$\frac{\sigma}{\sigma_{c0}}=\frac{\varepsilon}{\varepsilon_{c0}}\left(2.0-\frac{\varepsilon}{\varepsilon_{c0}}\right) \quad [7.5]$$

For $\varepsilon_{c0}<\varepsilon$,

$$\frac{\sigma}{\sigma_{c0}}=1+Z\left(\varepsilon-\varepsilon_{c0}\right) \quad [7.6]$$

On the plateau,

$$\sigma=\sigma_{pt} \quad [7.7]$$

where σ_{c0} is the peak stress, ε_{c0} is the peak strain, Z is the slope after the peak and σ_{pt} represents the residual stress.

Under cyclic loading, the compression non-linearity generally comes with stiffness degradation and the appearance of plastic strains.

In traction, concrete has strength, yet after cracking, its behavior is brittle. The law is linear for each traction part (Figure 7.6) and:

– up to the peak, the behavior is linear;

– softening is linear up to the ultimate strain.

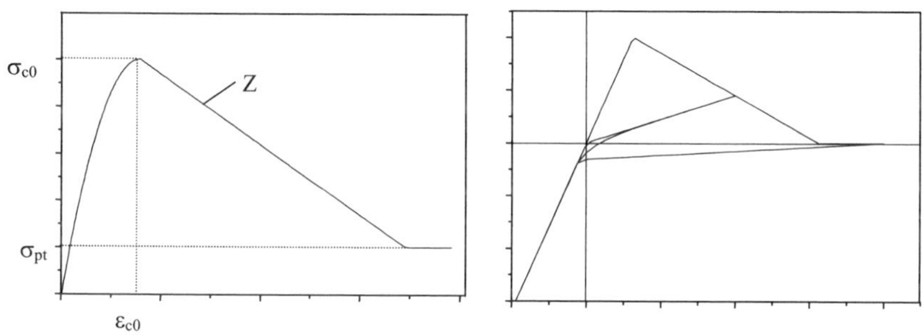

a. Compression *b. Traction*

Figure 7.6. *Example of compression and traction constitutive laws for concrete*

7.3.2. *Damage model*

Based on damage mechanics [LEM 90], [MAZ 86], the model [LAB 91] fits the description of the damageable behavior generated by the creation of micro-cracks (decrease of stiffness). During the cycles, the operation is linked to micro-crack re-closing. Two scalar damage variables are used, one in traction D_1 and the other in compression D_2, acting on the stiffness of the material and generating permanent strains. The 3D constitutive laws are given below:

$\underline{\varepsilon} = \underline{\varepsilon}^e + \underline{\varepsilon}^p$: partition of the plastic and permanent strain tensors [7.8]

$$\underline{\varepsilon}^e = \frac{\underline{\sigma}^+}{E_0(1-D_1)} + \frac{\underline{\sigma}^-}{E_0(1-D_2)} + \frac{\nu}{E_0}(\underline{\sigma} - Tr\,\underline{\sigma}\underline{1}) \quad [7.9]$$

$$\underline{\varepsilon}^p = \frac{\beta_1 D_1}{E_0(1-D_1)} + \frac{\partial f(\underline{\sigma})}{\partial \underline{\sigma}} + \frac{\beta_2 D_2}{E_0(1-D_2)}\underline{1} \quad [7.10]$$

where E_0 is the Young's modulus for the healthy material and ν is the Poisson's coefficient.

$\underline{\sigma} = \underline{\sigma}^+ + \underline{\sigma}^-$, with $\underline{\sigma}^+$ and $\underline{\sigma}^-$ are respectively the "traction stress" and the "compression stress"; D_1 and D_2 are respectively the traction and compression damage variables. Their evolution beyond a specific threshold, between 0 (healthy material) and 1 (broken material), is linked to the evolution of the local elastic energy. β_1 and β_2 are constants whose values calibrate the evolution of the permanent strains, and $f(\underline{\sigma})$ is the function of the crack opening-closing up adjustment.

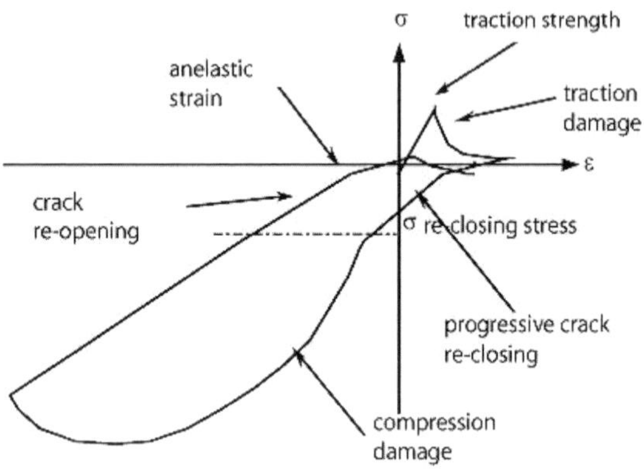

Figure 7.7. *Uniaxial response of the model with unilateral damage (from [LAB 91])*

Figure 7.7 gives the uniaxial response of the model with a compression-traction alternate loading. Beyond the peak in traction, we can see that damage (D_1) evolves, unloading reveals a Young's modulus affected by crack opening, E_0 ($1-D_1$), and a residual strain gradually gets cancelled out (in equation [7.10] $0<\mathrm{d}f(\underline{\sigma})/\underline{\sigma}<1$) until the cracks in compression close up ($\mathrm{d}f(\underline{\sigma})/\underline{\sigma}=0$ for the closing-up stress σ_f). Beyond a compression threshold, the damage (D_2) evolves, which creates non-linearity, reducing the modulus, E_0 (1-D_2), and definitive permanent strain. When σ_f is reached, the cracks open again, which reactivates the effects of D_1 ($0<\mathrm{d}f(\underline{\sigma})/\underline{\sigma}<1$). For further details refer to [LAB 91].

7.3.3. *Plasticity model for concrete*

Plasticity models are more common in the world of digital calculation of structures, mainly because they were developed for metals. Adaptations to the specific behaviors of materials like concrete have been recent developments, but give effective models that agree with the damage model principles and can describe the appearance of cracks and the effects of their opening and closing. Here, we present the concrete model in detail, which were developed at the INSA in Lyons [MAR 99], within the framework of plasticity theory. The failure surface is described by two criteria: Nadai in compression and bi-compression, and Rankine for traction cracking. The flow behavior law is associated and the strain-hardening rule (whether positive or negative) is isotropic. The behavior of cracked concrete is described using the distributed cracking concept, which considers the cracked material as a continuous medium. When the failure surface in the traction field has been reached, the biaxial plasticity is given up, and an orthotropic law is activated. Thereafter, the process of describing cracking involves three independent relationships between stress and strain, as defined in a system of coordinates parallel and perpendicular to cracking direction. The unilateral feature of concrete cracked during cyclic loading is dealt with by the restoration of the corresponding stiffness.

7.3.3.1. *Uncracked concrete*

Two failure surfaces are defined for the compression and traction domains (see Figure 7.8). Nadai's 2-parameter failure surface, of the Drucker-Prager type, is defined using a linear relationship between the octahedral stresses τ_{oct} and σ_{oct}. The mathematical expression of the surface for the compression domain is defined by:

$$f_{comp}\left(\sigma_{oct},\tau_{oct}\right)=\frac{\tau_{oct}+a\sigma_{oct}}{b}-f_{c'}=0,\ \sigma_1<0 \text{ and } \sigma_2<0 \qquad [7.11]$$

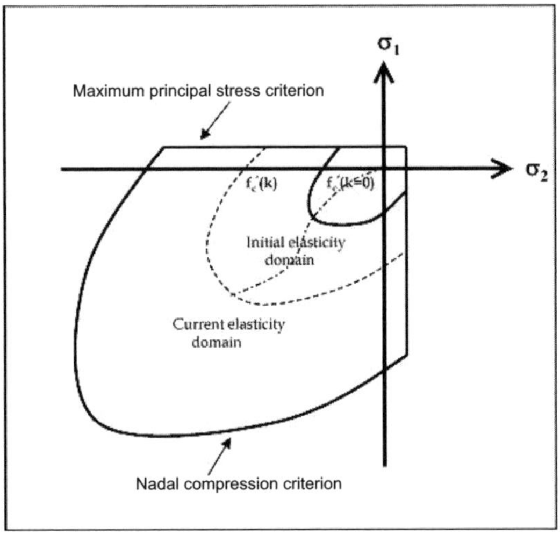

Figure 7.8. *Failure and loading domains in the main stress plane*

For the traction domain (pure traction or compression-traction), the failure area is expressed as follows:

$$f_{trac} - f_{t'} = 0, \sigma_1 < 0 \text{ and/or } \sigma_2 > 0 \qquad [7.12]$$

where:

$$\sigma_{oct} = I_1 / 3 \text{ and } \tau_{oct} = \sqrt{\frac{2J_2}{3}} = \sqrt{\frac{2}{9}} \sigma_{eq} \qquad [7.13]$$

σ_{eq} is the Von Mises equivalent stress:

$$a = \sqrt{2} \frac{\beta - 1}{2\beta - 1}, b = \frac{\sqrt{2}}{3} \frac{\beta}{2\beta - 1} \text{ and } \beta = f_{c'} / f_{c'c} \qquad [7.14]$$

where:
- $f_t^{'}$ = concrete strength in uniaxial traction;
- $f_c^{'}$ = concrete strength in uniaxial compression;

– f'_{cc} = concrete strength in biaxial compression for $\sigma_{xx} / \sigma_{yy} = 1$ and $\tau_{xy} = 0$;

– I_1 = first invariant of the stress tensor;

– J_2 = second invariant of the stress deviator.

The parameters of criterion a and b, functions of α and β, are derived from three tests: uniaxial compression f'_c, uniaxial traction f_t' and biaxial compression f'_{cc}, ($\sigma_{xx} / \sigma_{yy} = 1$ and $\tau_{xy} = 0$).

7.3.3.2. *Cracked concrete*

When the failure area in the traction domain has been reached, the behavior of the material is considered as uniaxial in each orthotropic direction, given by the directions parallel and perpendicular to the crack. The direction of the crack is perpendicular to the direction of the traction maximum main stress (Figure 7.9); it is traditionally determined owing to axis rotating matrices.

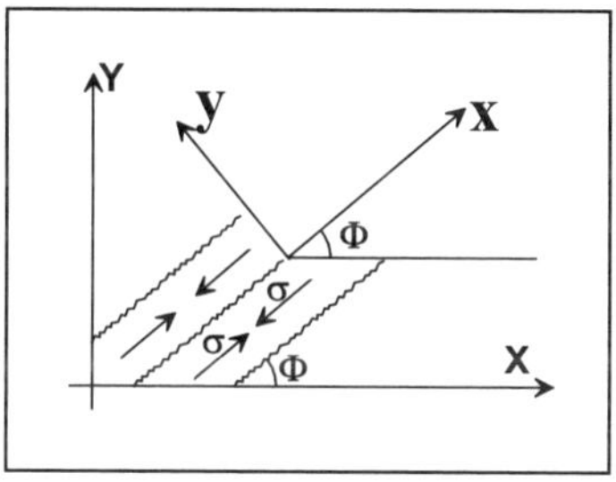

Figure 7.9. *Axes linked to the crack*

During the loading increment that causes cracking in a given material point, passing from a biaxial elastoplastic model to an uncoupled orthotropic model requires specific treatment, and several connecting rules are used to ensure continuity between both models. The behavior of cracked concrete is then described by a cyclic uniaxial law in all directions. The stress tensor within the local reference axis is complemented by the shear stress and elastically calculated with a restricted modulus of transverse elasticity to account for the meshing effect of the crack that corresponds to friction at both surfaces of the crack lips. This effect is implicitly taken into account in the law by arbitrarily decreasing the initial modulus of elasticity in shear using a shear transfer factor η. Best values depend on the kind of problem, but an initial value of factor η ranging from 0.2 to 0.4 seems capable of giving correct results [ILE 00]. This parameter can be adjusted. Whenever there is a pronounced rotation of the main directions, a second series of cracks is created perpendicular to the existing one.

7.3.3.3. *Cyclic uniaxial law*

The uniaxial law implemented allows us to account for the main phenomena observed when loading consists of a few alternating cycles.

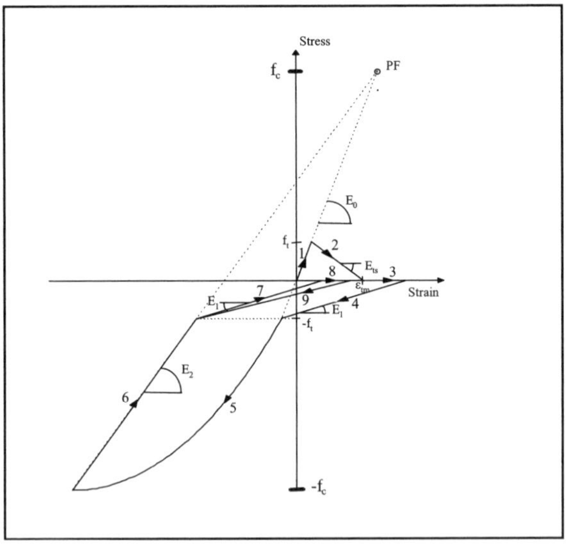

Figure 7.10. *Response of the model to a traction-compression cycle*

Let us examine (Figure 7.10) the behavior of a point initially in traction which cracks completely before being subjected to opposite compression reloading. The concrete is elastic until the strength in simple traction is reached: f_t (path 1), then it cracks according to negative stiffness (slope E_{ts}, path 2), up to a $\varepsilon_{tm} = \varepsilon_{rupture}^{trac}$ strain. Beyond this, crack opening occurs without any stress (path 3). When the stress changes its direction, an increasing compression stress is necessary for the crack to gradually close up (slope $E_1 \neq E_0$, path 4). The latter is considered to be completely closed for a stress lower than $-f_t$, the value at which stiffness is completely restored (path 5). The description of path 4 is based on tests that show the opposite lips of a crack do not coincide, that they become strained under the effect of a stress which tends to close the crack up, and that the stiffness of the uncracked concrete is only restored with complete closing. Nevertheless, the crack closes up with a stress equal to zero, provided the strain is higher than a specific threshold: 3*ε_{tm}. Path 5 follows the non-linear law of concrete in simple compression to a new inversion of loading, which involves a discharge along the E2 slope straight line (path 6) passing a focal point (f_c; ε_0), with $\varepsilon_0 = f_c / E_0$, as suggested by Mander *et al.* [MAN 88] and Park [PAR 90].

A concept similar to the damage concept affecting the modulus can also be found with stress intensity. When the stress exceeds -f_t (path 7), the E_1 modulus corresponding to the crack closing up is still found. Paths 8 and 9 obey the same rules as paths 3 and 4. The behavior of an initially compressed point (or a spot that has not completely cracked due to a reversed loading) is described by similar laws, presented in detail elsewhere [ILE 00].

7.3.4. *Cyclic models for steel*

Because of its geometry, the reinforcement can only be modeled with uniaxial laws, even for calculations involving 2D or 3D problems.

As a rule, the laws used are elasto-plastic, with or without hardening. The most exhaustive laws [MEN 73] successfully reproduce phenomena including non-linear strain hardening (Figure 7.11a), Bauschinger's effect with cyclic loading (Figure 7.11b) and compression bar buckling when frames are not close enough (Figure 7.11c).

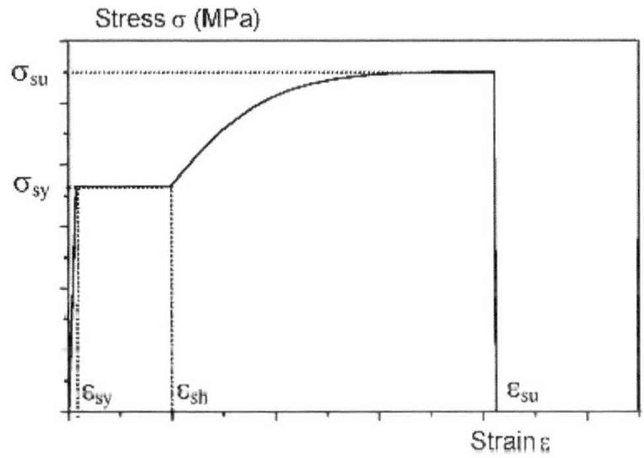

a) Behavior with monotonous loading

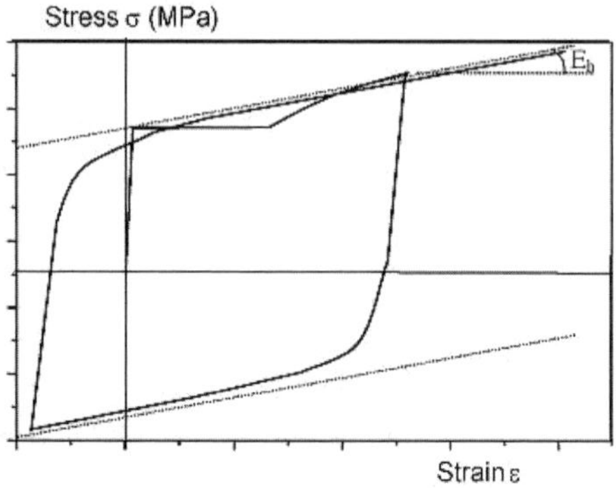

b) Behavior with cyclic loading

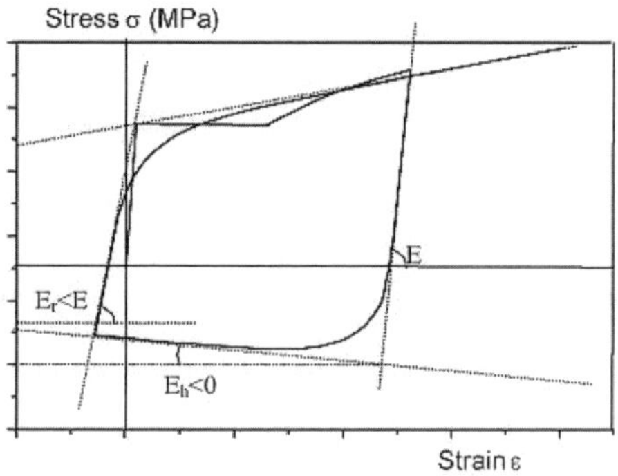

c) Taking buckling into account

Figure 7.11. *UNI-STEEL behavior laws (from [MEN 73])*

7.3.5. *Taking construction layouts and second-order phenomena into account*

Beyond material behavior, the quality of the model involves taking into account some phenomena that develop at the interfaces between materials (between steel and concrete, between soil and structure) or phenomena that are related to the severity of calculation assumptions (such as the effects associated with the transverse load or the effects of the geometric second order for slender elements). A few indications to help take them into account are given below.

7.3.5.1. *Influence of the transverse reinforcement (stirrups)*

Due to the formalism of multi-layer and multi-fiber models, some building arrangements can directly influence material laws. In this way, stirrups increase the ductility of an element and act on its strength. They are taken into account by increasing, in the behavior law of concrete, stress at the peak and/or by raising the post-peak phase to improve the failure strain when the volume ratio of the stirrups increases. As far as column-type or beam-type elements are concerned, stirrups delay the buckling of steel under alternating loading.

7.3.5.2. *Anchoring and overlapping of the reinforcements*

The phenomena linked to the steel/concrete bond (lack of overlapping or lack of anchoring) that tend to soften the structure and increase displacements can be modeled by specific non-linear and non-dissipative laws. Taking these phenomena into account can become really important in old reinforced concrete structures that do not respect the latest construction layouts (generally having inadequate overlapping or anchoring length). A modeling example ([CON 01], [MON 00] and [XIA 97]) involves drawing a parallel between a steel model and a steel-concrete interface model using an adapted law that splits the total strain into two parts: one related to the behavior of steel and the other describing sliding between steel and concrete [ELI 93].

7.3.5.3. *Taking transverse load non-linearity into account*

In Timoshenko beam elements, cross-sections do not stay orthogonal to the longitudinal axis. This can be taken into account by additional stresses linked to the shear diagonal cracking and enables verification of the transverse load collapse modes. The law describing layer behavior is no longer uniaxial, but should take the shear component forces into account [GHA 98]. As far as spatial problems (multi-fiber element) are concerned, the problem is somewhat more complex, but introducing a warping function is a good way to proceed [CAZ 03].

7.3.5.4. *P-Delta effects*

For very slender elements supporting an important axial load – a bridge pier for instance – a second-order bending effect due to the displacement at the top is added to the front-mass inertia effect. The digital implementation of this phenomenon for seismic conditions can be found in [GHA 98].

7.3.5.5. *Structure-support interaction*

This effect can be decisive. The support – typically the soil – has its own behavior and, at the interface, generates rotation and/or uplift effects. The main phenomena can be taken into account either using a 2D or 3D representation of the soil foundation or by using an integrating interface element, the softening of the support and the non-linear effects associated with the behavior of the support and the uplifts [CRE 01].

7.4. A few examples with their validation through experiments

Because of the complex phenomena they reproduce, and therefore the number of law parameters they use, using non-linear models involves permanent confrontation with results from shaking table, pseudo-dynamic and static tests. The experimental results used here come from tests carried out with the CEA AZALEE shaking table (Saclay-France) and the JRC ELSA reaction wall (Ispra-Italy) within the scope of two research programs dealing with reinforced concrete building structures, namely:

– the French CAMUS (Conception et Analyse des Murs sous Seisme) program; and

– the ECOEST-ICONS (Innovative Seismic Design Concepts for New and Existing Structures) European programs.

7.4.1. *Application of the semi-global method to a four-storey structure*

Within this context, validating non-linear calculation models first requires using the results from elementary tests on structure elements like columns, then applying these models to several-storey structures in which each element is submitted to series of complex and realistic loadings. The second stage notably allows the quality of the hypotheses linking the different elements to be checked. In the next section, we present a few results derived from the second phase of the study on an RC frame in a "multi-fiber beam" modeling context.

7.4.1.1. *Experimentation*

Within the framework of the ECOEST-ICONS European projects, two scale-1, 4-storey reinforced concrete frames were tested on the Ispra (Italy) JRC reaction wall using the pseudo-dynamic method. Both frames had the same dimensions and reinforcements, and differed only in the presence of masonry in filled walls (Figure 7.12 and [COM 96]). The reinforcement in these frame structures use building arrangements similar to those used from 1940 to 1970 in Mediterranean countries: smooth steel-bars, insufficient numbers of widely spaced stirrups with regard to current standards, and low reinforcements at nodes. The unfilled walls were made of perforated clay bricks.

	Frame without filling	Frame with filling
Seismic level	475 years – 2.18 m/s^2	
Displacement	60.8 mm	10.2 mm
Shear load at the basis	209 kN	754 kN
Seismic level	975 years – 2.88 m/s^2	
Displacement	116.7 mm$^{(1)}$	22.3 mm
Shear load at the basis	217 kN	846 kN
Seismic level	2000 years – 3.73 m/s^2	
Displacement	/	40.6 mm
Shear load at the basis	/	529.2 kN

Table 7.1. *Main experimental results of pseudo-dynamic tests on 4-storey frames carried out at the ELSA (Ispra, Italy). Tests stopping before the end of the signal*

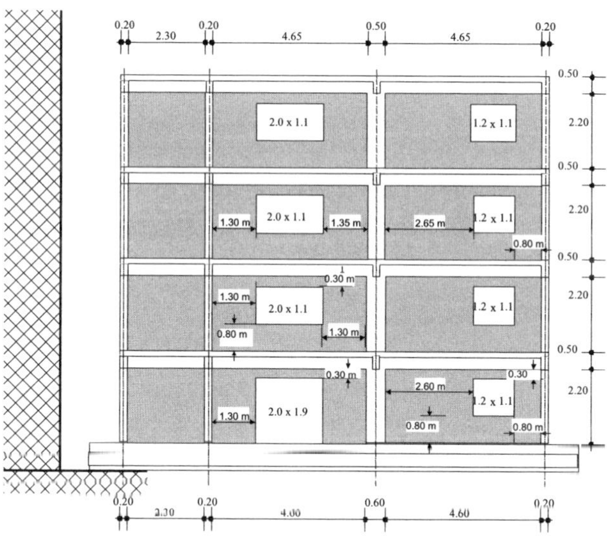

Figure 7.12. *4-storey frame tested at ELSA (CCR, Ispra, Italy)*

The artificial signals used for these high-level pseudo-dynamic tests corresponded to earthquakes with 475, 975 and 2,000 year return periods. Front displacements and shear loads at the basis of the structure measured during the tests are given in Figure 7.1 for the seismic levels applied.

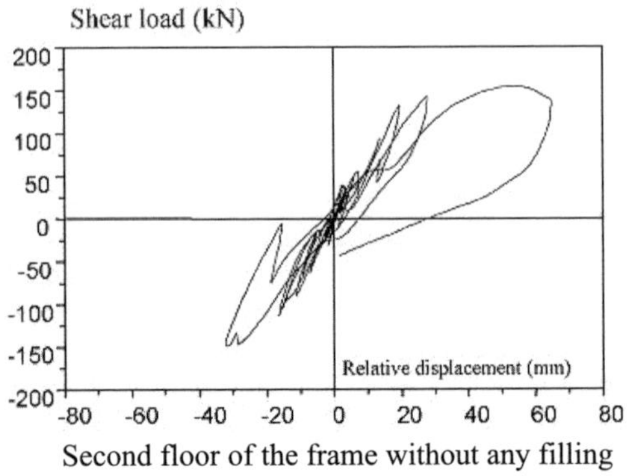

Figure 7.13. *Examples of the relative displacement-transverse load relationship between storeys for an earthquake with a 975 year return period*

For both structures, little damage could be observed for the 475-sequence period earthquake, whereas important displacement values (tests stopping before the end of the signal to allow the structure to be repaired) were reached for 975 (Figure 7.14) and 2,000 sequence period earthquakes respectively, for both the infilled frame and the frame without any infill.

The collapse mode is also different from one structure to the next: the damage is concentrated within the second-floor columns for the frame without any infill whereas the ground floor that has important openings (door and windows) constitutes the critical storey of in-filled frame.

7.4.1.2. *Modeling and comparison with experiments*

For time dynamic calculations, the reinforced concrete frame has been modeled with beam elements and the fiber model described previously. Each column and beam has been modeled in three parts: the two plastic hinges have been chosen to be non-linear, whereas the central part is considered as elastic with cracked stiffness. The masonry in filled walls has been modeled as diagonal truss rods (stiffness and strength) and has been determined with plane stress 2D refined calculations. These calculations allow us to estimate the drop in stiffness and strength caused by the appearance of openings.

Figure 7.14 shows a test-calculation comparison for the frame without infill for a high earthquake level (475-sequence period). Thus, non-linear models can assess the displacement and global strength of a structure. Furthermore, these calculations highlight the concentration of strains and damage on the second floor, as was the case during the tests. We should note that the damage mode with conventional linear calculations is hardly ever predictable.

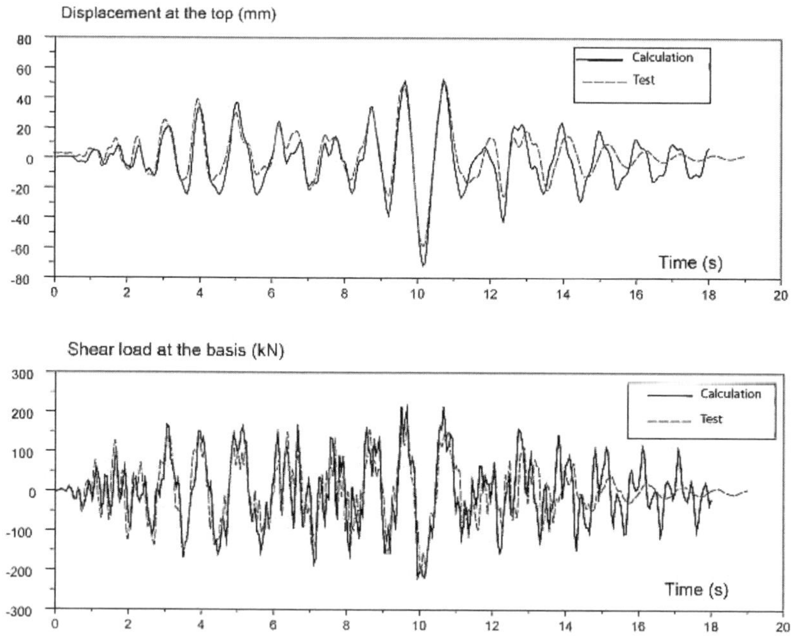

Figure 7.14. *Test-calculation comparison for the unfilled structure and the 475 sequence period earthquake*

Within the frame of para-seismic regulations, the justification of that structure with elastic calculation would involve using behavior coefficients respectively of 2.9 and 4.3 for earthquakes with 475 sequence periods (peak ground acceleration equal to 2.18 m/s^2) and 975 year sequence periods (peak ground acceleration equal to 2.88 m/s^2). For the first earthquake level, the steel must have plasticized but no damage could be observed. The 975 year sequence period test was stopped as initial crush had started on the second floor, and the structure had to be reinforced before being submitted to a further series of tests.

7.4.2. *Semi-global and local models applied to concrete walls*

Reinforced structure walls are quite widely used in France. Recent research has allowed us to work out their para-seismic efficiency better. We will deal below with work related to two specific studies:

– the case of weakly reinforced walls loaded in their plane, for which two modeling types are used: the “multi-layer” beam simplified finite element study or a more exhaustive 2D finite element study;

– the case of U-shaped walls loaded in two directions, the behavior of which is described using a 3D multilayer thin shell approach integrating the local plastic model.

7.4.2.1. *CAMUS experimental program*

Within the framework of the CAMUS program [COI 98], two 1/3-scale models (CAMUS I and II) consisting of two weakly reinforced slender walls have been tested on the AZALEE-CEA shaking table. The test layout has also been used for an additional model (CAMUS III) by the ICONS program that was dimensioned according to Eurocode 8 recommendations. The CAMUS model consists of two load bearing walls without any opening, linked together by six floors, and by a lower longitudinal girder anchored to the shaking table (Figure 7.15). Each wall is 5.10 m high, 1.70 m long and 6 cm thick. The walls are cast in two steps (first the longitudinal beam and levels 2 and 4, then levels 1, 3 and 5) in order to simulate the concrete work cast stages on each floor. Additional masses have been added to the higher and lower parts of each floor (except the ground floor) to impose a maximum load close to the vertical stress normally present on the foundation of conventional buildings (1.6 MPa).

The first two models have been designed to obtain a multi-storey operating mode with concrete cracking and steel plastification on several floors (“French-style” design). The third model – CAMUS III – had the same geometry as the first two models and was designed according to Eurocode 8 regulations so as to have an ultimate moment at the base of the wall next to the one in CAMUS I. The amount of reinforcement is given in Table 7.2. It is worth noting that the bending vertical reinforcement for the CAMUS II model is minimal (1ϕ 4.5 between the second and the fifth floor and 1ϕ 4.5 + 1ϕ 5 on the first floor). The amount of reinforcement in the CAMUS III model is definitely higher than the one used for the first model, and reinforcement discrepancies are obvious on the higher levels.

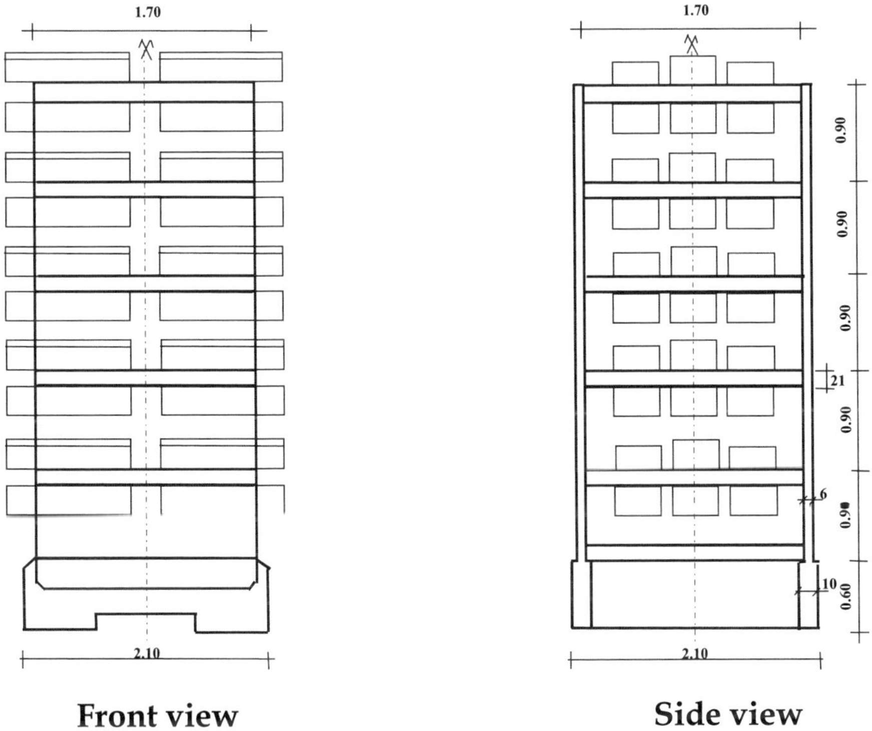

Figure 7.15. *CAMUS mock-up*

Most tests were conducted with an artificial acceleration meter deduced from the Nice S1 spectrum of PS92 regulations, with the timescale contracted by factor $\sqrt{1/3}$ to respect the similarity rules. CAMUS I and II models have been the subjects of additional tests to compare a proximal earthquake nearby (San Francisco or Melendy Ranch-type natural signal) with a distal earthquake (Nice-type artificial signal). The levels carried out during the test campaign were as follows:

– CAMUS I: Nice 0.24 g, San Francisco 1.11 g, Nice 0.24 g, Nice 0.40 g, Nice 0.71 g;

– CAMUS II: Nice 0.10 g, Nice 0.23 g, Nice 0.52 g, Nice 0.51 g;

– CAMUS III: Nice 0.42 g, Melendy Ranch 1.35 g, Nice 0.64 g, Nice 1 g.

CAMUS I		
Levels	At both ends of the walls	Central reinforcement
5th floor	1ϕ4.5 = 15.9 mm²	4ϕ5 = 78.4 mm²
4th floor	1ϕ6 = 28.2 mm²	4ϕ5 = 78.4 mm²
3rd floor	1ϕ6 + 1ϕ8 + 1ϕ4.5 = 94.4 mm²	4ϕ5 + 2ϕ 4.5 = 110 mm²
2nd floor	2ϕ6 + 2ϕ8 + 2ϕ4.5 = 189 mm²	4ϕ5 + 2ϕ 4.5 + 2ϕ6 = 138 mm²
1st floor	4ϕ8 + 2ϕ6 + 2ϕ4.5 = 289 mm²	4ϕ5 + 2ϕ 4.5 + 2ϕ6 = 138 mm²
CAMUS II		
Levels	At both ends of the walls	Central reinforcement
5th floor	1ϕ4.5 = 15.9 mm²	none
4th floor	1ϕ4.5 = 15.9 mm²	none
3rd floor	1ϕ4.5 = 15.9 mm²	none
2nd floor	1ϕ4.5 = 15.9 mm²	none
1st floor	1ϕ4.5 + 1ϕ5 = 35.5 mm²	none
CAMUS III		
Levels	At both ends of the walls	Central reinforcement
5th floor	2ϕ8 + 2ϕ4.5 = 132 mm²	2X5ϕ 4.5/200 = 159 mm²
4th floor	4ϕ8 + 2ϕ4.5 = 233 mm²	2X5ϕ 4.5/200 = 159 mm²
3rd floor	4ϕ8 + 2ϕ4.5 = 233 mm²	2X5ϕ 4.5/200 = 159 mm²
2nd floor	4ϕ8 + 2ϕ6 + 2ϕ4.5 = 289 mm²	2X5ϕ 4.5/200 = 159 mm²
1st floor	4ϕ8 + 2ϕ6 + 2ϕ4.5 = 289 mm²	2X5ϕ 4.5/200 = 159 mm²

Table 7.2. *Reinforcements of CAMUS models*

Model type	3D embedded	1D embedded	2D flexible support without interface	2D flexible support with interface	1D flexible support	Test
1st bending mode	10 Hz	10.3 Hz	8.5 Hz	7.3 Hz	7.4 Hz	7.3 Hz
1st pumping mode	45 Hz	40 Hz	28.3 Hz	22.3 Hz	19 Hz	20 Hz

Table 7.3. *Main natural modes, test-calculation comparisons*

7.4.2.1.1. Interest of simplified modeling for para-seismic studies

Importance of linking conditions at the support-structure interface

This analysis was conducted mainly on the CAMUS I model. For this analysis, the determination of natural modes was the subject of comparative analyses of calculations between the multi-layer model and a refined 3D model based on embedding conditions on a rigid foundation. The results show an excellent correlation between both calculations led to elasticity. Preliminary tests revealed that the two main modes were the first bending mode (in the plane of the walls) and the first vertical pumping mode, which was particularly visible, due to the flexibility of the table, a versatility not taken into account for the first calculations (Table 7.3). An accurate description, including the model, table, anchoring bars and the model-table interface mortar as an option was performed within a 2D EF model. For the two main modes, the results show the importance of the interface being softer than the remaining foundation), the taking into account of which allows the experimental results. As far as the "multi-layer" calculations are concerned, the flexibility of the table was adjusted to take the effects of the interface mortar into account. As we can see, the results compared with the experiment prove quite convincing.

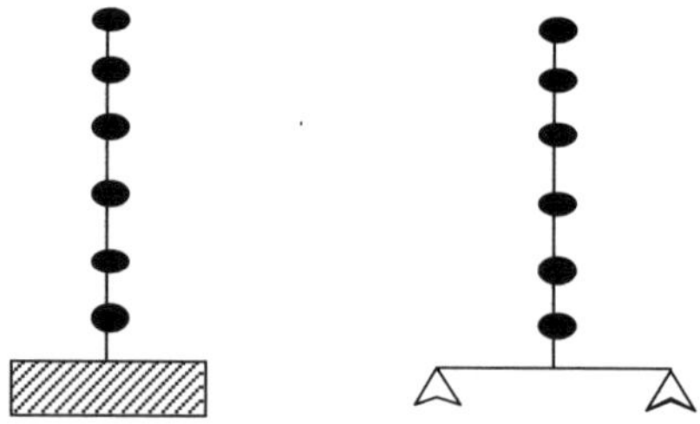

Figure 7.16. *CAMUS model: mechanical diagrams of the simplified model for stiff support and flexible support options; the dots represent the masses of the floors*

Behavior along the loading sequence (CAMUS I)

The entire loading sequence has been modeled on a multi-layer discretization basis, with a flexible support and the use of the damage model presented in section 7.4.2. As shown previously, two types of accelerograms have been used, one representative of an earthquake far away (Nice), the other one of an earthquake nearby (San Francisco), with a far shorter signal and higher acceleration peaks. The earthquakes applied have been successively: Nice 0.25 g, San Francisco 1.13 g, Nice 0.4 g, Nice 0.71 g (the quoted intensity value is the peak ground acceleration). Material parameters used for calculation were *concrete*: Young's modulus E = 30,000 MPa, maximum compressive strength, fc = 35 MPa, traction strength, ft = 3 MPa; *steel*: Young's modulus E = 200,000 MPa, elastic limit fe = 414 MPa, and strength at failure, fr = 480 MPa).

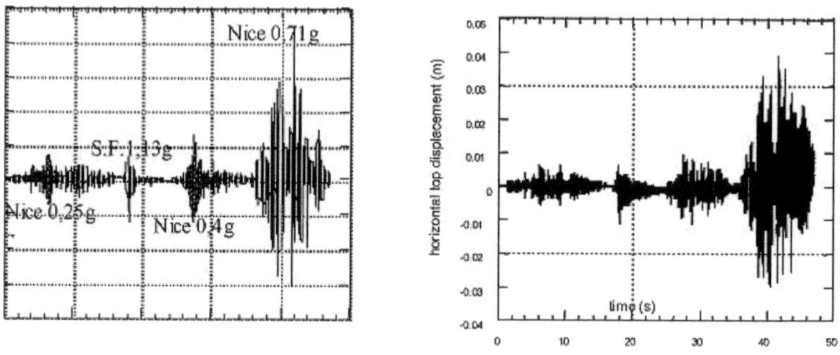

Figure 7.17. *Displacement at the top for the complete seismic sequence, test on the left, calculation on the right*

NOTE.– Some structural damping has been taken into account using Rayleigh's conventional description, both coefficients of which have been adjusted at 1% on the first mode and 2% on the second – advances have recently been made on the subject [RAG 99]. Finally, precautions have been taken as regards the finite element mesh to ensure objective results on a global scale, at least. Figure 7.17 compares the top displacement results obtained experimentally as well as by calculation for the whole loading sequence. Other results concerning load distribution throughout the damaged or plasticized areas have been obtained [MAZ 98]. All results are of good quality, which tends to lend credibility to such simplified methods.

Emphasizing the interaction between local behavior and overall function

The dialog between experiment and modeling groups has revealed the importance of interaction on local and global developments. During the test at the bottom of the wall we have been able to measure a dynamic variation of the vertical load with a frequency twice that of the bending moment and raising intensity. Analysis led to an appreciation of the link between the two phenomena: progressive cracking of the structure with cracks "breathing" (opening-closing) during loading, and the activation of the vertical pumping mode due to the flexibility of the table and the evolution of damage. Within this context, the damage model presented in section 7.4.2, which takes both effects into account, has revealed a direct connection between the closing-up stress (σ_f) and the intensity variation of the vertical load. A parametric study was conducted and Figure 7.18 shows this mutual influence for two particular values of σ_f, 3 MPa and 1.35 MPa, with the latter value leading to a connection with the vertical load variations measured.

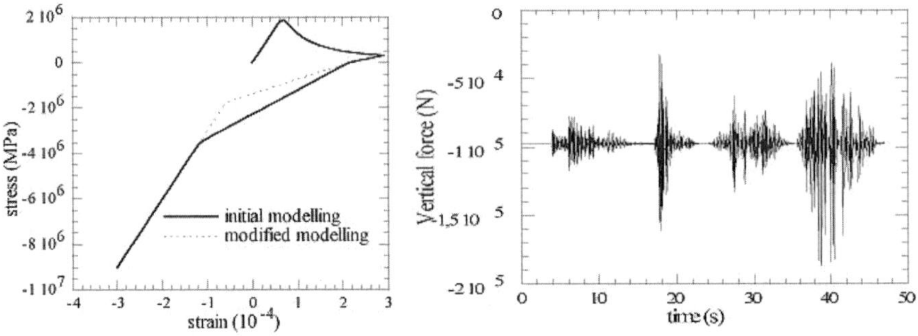

Vertical force	Initial model	Recalibrated model	Tests (kN)
Nice 0.25 g	115	119	138
S-F 1.13 g	150	160	198
Nice 0.40 g	119	132	146
Nice 0.71 g	140	190	248

Figure 7.18. *Connection between closing up stress and dynamic variation of the load at the bottom of the model. In the upper left hand corner, uniaxial response of the model for two values of σ_f (3.5 and 1.35 MPa). In the upper right hand corner, consequences on the variation of the vertical load at the bottom. Below, table of the maximum values of the compressive vertical load at the model-table interface*

7.4.2.1.2. Two-dimensional discretization and plasticity models

As the stress is horizontal and parallel to the planes of the walls, half of the structure is modeled using a biaxial local approach. An example of the complete FE mesh of the model is presented in Figure 7.19. The mesh of the wall was chosen so that the connection with the steel elements could be made in a position quite close to the exact position, in accordance with the reinforcement layout. The mesh allows the bar stoppers to be reproduced, as well as the concrete construction joints. Two-node bar elements have been used to represent the reinforcement.

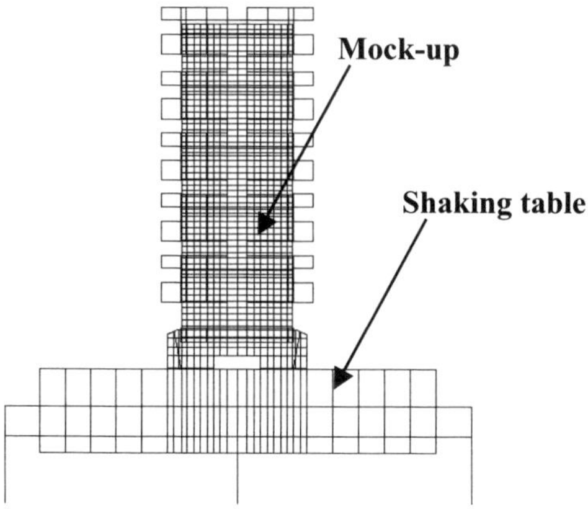

Figure 7.19. *2D meshwork of the CAMUS model*

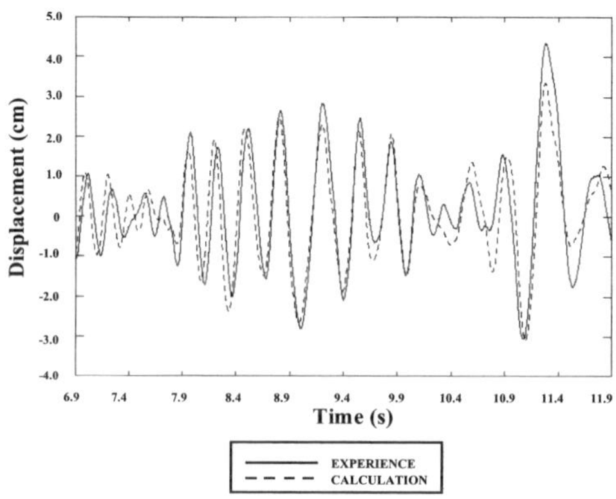

Figure 7.20. *Test-calculation comparison: CAMUS I model, Nice 0.71 g signal*

Figures 7.20, 7.21 and 7.22 show the calculation-test comparison of the front displacement for an earthquake level that caused important damage to each model. In each case, the non-linear plastic model is capable of estimating the overall behavior of the structure quite well. At the local level, the digital results highlight operating modes similar to the experimental modes – there lies the interest in such modeling (Figure 7.23).

In the case of the CAMUS I model, the behavior is clearly influenced by a strong localization of the strains at the level of the bar stoppers on the second floor, and by the appearance of inclined cracks due to the shear force. Such behavior is quite compatible with the operation observed during the last test: appearance of "fan-shaped" cracks resulting from a truss operation functioning followed by bending-predominant failure mechanism with bar failure at the level of the second-floor stoppers.

In the case of the CAMUS II model, the behavior is of the multi-block type, with horizontal cracks appearing at each concrete construction joint. Steel is plasticized at the bottom of the lower floors. This kind of behavior is well expressed in both the computed and experimental moment-curvature curves (Figures 7.24 and 7.25). The influence of the variation in vertical load dynamic is obvious in the plastic plateau area (when the cracks are wide open and the steel plasticized) because of the disruption brought about: the position of the plastic plateau depends on the vertical load value.

With the CAMUS III model, designed according to the Eurocode 8 regulations, the operating mode is quite different from the two modes previously described: a plastic hinge at the foot of the wall and a quasi-elastic behavior above the critical area. The large plastic rotations and dissipation of energy appeared mainly on the first floor (Figure 7.23).

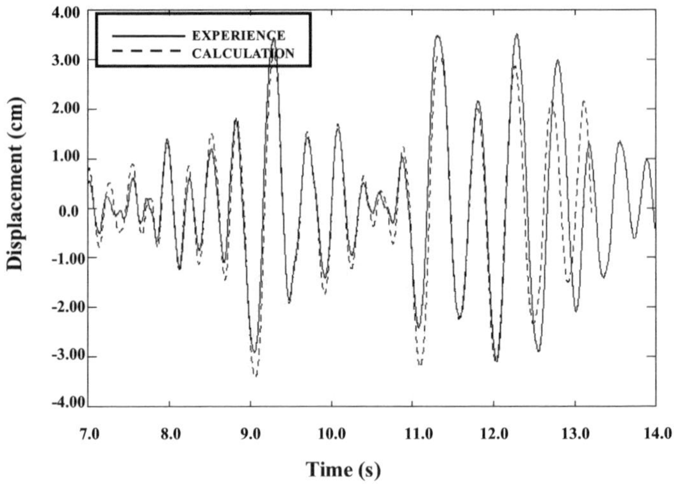

Figure 7.21. *Calculation-test comparison: CAMUS II model, Nice 0.51 g signal*

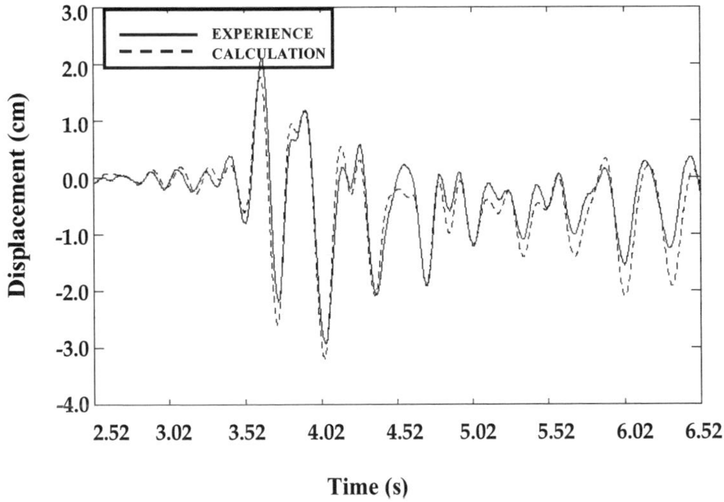

Figure 7.22. *Calculation-test comparison: CAMUS III model, Melandy Ranch 1.35 g signal*

Note that the collapse mode observed using the CAMUS I model could not be predicted by elastic calculation. Moreover, due to the disruption caused by the shear load, modeling based on a fiber model could not accurately reproduce this mode of behavior. The CAMUS tests, together with the digital analysis carried out, have shown the strong interaction that exists between the vertical and horizontal directions of the vibration.

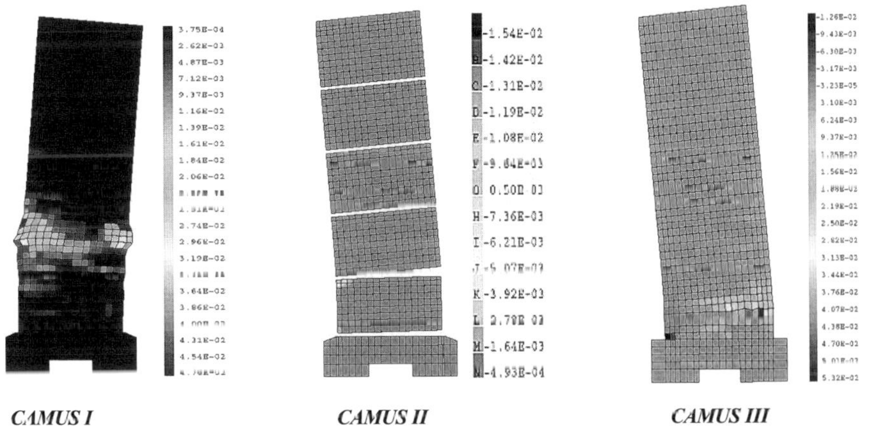

Figure 7.23. *Isovalues of the vertical strains in concrete*

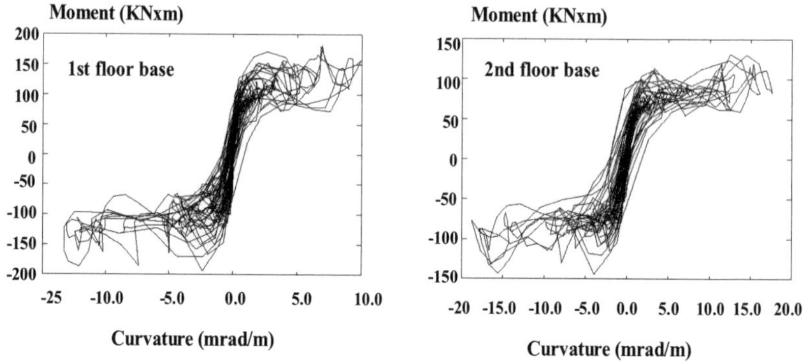

Figure 7.24. *Experimental moment-curvature relationships – CAMUS II – 0.51 g*

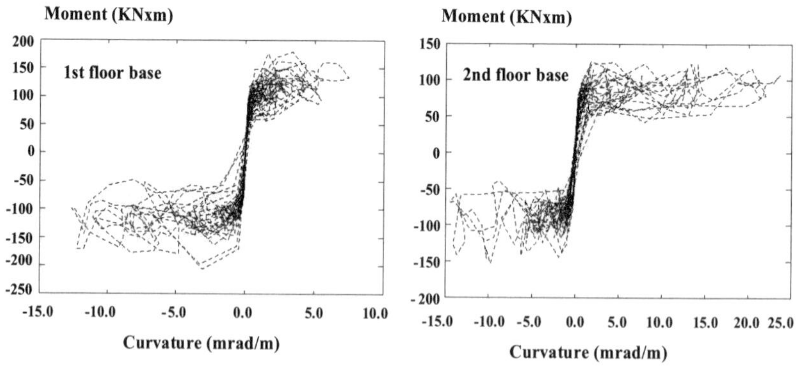

Figure 7.25. *Digital moment-curvature relationships – CAMUS II – 0.51 g*

The direct consequence of mass uplift is the appearance of a compression-traction dynamic vertical load, the magnitude of which can be considerable, depending on the values of the mass-weights, as seen before. It is important to note that the dynamic non-linear calculations are the only ones capable of estimating variation ranges of the dynamic vertical loads required in the designing stage.

7.4.2.2. *ICONS experimental program on u-shaped walls*

The structure was a 3.60 m high load-bearing wall, with a u-shaped constant section developed in accordance with Eurocode 8 regulations. The girder of the u-section was 1.50 m long, with flanges 1.25 m long and 25 cm thick. The wall was built on a foundation: a reinforced concrete square block 3.50 m long on each side and 1 m high. Another reinforced concrete block, 2.50 m long on each side and 60 cm high was placed at the head of the wall. The top and side views of the model are presented in Figure 7.26. The three specimens tested on the ELSA reaction wall are all identical.

The vertical reinforcement is ensured by steel bars 10 and 12 mm in diameter with a percentage ρ equal to 0.0056. The shear load reinforcement consisted of $\phi 8$ steel bars with 125 mm-wide spacing in the wings and 75 mm-wide spacing in the girder. The spacing of the diameter-8 stirrups was 90 mm wide. The wall section is shown in Figure 7.27.

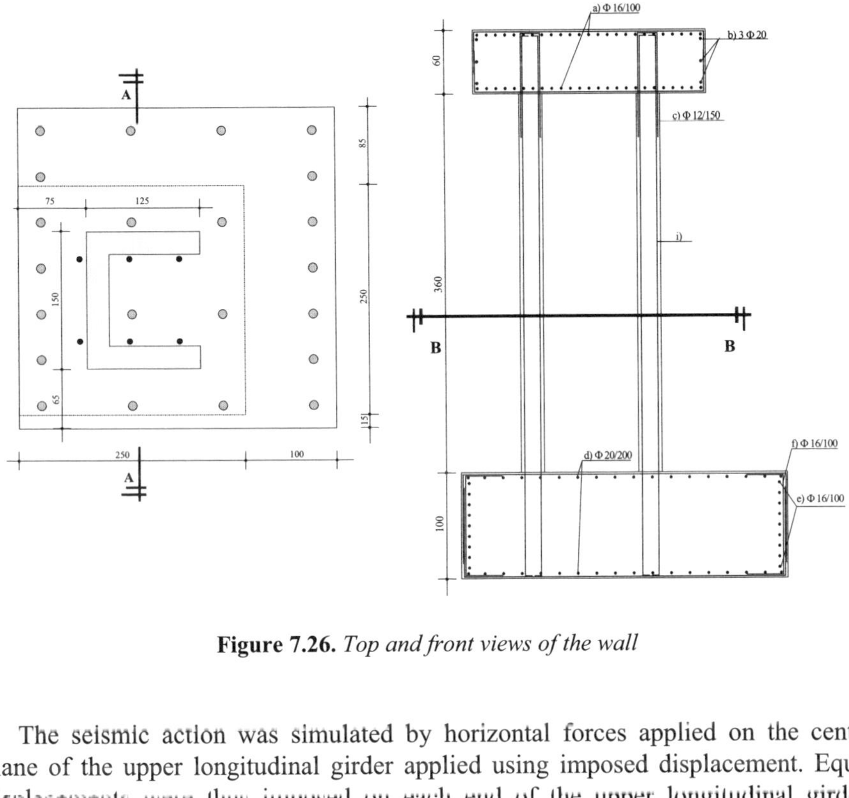

Figure 7.26. *Top and front views of the wall*

The seismic action was simulated by horizontal forces applied on the center plane of the upper longitudinal girder applied using imposed displacement. Equal displacements were thus imposed on each end of the upper longitudinal girder, therefore rotation within the horizontal plane of the girder was blocked. The experimental assembly plan and the axis positioning are shown in Figure 7.28.

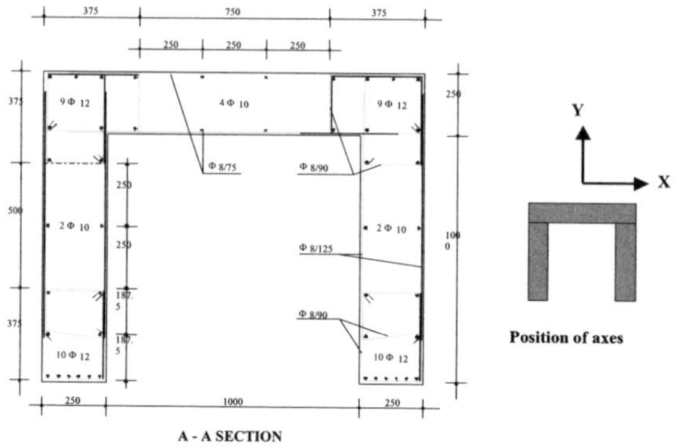

Figure 7.27. *Wall section*

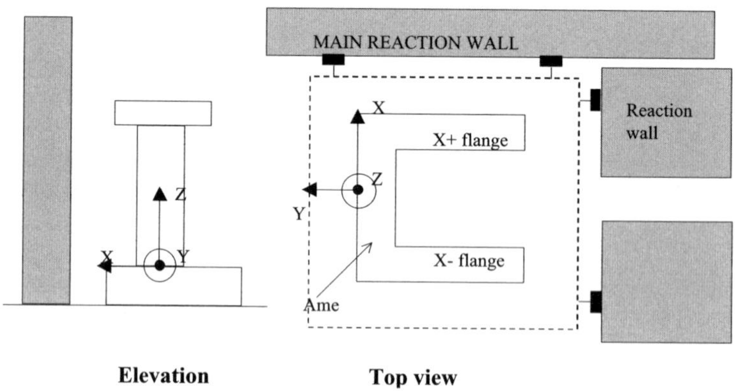

Figure 7.28. *Experimental assembly plan*

To simulate a vertical permanent load, before each cyclic test a 2,000 kN vertical load was applied via 6 cables arranged so that the result was close to the center of inertia of the area. Three different cyclic tests were carried out on the u-structure with the reaction wall: firstly, a USW1 test with horizontal loading along the Y-axis, then a USW2 test with horizontal loading along the X-axis, and finally, a USW3 test with horizontal loading along the X- and Y-axes.

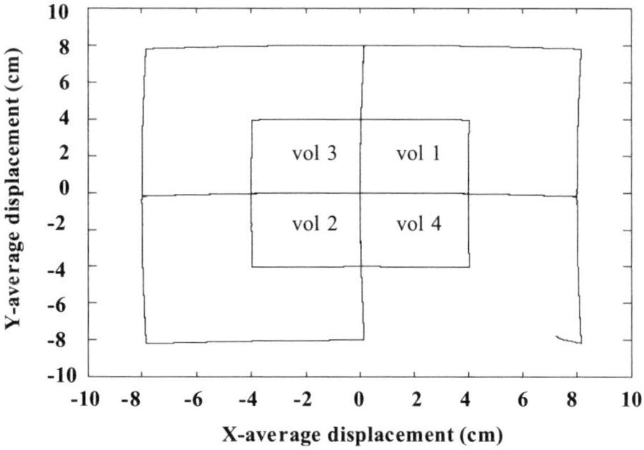

Figure 7.29. *Y-average displacement as a function of the X-average displacement in the USV3 test*

The displacements applied during the USW3 test are presented in Figure 7.29. Two program levels (± 4 cm and ± 8 cm) were involved. They were applied in a "butterfly-wing" manner: cycle n° 1 (vol 1 in the figure) carried out in the direct direction (clockwise), then cycle n° 2 (vol 2) in the indirect direction (anti-clockwise), then cycle n° 3 (vol 3) in the indirect direction and finally cycle n° 4 (vol 4) in a direct direction as well. Afterwards the same procedure was used for the upper level, which led to failure during cycle n° 4 as shown by Figure 7.29.

Modeling and comparison with experimentation

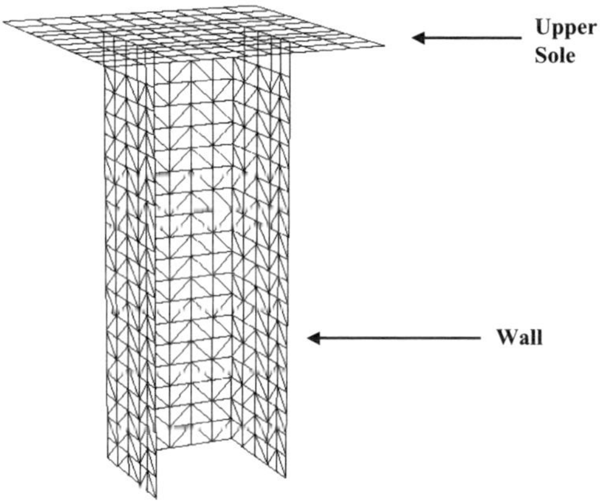

Figure 7.30. *Finite element mesh of the U-shaped non-rectangular wall*

A 3D model based on multi-layer thin shell modeling has been adopted to simulate the behavior of this structure. It should be noted that with such modeling, transverse shear is not taken into account, and the actual 3D behavior (transverse dilation) is not correctly described. Nevertheless, the approach allows the bi-bending of the different parts of the wall and the shear in the plane of the flanges and girder web to be represented quite well. The wall is modeled using integrated DKT-type elements, whereas four-node thin shell-type elements with linear elastic behavior are used to represent the upper sole. Vertical and horizontal steel reinforcements are modeled using off-center bar-type elements. The concrete-steel adherence is assumed to be perfect and the wall to be perfectly embedded at its bottom. The meshwork used for the wall is presented in Figure 7.30.

The behavior of the wall is ruled by shear in the flanges as well as in the girder web of the wall. As a matter of fact, the flanges change roles during the cyclic test, as most of the shear imposed in the Y-direction is taken up again by the compressed flange. During the 4 cm cycles, the steel reinforcements located at the flange-web junction tends to buckle and the concrete starts to exhibit spalling. During the 8 cm cycles, most of the steel at the ends of the flanges and the girder webs buckle, the concrete gets crushed and collapse takes place because of the shear in the compressed flange. The calculation-experiment comparison of the displacement-load cyclic curves recorded during the USW3 test is presented in Figure 7.31. Figure 7.32 shows the drawing of the loads in both directions: when the displacement in one direction changes, while keeping the displacement on the orthogonal direction practically constant, the load required in the latter direction to keep the corresponding displacement constant drops.

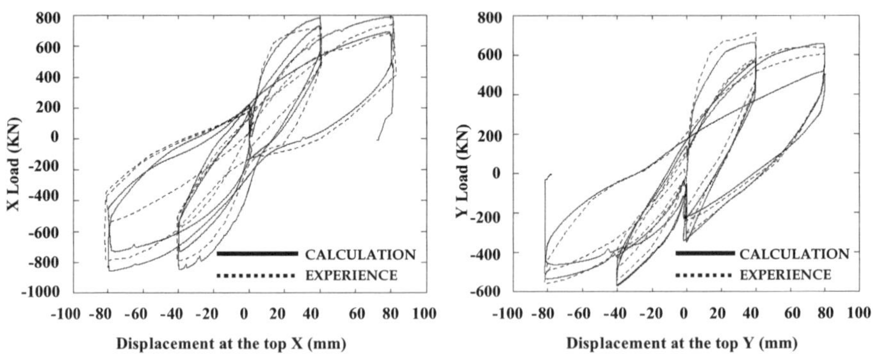

Figure 7.31. *USW3 test (XY dir.): comparison with cyclic calculation*

The effect of such coupling on the load-displacement diagram for both directions then consists of quasi-vertical unloading, which can be observed in Figure 7.31.

Thus, the hysteresis loops are broader and the dissipated energy higher. An important strength cyclic deterioration occurs during the cyclic loading. This can be seen in Figure 7.32, as collapse takes place during the last loop of loading for 8 cm displacement in the X direction and –8 cm displacement in the Y direction.

Comparing Figures 7.31 and 7.32 shows that the important phenomena observed during the tests (stiffness deterioration, strength cyclic deterioration) are correctly reproduced by the 3D model shell. Recent works have shown that for the same problem it was also possible to achieve good results with enriched Timoshenko multi-fiber elements [KOT 00].

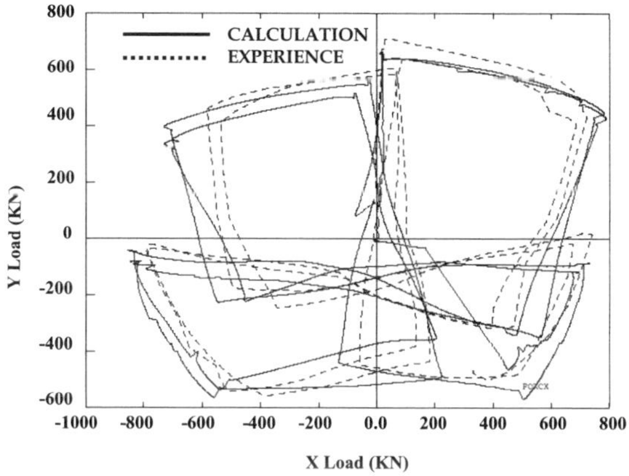

Figure 7.32. *USW3 test: graph of the load evolutions in X and Y directions during the test*

7.5. Conclusions

In this chapter, we showed some peculiarities of the functioning of RC structures under seismic loading and demonstrated the need to have efficient modeling tools to analyze their responses. This results in two major aspects to be considered:

– representation of the structure with its associated masses;

– relevance of the behavior models to use for both materials and interfaces (between the materials and between structure and support).

The choice should be made according to the expected accuracy of the results and the opportunity of expressing simplifying hypotheses.

With slender structures, frames and structural walls, the effectiveness of semi-global models (multi-fiber or multi-layer beams) has been proved, if the constitutive models are suited to the phenomena to emphasize (material models with damage and models for the interfaces that give rise to strong local non-linearity). Such models are especially recommended to access global and collapse-mode behaviors.

A good calculation experiment correlation should not leave aside the difficulty in finding local results (like crack opening or steel deformation) for important damage levels. Such difficulties are all the more awkward as the criteria that herald damage of the structure (and therefore its potential collapse) are generally local in para-seismic regulations.

To obtain more realistic apprehension of local values, a solution lies in more complete 2D and 3D discretization. We could state that the plastic model coupled with a correct description of the limit conditions and of loading allows us to justify the essential mechanisms of the behavior of a wall under cyclic loading. The main drawback of this local approach lies in the amount of calculation generated by the great number of degrees of freedom used, far more than with simplified models. Nevertheless, the plane stressed walls are quite suitable for 2D modeling, and from this point of view, current computer means are quite well adapted. The behavior of walls that have a U-shaped section and are loaded in bi-bending is modeled using a multi-layer thin shell 3D approach, based on the 2D local model. In the last case, although the amount of calculation generated is higher than the 2D approach, it is still quite reasonable and much lower than massive 3D modeling.

Whilst highlighting these model's possibilities, we should not conceal their limitations and it is important to note that the local phenomena that take place at the concrete-steel interface can only be represented approximately according to the concrete model used. The shear transfer factor η allows some of the transmission of the shear stresses in the cracking plane to be taken into account, but it cannot correctly represent the mode II cyclic energy dissipation.

Furthermore, for all the presented models, the modeling quality is affected by the parameter identification procedures used. Some experimental procedure regulations for the commonly admitted characterization of materials could prove quite useful.

7.6. Bibliography

[CAZ 03] CAZAUX G., Modélisation tridimensionnelle du comportement sismique d'ouvrages en béton armé – Développement de méthodes simplifiées, PhD Thesis, LMT-ENS de Cachan, 2003.

[COI 98] COIN A. *et al.*, "Recherche CAMUS, Rapport final, Présentation générale, Conclusions et perspectives", *Ministère de l'Equipement, Plan Génie Civil, Convention 9670011*, 10 pages, 1998.

[COM 96] COMBESCURE D., Comportement sismique des portiques en béton armé comportant des murs de remplissage en maçonnerie, PhD Thesis, Ecole Centrale Paris, 1996.

[COM 01] COMBESCURE D., SOLLOGOUB P., "Validation of non linear constitutive laws used for the seismic evaluation of existing industrial facilities", *4th International Conference on Fracture Mechanics of Concrete Structures*, Cachan, France, 2001.

[CRE 01] CREMER C., Modélisation du comportement non linéaire des fondations superficielles sous séisme. Macro-élément d'interaction sol-structure, PhD Thesis, LMT – ENS Cachan, 2001.

[ELI 93] ELIGEHAUSEN R., BALAZS G.L. "Bond and detailing", in *Bulletin d'Information CEB*, N° 217, Comité Euro-International du Béton, 1993.

[GHA 98] GHAVAMIAN S., Méthode simplifiée pour la simulation du comportement sismique des structures en béton armé. Traitement des effets de l'élancement et estimateur d'erreurs, PhD Thesis, LMT-ENS Cachan, 1998.

[GUE 97] GUEDES J., Seismic behavior of reinforced concrete bridges. Modelling, numerical analysis and experimental assessment, PhD Thesis Dissertation, University of Porto, Portugal, 1997.

[HOG 51] HOGNESTAD E., "A study of combined bending and axial load in reinforced concrete", *Bulletin Series 339*, University of Illinois, USA, 1951.

[ILE 00] ILE N., Contribution à la compréhension du fonctionnement des voiles en béton armé sous sollicitation sismique: apport de l'expérimentation et de la modélisation à la conception, PhD Thesis, INSA-Lyon, 262 pages, 2000.

[KOT 00] KOTRONIS, P., Cisaillement dynamique de murs en béton armé. Modèles simplifiés 2D et 3D, PhD Thesis, LMT-ENS Cachan, 2000.

[LAB 91] LABORDERIE C., Phénomènes unilatéraux dans un matériau endommageable: modélisation et application à l'analyse des structures en béton, PhD Thesis, University of Paris 6, 1991.

[LEM 90] LEMAITRE J., CHABOCHE J.L., *Mechanics of Materials*, Cambridge University Press, Cambridge, UK, 556 pages, 1990.

[MAN 88] MANDER, J.B., PRIESTLEY M.J.N, PARK R., "Theoretical stress-strain model for confined concrete", *Journal of Structural Engineering*, Vol. 114, N° 8, pp. 1804–1825, 1988.

[MAZ 86] MAZARS J., "A description of micro- and macroscale damage of concrete structures", *J. of Engineering Fracture Mechanics*, Vol. 25, N° 5/6, pp. 729–737, 1986.

[MAZ 98] MAZARS J., "French advanced research on structural walls: an overview on recent seismic programs", *Proceedings of the 11th European Conference on Earthquake Engineering*, Invited lecture, Paris CD-Rom, 1998.

[MEN 73] MENEGOTTO M., PINTO P., "Method of analysis for cyclically loaded reinforced concrete plane frames including changes in geometry and non-elastic behavior of elements under combined normal force and bending", *IABSE Symposium on Resistance and Ultimate Deformability of Structures Acted on by Well-defined Repeated Loads*, Lisbon, Portugal, 1973.

[MER 99] MERABET O., REYNOUARD J.M., "Formulation d'un modèle élasto-plastique fissurable pour le béton sous chargements cycliques", *Contrat EDF/DER, Rapport final*, 84 pages, N° INSA-1/943/002, 1999.

[MON 00] MONTI, G., "Consistent insertion of bond-slip into beam fiber elements for biaxial bending", *12th World Conference on Earthquake Engineering*, Auckland, New Zealand, 2000.

[PAR 90] PARK Y.S., "Fatigue of concrete under random loadings", *Journal of Structural Engineering*, Vol. 116, N° 11, pp. 3228–3235, 1990.

[PEG 93] PEGON P., "A Timoshenko simple beam element in Castem 2000", *JRC Technical Note*, N° I.93.05, Ispra, Italy, European Commission, 1993.

[RAG 99] RAGUENEAU F., Fonctionnement dynamique des structures en béton. Influence des comportement hystérétiques locaux, PhD Thesis, LMT-ENS Cachan, 1999.

[XIA 97] XIAO Y., MA R., "Seismic retrofit of RC circular columns using prefabricated composite jacketing", *Journal of Structural Engineering*, Vol. 123, N° 10, ASCE, 1997.

Chapter 8

Seismic Analysis of Structures: Improvements Due to Probabilistic Concepts

8.1. Introduction

Structures, including buildings, construction works and industrial facilities, have to be designed to resist earthquakes. To ensure this, engineers obtain data about the open field motions where the structure is to be built, and employ modeling and calculation methods to enable them to determine its likely response, in order to verify the design is acceptable.

The degrees of sophistication of the methods used will depend quite heavily on the importance of the structure and on the costs incurred if it collapses.

As a consequence, when designing large construction works or nuclear facilities, the most elaborate methods available are implemented. For example, we will care about the correlation of open field seismic motions within the considered soil area, and of the influence of structural motions on those of the soil (soil-foundation interaction). We will take non-linear behaviors into account. We will also characterize the motions of the components' anchoring points in order to carry out separate detailed calculation, etc.

The common denominator in all conventionally used methods is what is usually called the *modal method*. The method, based on the linear behavior of the soil and

---

Chapter written by Jean-René GIBERT.

the structures, is presented in a quite assertive *deterministic* light, through the seismic source data, through the whole approach for obtaining the response and the conclusion as far as strength or collapse are concerned.

The current chapter aims to show that this deterministic approach can sometimes lead to quite misleading results and that, in order to control things and to improve the predictions, it is necessary to become aware of the probabilistic feature of the problem raised and of the more or less justified hypotheses made. Thus, we are led to ask ourselves questions such as:

– What model should be chosen to represent the motion of the soil?

– Is the conventionally-used "oscillator response spectrum" data still relevant?

– How can we preserve the simple feature of the new methods proposed while keeping them consistent with regulations, especially in the case of non-linear behavior?

8.2. The modal method

For further details see [DAV 85], [DAV 88] and [GIB 88].

8.2.1. *Data about the seismic source*

Open field motion is a rather complex acceleration time signal (Figure 8.1) characterized by a level variable, the γ_{smax} maximum acceleration and frequency content.

In the deterministic approach one or several earthquake levels (operation and safety levels, for example) corresponding to different structure strength requirements are associated with a site.

The frequency content results from various compilations and averages of measured seismic signals. As we do not usually have any recordings of the concerned site, we often just use general standardized data such as the "Regulatory Guide 1-60" spectra (Figure 8.2), possibly corrected with the "site effects".

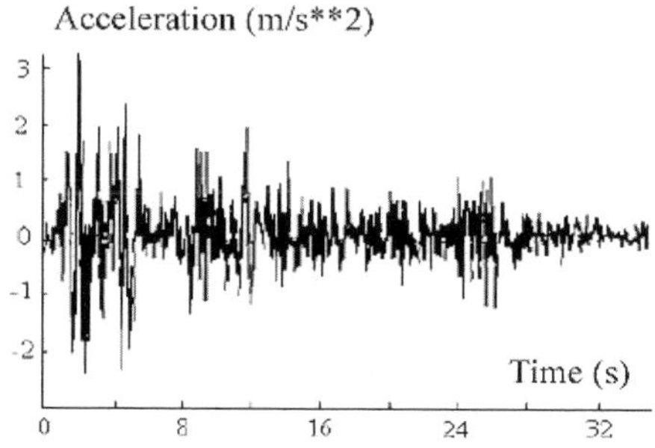

Figure 8.1. *Soil acceleration as a function of time during an earthquake: open field measurements*

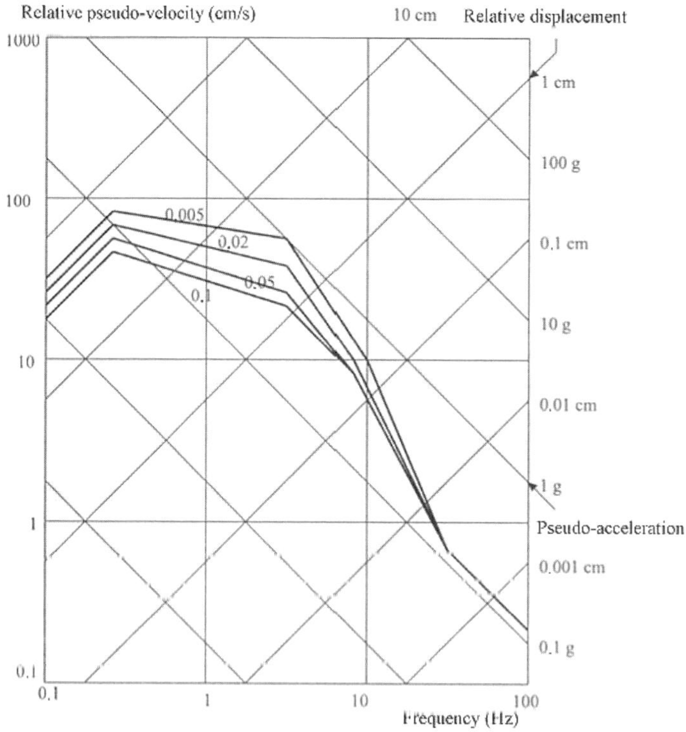

Figure 8.2. *Standardized spectra (at 0.15 g) of the "Regulatory Guide 1-60"*

Thus, this results in data that are unlikely to represent a real motion but, considering the way to characterize the frequency content and the modal method using the data, which constitutes a consistent set.

Let us examine the situation more accurately. The frequency content of the seismic signal is characterized by the oscillator response spectra (ORS), the definition of which is determined by the following:

– given a high degree of freedom oscillator (mass, spring, damper system) with imposed movement at anchoring points (Figure 8.3): $x_{soil}(t)$. The oscillator is characterized by its resonance frequency f and damping ratio ε;

– given x(t) the relative motion of the oscillator mass ($x(t) = x_{struct}(t) - x_{soil}(t)$), the maximum value reached by the absolute value of x(t) will be called S (f,ε).

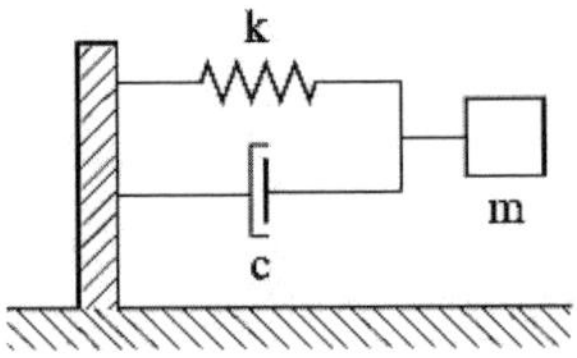

Figure 8.3. *Diagram of the seismically excited harmonic oscillator*

By varying f and ε, we obtain a curve network that is the ORS characterizing the $x_{soil}(t)$ signal.

The ORS are typically represented in the form of:

– pseudo-velocities $S_{pv}(f,\varepsilon) = 2\pi f\, S(f,\varepsilon)$;

– pseudo-accelerations $S_{pa}(f,\varepsilon) = (2\pi f)^2\, S(f,\varepsilon)$.

The ORS include 3 frequency bands (Figure 8.4):

– the very low frequency band (< 0.1 Hz) for which we can observe an asymptotic behavior in $S(f,\varepsilon) = (x_{soil})_{max}$;

– the frequency band where the seismic energy is located and where we can observe an amplification effect all the wider the smaller ε is;

– the high frequency band (> 20 to 30 Hz) where we can observe a behavior in $S(f,\varepsilon) = \gamma_{smax}/(2\pi f)^2$.

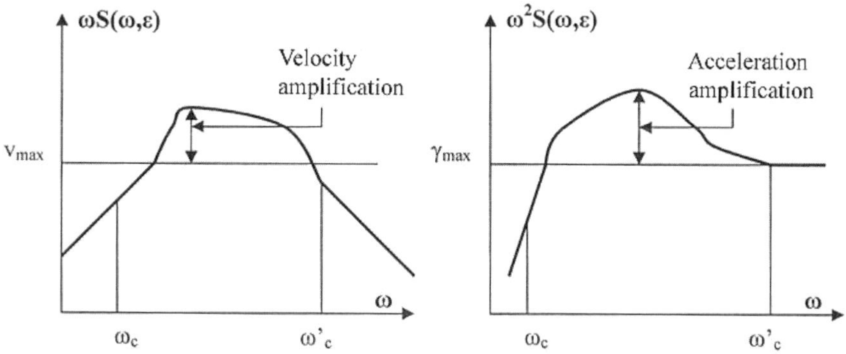

Figure 8.4. *The three frequency ranges of an oscillator response spectrum*

8.2.2. *Calculation of structural responses using the modal method*

Let us consider a *linear behavior* structure, with the $x_{soil}(t)$ motion imposed on a specified number of its points (for instance, structures with stiff soil foundation for which we can suppose that the anchoring points within the soil follow the open field motion).

The x(r, t) relative motions with regard to the soil confirm the system (r = space variable):

$$M\ \delta^2 x/\delta t^2 + A\ \delta x/\delta t + Kx = -M\ U\ \gamma_{soil}(t) \qquad [8.1]$$

with M, A, K being the operators of inertia, viscous damping and stiffness for the structure blocked at its anchoring points. U represents the space function "unit translation in the earthquake direction", and $\gamma_{soil}(t)$ is the soil acceleration (seismic signal).

Φ_n (r), ω_n and m_n being the modal elastic lines, resonance pulsations and generalized masses of the non-dampened structure with displacement limit conditions equals zero at the anchoring points. The $a_n(t)$ modal contributions confirm the following uncoupled system (when supposing that the damping operator becomes diagonal):

$$d^2a_n / dt^2 + 2\ \varepsilon_n\ \omega_n\ da_n / dt + \omega_n{}^2\ a_n = -(q_n / m_n)\gamma_{soil}(t) \qquad [8.2]$$

where ε_n and q_n are respectively the modal damping and the seismic participation factor (the projection of the MU force of inertia on mode $q_n = (\Phi_n, M\,U)$).

The ORS data allows the maxima reached by the absolute values of modal contributions to be obtained without calculation from:

$$a_{n\,max} = \left[|q_n| / \left(m_n \omega_n^2\right)\right] S_{pa}\left(f_n, \varepsilon_n\right) \qquad [8.3]$$

where $f_n = \omega_n / 2\pi$ = resonance frequency.

Obtaining the variables useful for strength diagnostics (relative displacements and stresses in certain points of the structure) requires an assumption.

Actually, if we are interested in the relative displacement at a specific point r, for instance, the latter will be a linear combination of the modal contributions:

$$x(r, t) = \sum_n a_n(t) \Phi_n(r) \qquad [8.4]$$

To obtain the absolute value of x, the so-called *simple quadratic combination* rule is applied:

$$X_{max}^2(r) = \sum_n a_{n\,max^2} \Phi_n^2(r) \qquad [8.5]$$

This rules implies, as will be shown later, the statistical independence of the different terms of the sum. In particular, the previous formula cannot be applied with modes whose frequencies are beyond the seismic range.

The summation is then restricted to the first N modes (the higher limit of the seismic range) and the contribution of the higher modes is represented by a "pseudo-mode", determined by virtue of an associated static solution:

$$x_{max^2}(r) = \sum_{n-1}^{N} a_{n\,max^2} \Phi_{n^2}(r) + \gamma_{s\,max} \left\{ X_{\underset{n=1}{s}}(r) + \sum_{n=1}^{N} \left[q_n / \left(m_n \omega_{n^2}\right)\right] \Phi_n(r) \right\}^2 \qquad [8.6]$$

with the X_s (r) static solution given by $K\,X_s = -\,M\,U$.

REMARK 8.1.– The coherence of the method is quite obvious. An earthquake is defined due to the response maxima of oscillators, with the maxima being precisely the values that enable us to rule about the strength of the structure. As a matter of fact, this implies that collapse is determined by a threshold that is exceeded once (collapse from excessive strain and not from oligo-cyclic fatigue).

8.3. Criticism of the modal method

The drawbacks of the modal method are numerous and include:

– the simple quadratic combination hypothesis cannot always be verified;

– knowing the response maxima is not always sufficient. For example, calculating a complex installation must be performed in several steps. The calculation for buildings is effected, with the main elements represented through their mass. To calculate the equipment, the motions of their anchoring points have to be characterized. This is called the *floor spectrum* problem;

– collapse can be caused by combination of several cycles, which therefore involve several relative maxima (oligo-cyclic fatigue, cumulative plastic strain);

– if we want to take a highly non-linear behavior into account, the modal method is ineffective.

Therefore, other approaches have to be used. For instance, we might think of resorting to time calculations. Carrying out shaking table tests implies a time approach as well. The obvious question that must then be asked deals with associating a temporal seismic signal with the ORS data.

In order to bring solutions to this set of questions, it is necessary to revert to the definition of a seismic signal.

Observation of the seismic motions on a same site shows noticeable time rate differences from one signal to the next. Such differences result from the complex phenomena involved, and it would be mistaken to believe that it is possible to give a detailed explanation of them.

The only solution involves representing the seismic motion of a site through a random process that will be characterized by means of average values. The previously defined ORS can be considered as such values. Are they relevant ones, that is to say, is the ORS data enough to determine the average values associated with the response of the structure and which are necessary to predict the collapse probability?

In order to answer that question, we will first recap a few ideas about random processes.

8.4. A few reminders about random processes

For more information, see [GIB 88], [PRE 90] and [KRE 83], though these are more theoretical.

8.4.1. *Definition and characterization of a time random process*

Consider a set of time instants arranged in ascending order. Each is associated with an X_n random variable. The set of these variables constitutes a random process.

The process will thus be completely characterized by the joint repartition functions and probability densities of the X_n set. The discrete process defined above can be more generally applied to a continuous process, denoted X(t), by having the between t_n intervals tend to zero. In that case, complete characterization obviously requires an infinite amount of information.

From a practical point of view, we will only be able to consider a limited number of joint probabilities, and therefore the characterization will be imperfect.

In most cases, considering only the p_2 $(x_1,t_1;\ x_2,t_2)$ joint probability densities or two X_n variables will be enough. If the probability density completely characterizes the process, the latter is said to be *Markovian*.

8.4.2. *Second order characterization*

Rather than using the p_2 function, we will consider the associated moment function, limiting ourselves to order 2 (which, in the case of a *Gaussian* process, is enough to characterize it). Hence if (D) is the set of values X can assume, we can define:

(i) the process average;

$$m(t) = \int_D x \, p_1(x, t) \, dx \qquad [8.7]$$

(ii) the correlation function of the process;

$$\rho(t_1, t_2) = \int_D \int_D x_1 \, x_2 \, p_2(x_1, t_1; x_2, t_2) \, dx_1 \, dx_2 \qquad [8.8]$$

The process is said to be stationary if these characteristics stay invariant through time translation. Thus, we will obtain:

$$m = \int_D x \, p_1(x) \, dx \text{ and } \rho(\tau) = \int_D \int_D x_1 x_2 p_2(x_1, x_2, \tau) \, dx_1 \, dx_2 \qquad [8.9]$$

with $\tau = t_2 - t_1$.

It is interesting to mention that for stationary processes, the "ergodicity" quality allows us to determine m and $\rho(\tau)$ owing to time averages carried out on an x(t) realization:

$$m = \lim_{T \to \infty} 1/T \int_{-T/2}^{T/2} x(t) \, dt \text{ and } \rho(\tau) = \lim_{T \to \infty} 1/T \int_{-T/2}^{T/2} x(t) \, x(t+\tau) \, dt \qquad [8.10]$$

In the case of stationary processes, the *power spectral density (PSD),* which is the Fourier transform of $\rho(\tau)$, is quite commonly used. We can note:

$$S(f) = 2 \int_{-\infty}^{\infty} \rho(\tau) \exp(-2i\pi f\tau) d\tau \qquad [8.11]$$

Quite often the signal will have an average equal to zero (the seismic signal for example) and we will obtain:

$$\text{Variance of X (t)} = \sigma^2 = \rho(0) = \int_0^{\infty} S\ (f)\ df = m_0 = 0 \text{ order spectral moment}$$

Physically, for the signal, the PSD represents the energy distribution according to the frequency. The notion of dX/dt(t) derivative process will also be used; it is defined as a quadratic mean by:

$$\text{Expected value: } E\left\{\left[\left(X(t+\varepsilon) - X(t)\right)/\varepsilon - dX/dt\ (t)\right]^2\right\} \to 0$$

$$\varepsilon \to 0$$

For a stationary process, the derivative process exists if $d^2\rho/d\tau^2$ exists for $\tau = 0$.

The derivative process is "orthogonal" to the initial process for:

$$E\left[X\ (t)\ dX/dt\ (t)\right] = d\rho/d\tau\ (0) = 0, \text{ and its PSD is equal to } (2\pi f)^2 S\ (f).$$

Thus, the variance of the derivative process for a zero- average signal is:

$$\sigma^2_{\text{der}} = (2\pi)^2 \int_0 f^2 S(f) df = (2\pi)^2 m_2 \qquad [8.12]$$

where m_2 = second order spectral moment.

8.4.3. *Response of a linear system to random stress*

Generally we know how to express the second order characteristics of the response of a linear system characterized by its transfer functions to stress in the shape of a space-time random field that is also second order characterized.

In order not to make this chapter too heavy, we will develop the formulation for the case described previously of a structure loaded with an imposed acceleration of its anchoring.

The imposed acceleration will be considered as a zero-average stationary *time random process* Γ(t), characterized by its PSD S(f). The structure is characterized by its by-pulse response G (r, τ) (response at point r and instant τ to an anchoring acceleration corresponding to a unit pulse exerted at instant τ = 0).

Let us consider the resting structure at τ = 0 and let us exert the Γ(t) process from that moment. The relative displacement of the structure at point r and instant τ can be considered as a stationary random process of time t: X (r, τ, t), the PSD of which is expressed by:

$$S_x(r, \tau, f) = \left|H(r,\tau,f)\right|^2 S(f) \quad [8.13]$$

H (r, τ, f) has the meaning of a transfer function and is given by:

$$H(r,\tau,f) = \int_0 G(r, u)\ \exp\ (-2i\pi fu)\ du \quad [8.14]$$

If τ → ∞, the transfer function tends to the conventional expression of the Fourier transform of the by-pulse response, written h(r, f).

Formula [8.13] then gives:

$$S_x(r, f) = \left|H(r, f)\right|^2\ S(f) \quad [8.15]$$

which represents the PSD of the structure response set in r.

In [8.15] we can express the H function owing to the natural mode basis of the structure:

$$H(r, f) = \sum_n H_n(f)\Phi_n(r) \quad [8.16]$$

with $H_n(f) = -(q_n / m_n)/\left(-\omega^2 + 2i\varepsilon_n\omega_n\omega + \omega_n^2\right)(\omega = 2\pi f)$.

Hence:

$$S_x(r, f) = \sum_n \sum_m H_n(f)\, H_m^*(f) \Phi_n(r) \Phi_m(r)\; S(f) \qquad [8.17]$$

For a wide frequency band source signal (quasi-random noise), it is possible to integrate [8.17] into the frequency domain. For example, if we take S (f) = 1, we obtain the response variance:

$$\sigma_X{}^2(r) = \sum_n \sum_m \sigma_{nm^2} \Phi_n(r)\, \Phi_m(r) \qquad [8.18]$$

with $\sigma_{nm}{}^2 = 1/[2\omega_n\omega_m (\varepsilon_n\omega_n + \varepsilon_m\omega_m)(1+\lambda_{nm}{}^2)]$ and $\lambda_{nm} = | \omega_n - \omega_m | / (\varepsilon_n\omega_n + \varepsilon_m\omega_m)$.

In the case of quite distinct modes ($\lambda_{nm} >> 1$ for n different from m), the non-diagonal terms in the double summation can be neglected:

$$\sigma_{X^2}(r) = \sum_n \sigma_{n^2} \Phi_{n^2}(r) \qquad [8.19]$$

with $\sigma_n{}^2 = 1 / (4\, \varepsilon_n\omega_n{}^3)$.

Equations [8.18] and [8.19] can be interpreted in the following way.

Inside the natural mode basis of the structure, modal contributions are the motion variables. Their evolution over time results from the random stimulation (here the unit random noise). Thus, they are $A_n(t)$ random processes and we obtain:

$$X(r, t) = \sum_n A_n(t) \Phi_n(r) \qquad [8.20]$$

The $\sigma_{nm}{}^2$ values represent the correlation coefficients of the $A_n(t)$ values with one another (for the same value of t), whilst $\sigma_n{}^2$ values represent their variances. Equation [8.19] shows that if the modes are very distinct, the $A_n(t)$ will be statistically "decorrelated".

8.4.4. *Using stochastic equations*

After these basic reminders, it seems timely to say a few words about stochastic equations, as they represent an important investigative tool as far as the calculation of randomly stimulated structures is concerned. Although engineers seldom use them, for simple cases, they can provide us with quite useful solutions for approximate methods and the interpretation of digital simulations.

Consider a not necessarily linear N dof (degree of freedom) structure, stimulated by a stationary *Gaussian* random noise ("no memory" random process). Its response is a "one memory step" (*Markov*) 2N dimension vectorial process (the dofs and their first derivatives). Such a process is completely characterized, as we have already said, by its second order joint probability or, which is equivalent, by its transition probability p_2 (y, t / y_0, t_0), with $t > t_0$, and y represents a 2N-component vector.

We can show that such probability verifies a so-called *Fokker-Planck* equation, with partial derivatives according to y and t, the terms of which can be explained from those of the motion equations of the structure.

Let us take as a simple example a 1-dof dampened oscillator, with a stiffness derived from a potential:

$$d^2x/dt^2 + 2\varepsilon\omega\, dx/dt + d\left(U(x)\right)/dx = f(t) \qquad [8.21]$$

Initially the oscillator is at rest. We will take $y_0 = 0$ and $t_0 = 0$. Furthermore, we will only be interested in its set response. Then the transition probability does not depend on t time and it will be written as q (x, v), with v = dx/dt. The source process F(t) is a S_0 level random noise (its correlation function will be $S_0\,\delta(\tau)$, with $\delta(\tau)$ = delta functional).

The *Fokker-Planck* equation confirmed by q (x, v) is:

$$v\,\partial q/\partial x - \partial/\partial\left\lfloor\left(2\varepsilon\omega v + dU/dx\right)q\right\rceil - \left(S_0/2\right)\partial^2 q/\partial^2 v = 0 \qquad [8.22]$$

the solution to which is.

$$q(x, v) = C \exp\left[-(4\varepsilon\omega/S_0)\left(v^2/2 + U(x)\right)\right] \quad [8.23]$$

being a constant such that $\iint q \, dx \, dv = 1$.

8.4.5. *Extrema statistics in a stationary process*

Second order representation leads to simple formulation for linear systems at least, but it is not enough as far as seismic analysis is concerned. As a matter of fact, the values handled are the maxima reached instead of the average quadratic values.

This section will present the main formulations at our disposal, while keeping in mind the fact that a definitely more complex problem is involved.

8.4.5.1. *Wide-band processes and narrow-band processes*

Extrema statistics depend greatly on the bandwidth of the process being investigated.

Typically we distinguish between the following:

– wide-band processes, the PSD of which covers a wide frequency range. The soil's seismic motions without any site effect belong to this category;

– narrow-band processes, the PSD of which corresponds to a narrow peak (the whole energy is located around a middle frequency f_0). The response of a weakly dampened harmonic oscillator with wide-band noise belongs to this category.

– intermediate processes consisting of a low number of well-separated narrow peaks. The response of a structure on its first natural modes belongs to this category.

Each category corresponds to a quite characteristic rate of the signal. In particular, wide-band noise presents isolated extremas, whereas narrow-band noise consists of "extrema packages" (Figure 8.5).

The bandwidth is characterized by the parameter:

$$\delta = \left(1 - m_{1^2}/m_0 m_2\right)^{1/2} \quad [8.24]$$

where m_0, m_1 and m_2 are order 0, 1 and 2 spectral moments.

For example, the bandwidth of the response of a one-random noise harmonic oscillator is:

$$\delta = \left(4\varepsilon/\pi\right)^{1/2} \quad [8.25]$$

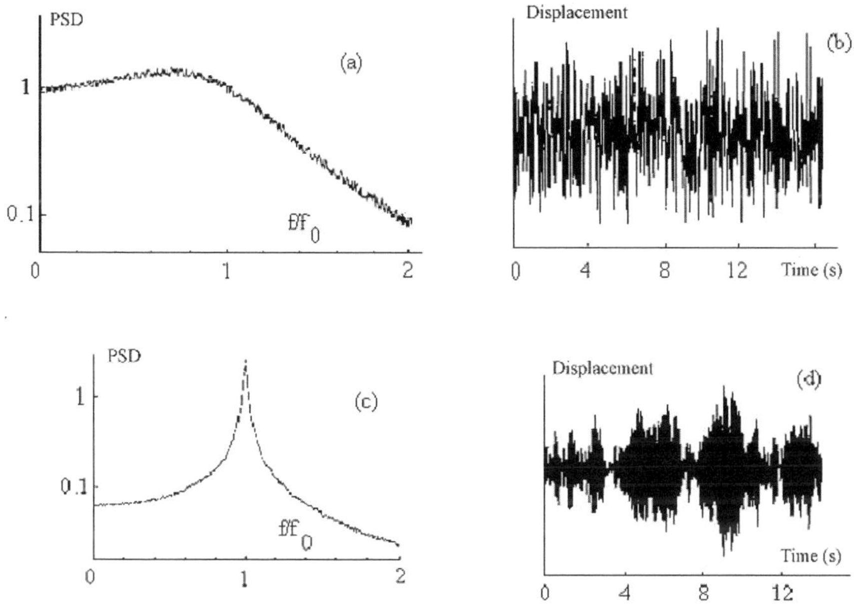

Figure 8.5. *Wide-band (a and b) and narrow-band (c and d) processes PSD and time rate*

8.4.5.2. *Envelope of a narrow-band process*

A narrow-band process consists of both amplitude and frequency-modulated frequency f_0 oscillations. The amplitude modulation gives conditions for the analysis of extremas. It can be represented by a new process, whose envelope can be defined by:

$$E^2(t) = X^2(t) + \left[dX/dt(t)\right]^2 / \left(2\pi f_0\right)^2 \quad [8.26]$$

The envelope process is wider band. If X(t) is *Gaussian*, E(t) obeys a *Rayleigh* law.

8.4.5.3. *Statistical study of the absolute maximum reached by a zero-average process in a given time period*

This section will be directly useful for the seismic analyses described above. We will present below a model that can be applied to wide-band Gaussian processes before trying to extend it to narrow-band processes.

8.4.5.3.1. *Average number of threshold crossings per time unit*

We consider a positive threshold x_m. We will deal with the crossings of that threshold owing to the process:

$$X(t) < -x_m \text{ or } X(t) > x_m$$

The crossing average frequency is given by:

$$f(x_m) = \int_{-\infty}^{\infty} |v| \, p(x_m, v) \, dv \qquad [8.27]$$

where p (x,v) are the joint probability density of the process and its derivative.

For a Gaussian process we have:

$$p(x, v) = (1/2\pi\sigma\sigma_{der}) \exp\left\{-\tfrac{1}{2}\left(x^2/\sigma^2 + v^2/\sigma_{der^2}\right)\right\} \qquad [8.28]$$

Once the calculations are over, by using [8.12] and introducing the a-dimensional threshold $r = x_m/\sigma$ we obtain:

$$f(r) = 2(m_2/m_0)^{1/2} \exp\left(-r^2/2\right) \qquad [8.29]$$

REMARK.– $f(0)/2 = (m_2/m_0)^{1/2}$ represents the average frequency of the process according to the *Rice* formula.

8.4.5.3.2. Determination of the reliability function and the peak factor

Reliability is defined as the probability for the process *not* to cross the threshold in a given T time period, which is usually chosen to be non-dimensional.

N = f (0) T represents the average number of semi-cycles performed by the process during the T time period. The reliability function is noted: W (r, N).

For wide-band processes and for relatively high thresholds ($r \geq 2.5$ to 3), the "threshold-crossing" occurrence is both rare and isolated.

Therefore, we can represent its occurrence owing to a *Poisson* distribution:

Probability (n occurrences in the T time period) =

$$\left\{\left[f(r)T\right]^n / n!\right\} \exp\left[-f(r)T\right]$$

Reliability involves a zero occurrence probability. Hence, by replacing f(r) with its [8.29] expression:

$$W(r, N) = \exp\left[-N \exp\left(-r^2/2\right)\right] \qquad [8.30]$$

we can recognize *Gumbel*'s asymptotic law.

From reliability, we can define the absolute maximum reached by the process during the time period T. This is a stochastic variable the probability density of which is:

$$p_{max}(r, N) = \partial W/\partial_r \qquad [8.31]$$

The most frequently used average value is the *average maximum*, which is also called *peak factor* μ (N) when it is referred to the mean square deviation.

$$\mu(N) = \int_0^\infty r \partial W/\partial r \, dr \qquad [8.32]$$

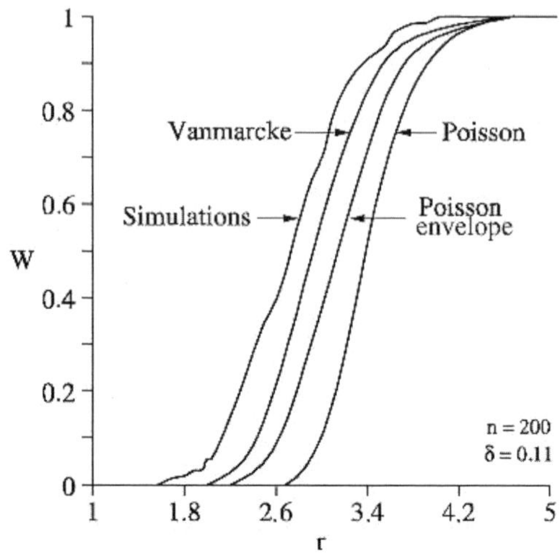

Figure 8.6. *Rate of the reliability function – comparison between the different estimations for a narrow-band process*

The function W varies from 0 to 1 when r varies from 0 to infinity. The crossing occurs quite suddenly around such r-values as W(r) = ½, which is an increasing function of N (Figure 8.6).

Thus, we can explain integral [8.30] owing to expression [8.32]. If N is not too large ($N \geq 100$), we have:

$$\mu(N) = \left[2 \text{ Ln}(N)\right]^{1/2} \qquad [8.33]$$

It is also possible to explain the maximum standard deviation (related to the process standard deviation):

$$\sigma_{max}(N) = \pi(6)^{-1/2}\left[2 \text{ Ln }(N)\right]^{-1/2} \qquad [8.34]$$

8.4.5.3.3. Extension to a narrow-band process

As the maxima of a narrow-band process come in the shape of packages, the *Poisson* model for the calculation of W is no longer valid. Instead of applying the model to the process itself, we could try to apply it to its envelope.

For a *Gaussian process,* the average number of maxima per package is given by:

$$E\left[N_{maxi}(r)\right] = \left[1 - \exp\left((\pi/2)^{1/2}\,\delta r\right)\right]^{-1} \qquad [8.35]$$

where the bandwidth parameter δ appears.

Thus, we can derive from it a threshold crossing average frequency by the envelope:

$$f(r)/E\left[N_{maxi}(r)\right] \qquad [8.36]$$

and use this frequency as previously.

It will also be possible to adjust semi-empirical formulas inspired by what comes before, the most sophisticated of which is *Vanmarcke*'s formula:

$$W(r, N) = \left[1 - \exp\left(-\left(r^2/2\right)\mathrm{Ln}2\right)\right]\exp\Big\{-N\exp\left(-r^2/2\right)$$

$$\left[1 - \exp\left(-(\pi/2)^{1/2}\,\delta^{1.2}\,r\right)\right]/\left[1 - \exp\left(-r^2/2\right)\right]\Big\} \qquad [8.37]$$

Figure 8.6 shows a comparison between different models and simulation results carried out with a reduced 0.01 (δ – 0.11) damping harmonic oscillator.

Figure 8.7 shows the results of digital simulations concerning the evolution of the peak factor and the maximum standard deviation as a function of N and δ.

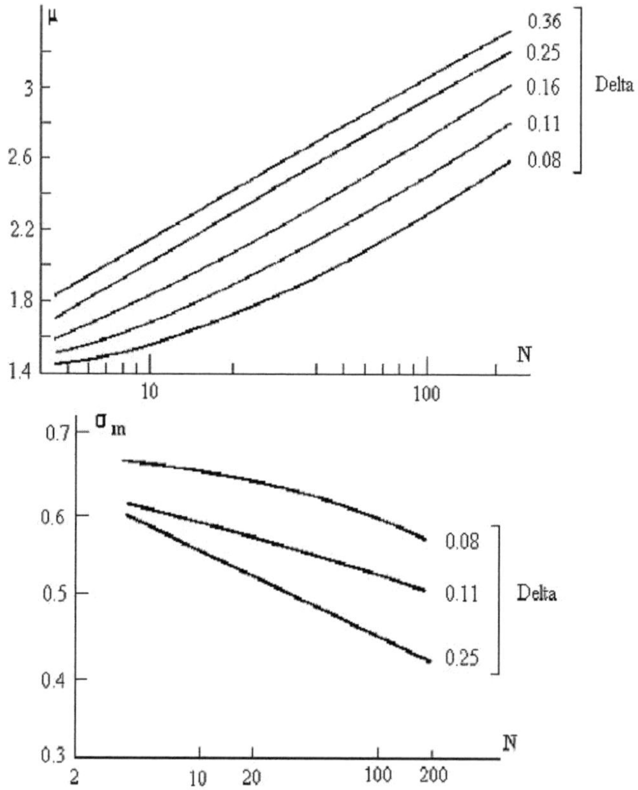

Figure 8.7. *Abacuses to determine the peak factor of a narrow-band process (numerical calculation results of a harmonic oscillator stimulated by random noise)*

8.5. Improvements to the modal method

Let us return to the seismic analysis problem.

What we said before will be used to rephrase the definition of seismic data, namely the oscillator response spectra (ORS).

The seismic signal should be considered as a random process characterized by average values suited to the response we are looking for, typically predicting the collapse of the structure. According to the regulations and practice, the ORS characterization is used, although in some cases it is not the best suited, as we will show later.

Therefore, the problem involves specifying the meaning of ORS in probabilistic terms. As we recalled at the beginning, the ORS represents the maximum reached during the seismic stimulation by the response of a harmonic oscillator. The notion was analyzed in probabilistic terms in the previous section.

One ORS point represents a random variable. Calculating the response to a given seismic signal represents a carrying out of that random variable. Figure 8.8 shows such calculation results. We can observe quite fluctuating curves that reflect the straggling. Actually, the regulation curves used by engineers are much smoother. Though they result from various averages not necessarily consistent with a well-defined random process, they should be considered as a statistic average associated with a random process.

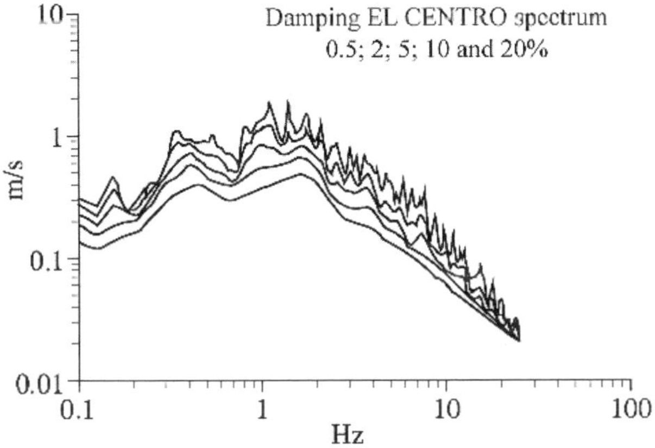

Figure 8.8. *OSR associated with a seismic signal (the higher the damping, the lower the statistical fluctuation)*

The first goal of probabilistic seismic analysis is therefore to adjust a random process to the regulation ORS data considered as the statistic average of the "maximum reached" random variable. In fact, the problem has several solutions that may involve quite different results in the case of very non-linear structure behaviors.

Let us return to the criticisms of the modal method listed above. The first concerned the simple quadratic combination hypothesis. As a matter of fact, there are actually two distinct causes of non-verification of this hypothesis.

8.5.1. *Complete quadratic combination*

The first cause is that modal contributions are statistically dependent processes. This is most obvious when two natural modes have neighboring frequencies.

The effect is obvious in the expression of the response variance of a structure represented by its natural modes (equation [8.18]). The parameter characterizing the modal contribution-coupling is written:

$$\lambda_{nm} = \left|\omega_n - \omega_m\right| / \left(\varepsilon_n \omega_n + \varepsilon_m \omega_m\right) \qquad [8.38]$$

We can give an expression on the equation [8.18] model for the maximum. Thus we obtain the *complete quadratic combination* formula:

$$x_{max^2}(r) = \sum_{n=1}^{N} a_{nmax^2} \Phi_{n^2}(r) + 2 \sum_{\text{couples } n,m(n \neq m)} a_{nmax} a_{mmax} \Phi_n(r) \Phi_m(r) / \left(1 + \lambda_{nm^2}\right)$$

$$+ \gamma_{smax^2} \left\{ X_s(r) + \sum_{n=1}^{N} q_n / \left(m_n \omega_{n^2}\right) \Phi_n(r) \right\}^2 \qquad [8.39]$$

Furthermore, we can improve the formula by taking the transient feature of the seismic stimulation into account in introducing, instead of the modal damping, equivalent modal damping given by:

$$\left(\varepsilon_n\right)_{equ} = \varepsilon_n \left[1 - \exp\left(-2\varepsilon_n \omega_n T\right)\right] / \left[1 - \exp\left(-\varepsilon_n \omega_n T\right)\right]^2 \qquad [8.40]$$

where T is the earthquake duration.

Formula [8.40] takes into account the fact that when the earthquake takes place, the structure at rest includes in its response a transient phase all the more important, compared with the T duration, the smaller the damping ε.

8.5.2. *Peak factor effect*

The second error cause lies in the "peak factor effect": the transposition of formulae that can be applied to a variance calculation to maxima is generally not valid. To understand such an effect, we will consider the special case of a structure that responds according two almost similar contribution modes, the resonance frequencies of which are quite separate ($\lambda_{12} >> 1$).

In this case, the simple quadratic combination formula well applies to the variance calculation:

$$\sigma_{x^2} = \sum_n \alpha_n \sigma_n{}^2 \qquad [8.41]$$

Transposition to maxima would give:

$$x_{max^2} = \sum_n \alpha_n a_{n\,max^2} \qquad [8.42]$$

The latter formula is only correct if the μ_n peak factors associated with each $A_n(t)$ process are equal to each other and the $X(t)(\mu_X)$ as well. If both modes have neighboring damping ratios, the δ_n bandwidths will also be close, and providing that both resonance frequencies are of similar orders, both peak factors will not be very different either.

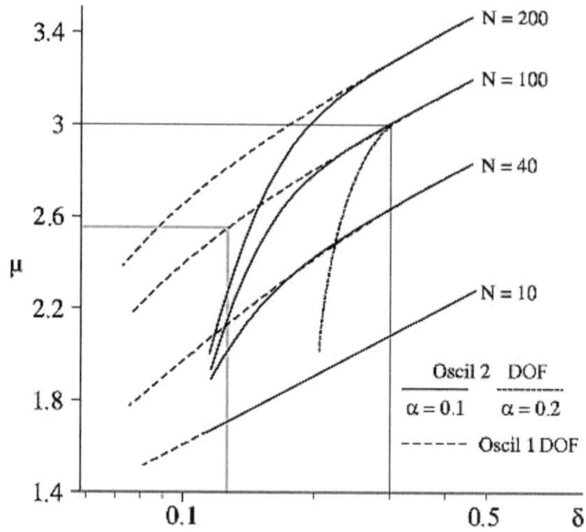

Figure 8.9. *Illustration of the peak factor effect*

On the other hand, the δ_X bandwidth of the X(t) double peak process will be definitely wider. It can be estimated owing to the formula:

$$\delta_X \approx \left[\alpha^2 + (8/\pi)\varepsilon\right]^{1/2} \text{ with: } \alpha = \left|\omega_n - \omega_m\right| / (\omega_n + \omega_m) \qquad [8.43]$$

valid in the case: $\varepsilon_1 = \varepsilon_2 = \varepsilon \rightarrow \delta_1 = \delta_2 = \delta = (4\varepsilon / \pi)^{1/2} << 1$.

For example: $f_1 = 4$ Hz, $f_2 = 6$ Hz, $\varepsilon = 0.01 \rightarrow \delta_X = 0.26$ and $\delta = 0.11$.

In Figure 8.9, we represented the evolution of the peak factor according to the bandwidth, for the response process of 1-dof and 2-dof oscillators. The curves allow us to determine the μ_1, μ_2 and μ_X peak factors. In the previous example, we have, for 100 half-cycles carried out with an average 5 Hz frequency, $\mu_1 \approx \mu_2 \approx 2.5$ and $\mu_X \approx 3.0$, that is, a 1.20 ratio. This ratio represents the error made when carelessly applying the maximum quadratic combination formula. This error always tends to be underestimated. In our example, it is equal to 20%.

There exist correction processes of the complete quadratic combination formula that consist of introducing a multiplying function determined from abacuses into the cross-terms of the sum (see for instance [GIB 88]).

REMARK.– Peak factor corrections are based on stationary process considerations. As a first approximation, it is possible to take the transient feature into account by replacing the damping ε with the equivalent damping given by [8.40].

8.6. Direct calculation of the floor spectra

The seismic calculation of a complex structure is performed in several steps:

– the first step involves the calculation of buildings, including the soil-foundation interaction effects and simplified modeling of large equipment;

– the second step involves detailed calculation of large equipment;

– in some cases a third step is necessary to study the secondary equipment, i.e. the pipework connecting the main equipment.

If we want to avoid time calculations and therefore continue to apply the modal method concept, then at the start of the second and third steps we will need the ORS associated with the motions of the floors on which the large equipment rests or to which the secondary equipment is anchored. However, the modal method such as it has been described so far does not allow us to obtain such information when it is applied to the first step.

The objective of the so-called "floor spectrum direct calculation" method is to answer that question. As in the previous section, it implies resorting to an accurate probabilistic definition of seismic loading. On this occasion we will be driven to study the transient feature of the loading and of the response of the structure, which has only been touched on so far.

8.6.1. *Representation of non-stationary processes*

If staying at a second order characterization, a general (non stationary) process fulfills a correlation function depending on two time variables:

$$\rho(t_1, t_2) \text{ or } \rho(t, \tau) \text{ with: } t = (t_1 + t_2)/2 \text{ and } \tau = t_2 - t_1$$

A PSD can be defined either by carrying a double Fourier transform of the correlation function or by performing a simple Fourier transform (with regard to τ).

Then we obtain either a PSD that is a function of two frequencies f_1 and f_2, or a PSD that is a function of frequency f and time t.

Obviously, adjusting a PSD to two variables during a series of observations or of experiments is far more problematic and often illusive. A way to simplify involves supposing that the ρ (t, τ) function evolves more slowly according to the t variable than according to the τ variable (which often proves to be true). It is then easier to carry out adjustments.

We can also assume that the t evolution occurs in a deterministic way, and thus adjust a frequency content varying in time according to some given law. We can be still more restrictive by considering an average, and therefore constant with time frequency content, and by making only the overall amplitude time-dependent. The latter assumption leads us to the *divisible model* of seismic stimulation:

$$\Gamma(t) = a(t)F(t) \qquad [8.44]$$

where a(t) is a deterministic envelope and F(t) is a S(f) PSD stationary random process.

As we will see, the latter representation is fairly easy to adjust to conventional seismic data. Generally a simple envelope (half-sine) is taken. Even a constant level slot gives satisfactory results because the most important effect to take into account is the start of the stimulation for t = 0 time and its stopping for time t = T.

8.6.2. *Adjusting a separable process from the ORS data*

We have seen that the seismic stimulation is given to an engineer in the shape of a set of ORS drawn up for several damping values. Thus, the problem involves adjusting an a(t) envelope and a S(f) PSD from $S_{pv}(f,\varepsilon)$ or $S_{pa}(f,\varepsilon)$ functions.

As we remember that the ORS have the meaning of maximum statistic averages, we can show that the relationship:

$$\left[S_{pv}(f, 0)\right]^2 = E\left|\Gamma(f)\right|^2 \qquad [8.45]$$

is quite well confirmed.

The second term of [8.45] represents the mathematical expectancy of the f parameter random variable:

$$\left|\Gamma(f)\right|^2 = \cdot \int_0^T\int_0^T \Gamma(t)\Gamma(t')\exp\left[-2i\pi f\left(t-t'\right)\right] dt\, dt' \qquad [8.46]$$

By using [8.46] we obtain, if we suppose a slow a(t) variation:

$$\left[S_{pv}(f,0)\right]^2 \approx \frac{1}{2} S(f) \int_0^T a^2(t)dt \qquad [8.47]$$

We take a standardized envelope with a given (half-sine) shape. The integral given by [8.47] only depends on the unknown T. In order to obtain T from the ORS, for example, for an f frequency centered within the seismic range and for two given damping values, we can consider the ratio:

$$R(f,\varepsilon,\varepsilon') = S_{pv}(f,\varepsilon')/S_{pv}(f,\varepsilon) \qquad [8.48]$$

This relationship characterizes the duration of the stimulation. Actually, if we consider the case when the duration of the transient phase of the response of the f frequency and the ε damping frequency oscillator where ε‘ is small with regard to T, for instance, R is given by:

$$R\ (f,\varepsilon,\varepsilon') = (\varepsilon/\varepsilon')^{1/2} \mu(N,\delta')^{o} / \mu(N,\delta) \qquad [8.49]$$

From Figure 8.7 abacuses, the latter relationship allows us to obtain N and therefore T. Then relation [8.47] gives S(t).

8.6.3. *Determination of the floor spectra*

As we now know the characteristics of the source separable process, it is possible to determine those of the structure response. To do so, all we have to do is apply equation [8.13] which gives the PSD of the response at point r and time τ after the beginning of the stimulation, in which the expression of the H transfer function is slightly modified to take the envelope of the source process into account:

$$S_X(r,\tau,f) = \left|H(r,\tau,f)\right|^2 S(f) \quad [8.50]$$

with:

$$H(r,\tau,f) = \int_0^{\tau} G(r,u)\ a\ (\tau-u)\ \exp\ (-2i\pi fu)\ du \quad [8.51]$$

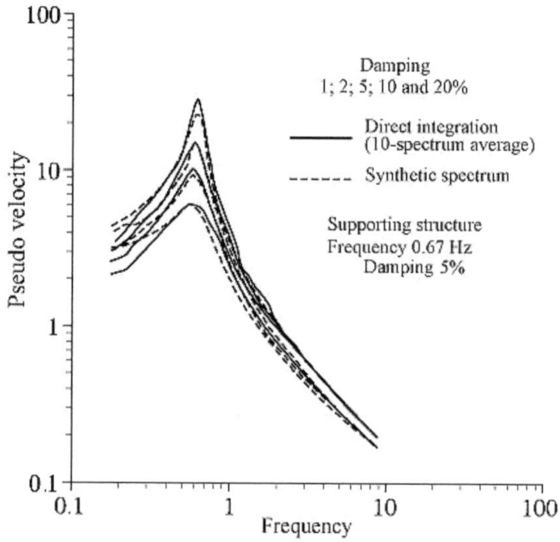

Figure 8.10. *Floor spectra: comparison between the proposed method (synthetic spectra) and direct time integration calculation (10 realizations)*

The last step involves passing from S_X (r, τ, f) (r being the position of the floor or the anchoring considered) to the associated ORS that will be the "floor spectra" sought. To obtain this, we also go through a separable process representation.

On the one hand, we consider the average according to τ of S_X (r, τ, f) that we take as PSD: S_{floor} (f), and on the other hand the integral according to f of S_x (r, τ, f), which represents the square of the standard deviation of the response σ_X (r, τ)), which is taken, after being standardized, as an envelope: $a_{floor}(\tau)$.

Finally, we can reconstruct the oscillator spectra $[S_{pv]\ floor}$ (f, ε)] using equations [8.47] and [8.49]. An illustration of the floor spectrum calculation for which the carrying structure behaves as a 1-dof oscillator is presented in Figure 8.10.

8.7. Creation of synthetic signals and direct numerical integration

So far we have examined seismic analysis methods that do not require any direct time integration of the movement equations. The main interest of such methods lies in them being more "physical", far lighter and far more flexible to use (depending on the problem to deal with we adjust the number of modes used), and in them being perfectly consistent with the regulation data of the seismic source.

Nevertheless, they cannot be applied in two cases:

– when we want to carry out an experimental analysis, in the case of the shaking table assessment of equipment, the good behavior of which has to be guaranteed under seismic loading defined by a given ORS, for instance. The jacks of the shaking table then need a load time law;

– when the structure has to be represented by a highly non-linear model. Then the modal theory cannot be applied.

We could also prefer the time method for some linear but complex structures, the modal core truncation criteria of which are not really clear or involve a very high number of modes. As we said at the beginning of this chapter, the seismic data associated with a site is seldom of the time kind because of the absence of measurement data, of the "envelope" feature that we mean to give it, etc.

Therefore, the problem is as follows: from an ORS set, how can we define a random process, the ORS of which will represent the average maxima like that defined above?

Such a problem is not well put. Is there one solution to it? Are there several? Do these solutions have any relationship with the physical phenomena brought into play?

As far as we know, these questions have not been satisfactorily answered yet.

In practice, various methods allow us to make random processes, the average characteristics of which are adjusted on an ORS family within an eigenfrequency and damping range, and thus to produce time creations that can be used for both calculation and experimentation.

These methods even include the correction of the most obvious non-physical characteristics (for instance, time averages not equal to zero for acceleration signals).

The nature of the generated signals can be quite different from one method to the next. It would not be that important for a linear behavior structure because in this case the ORS represent relevant values as far as the response maximum prediction is concerned. Besides, that is the reason why the adjustment of a separable process involves satisfying results for the direct calculation of the linear floor spectra.

In the case of a marked non-linear behavior (modeled or real structure), such relevance is not at all obvious. Thus, from response average maxima, we can obtain quite different results according to the method used (with adjustment on the same ORS).

As an illustration we will hereafter give the principles of the quite commonly used "random phase harmonic sinusoid" method.

Generally, the adjustment is carried out on a given S (f, ε) ORS (ε being given). The probabilistic model used is the separable type;

$$\Gamma(t) = a(t)F(t) \qquad [8.52]$$

The stationary process F(t) results from a sum of sinusoids with deterministic A_i amplitudes and φ_i random phases that are independent and equiprobable within the $[-\pi, \pi]$ interval:

$$F(t) = \sum_{i=1}^{N} A_i \sin(2\pi f_i t + \varphi_i) \qquad [8.53]$$

The f_i values are a set of discrete frequencies that describe the studied range. We often take:

$$f_i = i\Delta f \text{ with } \Delta f = 1/T \qquad [8.54]$$

T being the duration of the signal. According to the *Shannon* theorem, the studied $[0, f_{max}]$ frequency range is conditioned by the choice of the time discretization slot $1/2f_{max} = T/2N$ with $f_{max} = N\,\Delta f$.

The problem involves determining the A_i values and T from the ORS. For this we can show that the PSD discretized (with the Δf slot) corresponding to the F(t) process is given by:

$$S(i\Delta f) = A_i^{\,2} T / 2 \qquad [8.55]$$

Thus, we are brought back to the problem in the previous section: the relationships expressed in [8.47] and [8.49] allow us to obtain S (f) and T from the ORS; equation [8.55] allows us to obtain the A_i values.

REMARK.– Regarding the time integration methods: the seismic analysis must result in obtaining statistical characteristics concerning the absolute maxima (averages and possibly standard deviation) of different physical values (displacements, accelerations, stresses). Thus, time calculation represents the realization of the corresponding random variables. In order to carry out their statistics with reasonable accuracy, many calculations have to be done. Therefore the time method is cumbersome and expensive, especially in the case of non-linear behaviors. We lay great stress on the fact that the result of one only calculation, however complex it may be, brings only little information. The potentially chaotic behaviors of the non-linear systems we will deal with in the next section only reinforce that assertion.

Let us notice however that if the calculation includes several quite isolated maxima, it will be possible to take advantage of all these maxima, not only of the most important of them, which will allow us to decrease the number of calculations to carry out for any given accuracy.

8.8. Seismic analysis of non-linear behavior structures

8.8.1. *Introduction*

Generally speaking, the dynamic study of non-linear systems is a mathematics topic that is quite extensive and complex with many remaining gray areas. Unlike linear cases with which we have been occupied so far, there is no theoretical frame strong enough to allow us to draw predictive methods engineers could use in an acceptable way. It is not surprising, as the non-linear feature does not correspond to the general case, whereas the linear feature corresponds to quite specific properties.

As a consequence, we will be able to propose valid predictive methods only if the problem becomes particular, i.e. if we can precisely know the nature of the non-linearities encountered, and those of the stimulation sources, as losing the linearity property forbids characterization of the structure as a "responsive system" independently of the source.

Let us note that the type of result we are interested in, regarding the response (here essentially the maxima), also plays an important part.

In this section we are not dealing with developing theories about non-linear systems but, after clarifying the main types of non-linearities commonly used in practice, with illustrating the associated dynamic (seismic) behaviors and describing the simplified methods used by engineers for direct prediction or interpretation and thus for a reasonable use of time digital simulations.

8.8.2. *Main non-linearities of seismically-loaded structures*

Non-linear behaviors can be generated by:

– geometry due to the important displacements of the structure (bridge stay-cables or high-voltage cables, for instance) or due to limit conditions (the impact of two neighboring structures like pipework with its supports, for example, or sliding at the level of anti-seismic bearing devices); or

– behavior laws of materials (for example, the plastic behavior of steel or of reinforced or pre-stressed concrete extensively illustrated in this book).

In dynamics, it is important to distinguish non-linearities leading to great energy dissipation (like plastic or visco-plastic behaviors, or friction) from those implying only a little dissipation, like great movement effects or moderate shocks that can be deemed elastic.

It is also necessary to distinguish the weak non-linearities which often appear at the level of imperfect linking and which do not question the modal structure of the system from the strong non-linearities that radically modify the mode structure and the behavior.

The former are taken into account through a slight modification of the resonance frequencies and mostly through the introduction of a relatively small damping term compared with the stiffness and inertia terms. In most such cases we give up an actual physical modeling to merely globally adjust a viscous dampening coefficient ε_n for each mode, which corresponds to linear modeling.

The latter can only be dealt with using specific non-linear models. In current practice, it does not prevent engineers from deriving *equivalent linear models* from these models, within restricted parametric ranges.

This last aspect will be the focus of our attention in the rest of the chapter.

We will not systematically study the different non-linear models used in seismic analysis, but we will rely on two characteristic examples:

– the *elasto-plastic* behavior model, which is the most often used, or the "according to *Coulomb*'s law adherence-sliding-type" behavior, which has the same nature;

– the "elastic shock-type" behavior model.

8.8.3. *Notion of "inelastic spectra"*

This notion is the most used method in seismic analysis practice for taking the elasto-plastic behavior of the structure into account.

Consider a perfect elasto-plastic m-mass 1-dof oscillator loaded by an $\gamma(t)$ imposed acceleration of its support. It is characterized by its resonance pulsation ω in its elastic phase, its limit elastic displacement x_e, its limit force $m\omega^2 x_e$ and its ductility coefficient μ (Figure 8.11, where $\alpha = 0$).

Suppose that $\gamma(t)$ consists of two opposed I intensity impulses in the shape of $\Delta\tau$ time slots and separated by a time period τ, with $\Delta\tau << \tau$ (Figure 8.12).

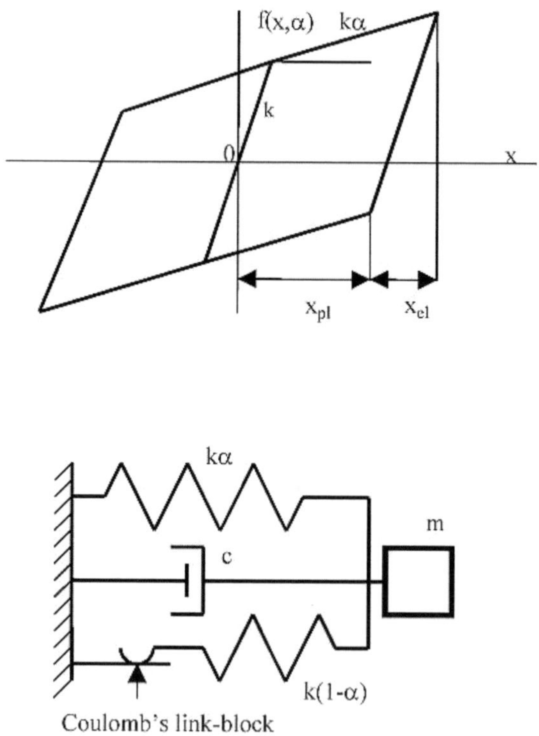

Figure 8.11. *Diagram of a bi-linear-strain hardening elasto-plastic oscillator*

The idea involves comparing the limit intensities of the impulses bringing about collapse that can be calculated thanks to:

– elastic analysis: we will write that the displacement remains lower than x_e;

– elasto-plastic analysis: we will write that the displacement remains lower than μx_e.

We can distinguish three frequency ranges:

(i) If $\omega\tau << 1$, it is the low frequency range of the ORS. The maximum relative displacement of the mass depends little on the stiffness law and it is given approximately by $x_{max} = I\tau$. A purely elastic analysis gives a limit intensity $I_{le} = x_e/\tau$.

An elasto-plastic analysis leads to $I_l = \mu\ x_e/\tau$. Therefore, taking plasticity into account involves admitting a higher limit intensity of a factor μ.

(ii) If $\omega\tau \approx 1$, it is the medium frequency range of the ORS. We can suppose that the maximum displacement of the mass occurs before the second impulse takes place. The elastic analysis will give $I_{le} = x_e\,\omega$.

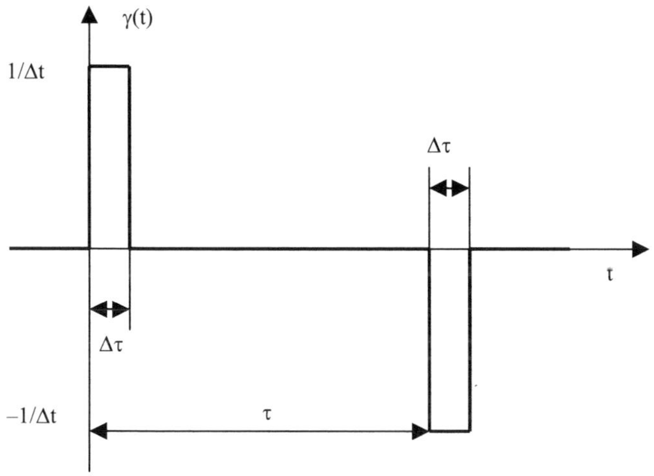

Figure 8.12. *2-impulse stimulation model*

When writing the preservation of energy, the elasto-plastic analysis will give:

$$I_{l^2} = (\omega\ x_e)^2 + 2\omega^2 x_e (\mu\ x_e - x_e) \qquad [8.56]$$

As a consequence, taking plasticity into account involves admitting a higher limit intensity of a factor $(2\mu - 1)^{1/2}$.

(iii) If $\omega\Delta\tau \leq 1$, it is the high frequency range of the ORS. The maximum displacement of the mass is conditioned by the application of the first acceleration step. The elastic and elasto-plastic analyses will give the same results as far as the limit intensity is concerned.

These results are extrapolated to the case of an average seismic stimulation:

– Constituting the "inelastic spectrum" involves applying the reduction factors defined above within the three frequency ranges to the original ORS. Then we will use the new spectrum to apply the modal method the way we would do with a linear behavior structure.

– In the para-seismic Eurocode, we use a more continuous reduction factor depending on the frequency but based on the same principles (Figure 8.13).

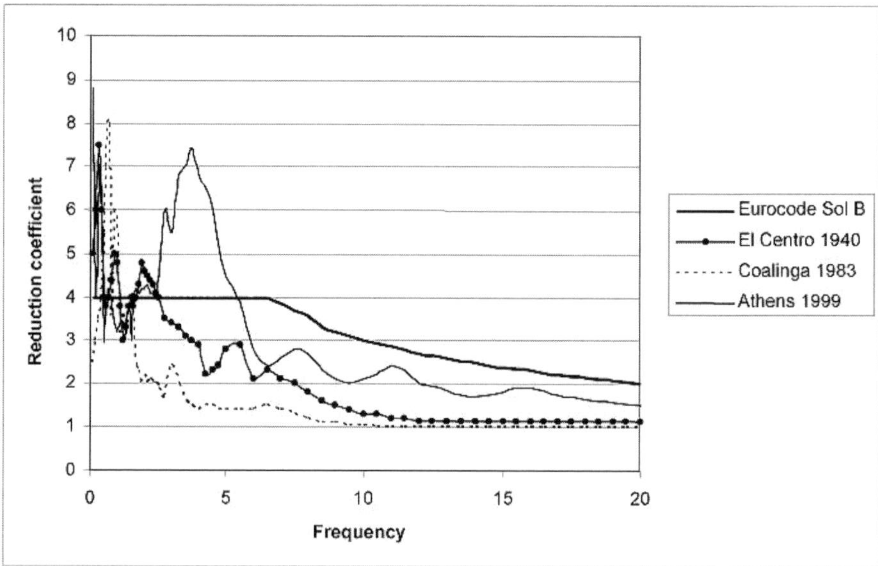

Figure 8.13. *Inelastic reduction coefficient for a 4-ductility factor dispersion effect between successive signals*

Applying the technique raises some problems:

– the most serious one lies in the application of a system that cannot be reduced to 1 dof. Actually it is difficult to define "basic solutions" (the equivalent of linear natural modes); moreover, the result we are trying to achieve generally cannot be represented in the shape of a sum of the contributions of these basic solutions. Applying the modal method starting from those hypotheses can bring about very serious errors;

– on the other hand, the behavior law is not generally the perfect elasto-plastic type (for example the strain-damping model in Figure 8.11);

– the nature of the source plays a part. As a matter of fact the model is based on the assumption of a seismic signal consisting of rare and isolated impulses that have the oscillator enter the plastic field. This is more or less achieved with wideband signals, yet it is completely wrong for narrow-band signals (floor motions).

This latter point can be illustrated by taking different seismic motions and by numerically calculating the reduction factors. Figure 8.13 shows their evolution according to ω impulses (compared with the average impulse of the seismic signal) for a given μ value.

REMARKS:

1) Besides the factors discussed above, the inelastic spectrum technique assumes that the highest impulse will cause collapse (notion of absolute maximum contained in the ORS). This is not always the case. In fact, the μx_e limit value can be reached by combining several intrusions into the plastic domain. *Vanmarcke* has proposed a calculation method (see [VAN 73]) that applies whenever such intrusions are rare and isolated. It is based on the following ideas:

– the plastic threshold overstepping statistics are little modified by the non-linear behavior. The considerations developed in the summaries about maximum statistics can therefore apply (average overstepping number, Poisson model etc.);

– the plastic displacement during an excursion is calculated according to the same hypotheses as for the inelastic spectra ($\omega\tau \approx 1$ case).

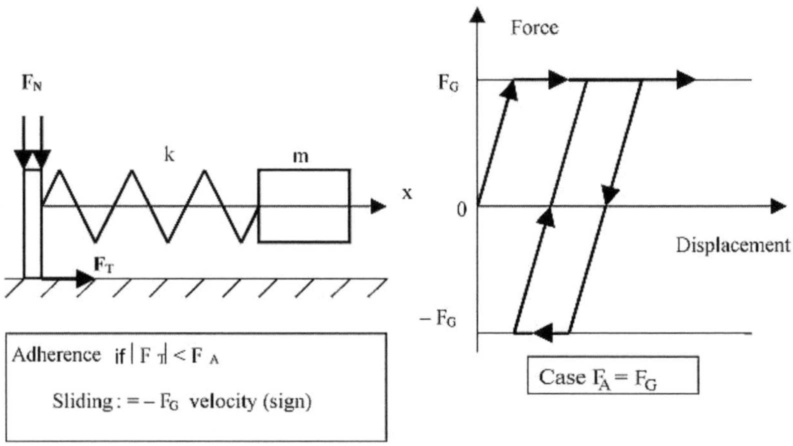

Figure 8.14. *Diagram of a sliding oscillator*

Thus, we can obtain the average value of the residual plastic displacement, the probability for the cumulative plastic displacement to exceed a specific value, etc.

2) Structures presenting a series of adherence and sliding phenomena that can be represented owing to a *Coulomb's* law belong to the above category. Thus, the 1-dof sliding oscillator represented in Figure 8.14 obeys the same behavior law as the perfect elasto-plastic oscillator when the adherence force equals the friction force. The sliding displacement plays the same part as the previous plastic displacement. We can mention studies on the sliding problem carried out with a technique similar to the reference inelastic spectrum technique (see [NED 99], [NOE 93] and [SAR 98]).

8.8.4. *Conventional method of stochastic linearization*

The flaws of the inelastic spectrum method result from the fact that the probabilistic nature of the seismic source is unknown. Thus, a solution may consist of coming back to the separable probabilistic model defined in the previous sections and trying to statistically analyze the response of the non-linear system.

The *stochastic linearization* method tries to solve the problem in a quite comprehensive way. We are going to present its principles while shedding light on its limitations. In order to simplify, we will use a 1-dof system for the presentation, as extending the method to N-dof systems does not raise any specific problem.

Given a non-linear oscillator with a motion equation:

$$d^2x / dt^2 + h(x, dx/dt) = -\gamma(t) \qquad [8.57]$$

$\gamma(t)$ being stationary, we are interested in the proven solution for [8.57], for example, we wish to replace this oscillator with an *equivalent linear oscillator* of an equation:

$$d^2y/dt^2 + 2\omega_{eq}\varepsilon_{eq} dy/dt + \omega_{eq}^2 y = -\gamma(t) \qquad [8.58]$$

The y(t) displacement, the solution to [8.58], does not verify [8.57], but we want to adjust ω_{eq} and ε_{eq} so that it is as close as possible to its solution x(t). The adjustment is done by minimizing, in the mean squares sense, the difference between the non-linear term and the equivalent linear term both applied to the exact solution of the problem (equation [8.57]). The ω_{eq} and ε_{eq} thus obtained are such that the variance of the response calculated with [8.58] corresponds quite well to the solution value of [8.57].

Let us call $\Gamma(t)$, X(t) and Y(t) the random processes associated with $\gamma(t)$, x(t) and y(t).

Unfortunately, the statistic characteristics of the X(t) "non-linear response" process are not known *a priori*. In practice, we will therefore try to minimize the e(t) gap between the non-linear term and its linear equivalent applied to the Y(t) "equivalent linear response" process. This approximation explains the low efficiency of the method in some cases:

$$e(t) = h(Y, dY/dt) - \left(2\omega_{eq}\varepsilon_{eq} dY/dt + \omega_{eq}^2 Y\right) \qquad [8.59]$$

ω_{eq} and ε_{eq} are constants given by the equations:

$$\partial E\left[e(t)^2\right] / \partial\left(\omega_{eq}^2\right) - \partial E\left[e(t)^2\right] / \partial\left(\varepsilon_{eq}\omega_{eq}\right) = 0 \qquad [8.60]$$

For a *Gaussian* stimulation (the *Gaussian linearization* method), the $E[e(t)^2]$ variance can be expressed according to the co-variance coefficients of Y(t) and Y/dt(t) which themselves are functions of ω_{eq} and ε_{eq}. The [8.60] system consists of 2 equations in 2 unknowns. Its solution will give the characteristics of the equivalent linear oscillator.

We can also show that:

$$\omega_{eq}{}^2 = E\left[\partial h\left(Y, dY/dt\right) / \partial Y\right] \qquad [8.61]$$

$$2\omega_{eq}\varepsilon_{eq} = E\left[\partial h\left(Y, dY/dt\right) / \partial\left(dY/dt\right)\right] \qquad [8.62]$$

The equivalent linear stiffness and damping represent the average statistics of the "tangent" stiffness and damping values associated with the different values of the response of the equivalent linear oscillator.

More generally, the method can apply in the non-stationary case of stimulation owing to a separable process. We are going to discuss the equivalent linearization method in two simple examples.

8.8.4.1. *Application to an elasto-plastic oscillator*

The formalism described above does not apply directly to elasto-plastic or adherence-sliding behavior types. As a matter of fact, conventional models (perfect elasto-plastic models with bilinear strain hardening) imply dependence on the history of movement that cannot be represented by a mere h (x, dx/dt). Thus, we have to resort to either simplified models representing the plastic cycles according to the movement maximum amplitude or to more sophisticated models that introduce an additional subsidiary variable confirming a differential equation (*Wen-Bouc*'s model [BOU 94], for example). Figure 8.15 shows the aspect of cycles obtained with Wen-Bouc's model for different values of its adjustable parameters.

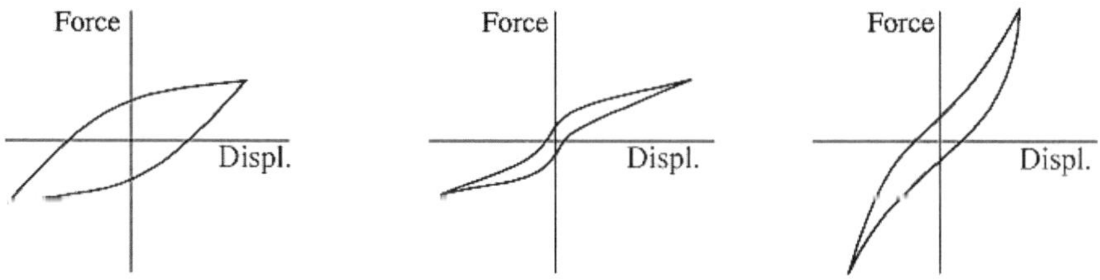

Figure 8.15. *Different cycles shapes obtained owing to Wen-Bouc's model*

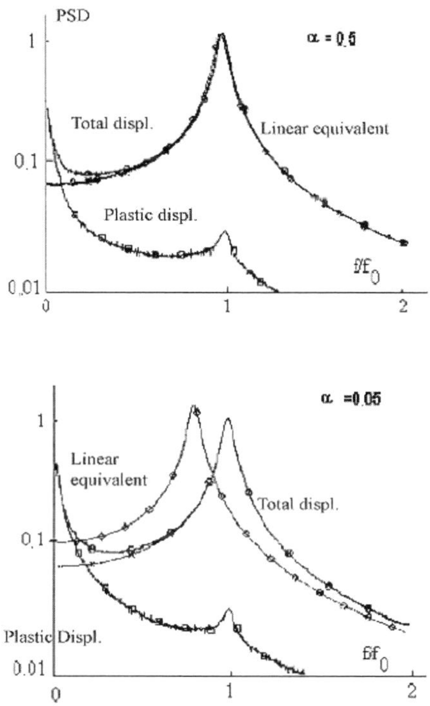

Figure 8.16. *Adjustment attempt of a linear oscillator equivalent to a bilinear strain-hardening elasto-plastic oscillator*

Such models give quite good results as far as the prediction of the variances and PSDs of the response, for linearities that are either weak or with strain-hardening slope/elastic slope ratios higher than 0.5 or so.

For strong non-linearities with slope ratios lower than 0.5, the non-linear calculation reveals the existence of two phenomena:

– a phenomenon at the elasto-plastic cycle frequency;

– a low-frequency phenomenon that corresponds to the cumulative effects of plastic displacements regarding the inelastic spectrum method referred to above.

A 1-dof linear oscillator cannot represent this "two-frequency" behavior, thus, in that case we can observe discrepancies as far as the variance estimate is concerned, as well as poor PSD results (Figure 8.16).

8.8.4.2. *Application to a 1-dof elastic shock oscillator*

Figure 8.17 shows the diagram of a 1-dof-*shock oscillator* and the aspect of the behavior law. K_c is the stop stiffness, which is typically high relative to that of the oscillator. This law is the f(x) type and it fits quite well with the stochastic linearization technique.

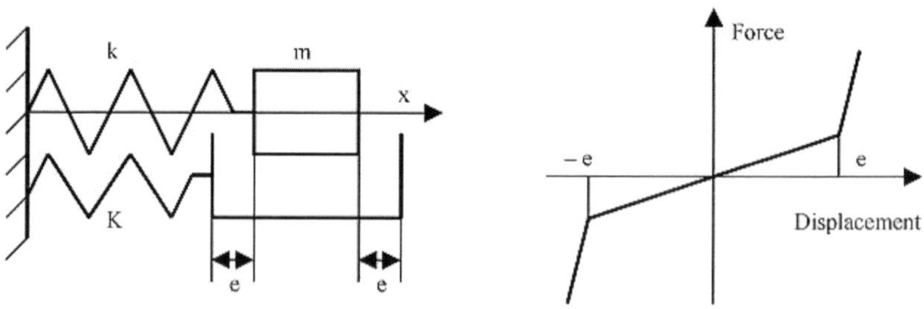

Figure 8.17. *Diagram and law of a 1-dof-shock oscillator*

However, applying this technique here gives very disappointing results:

– The graphs in Figure 8.18 extracted from [GUI 90] derive from non-linear digital simulations. They show the physical behavior of the shock oscillator. They represent the mass displacement PSDs for different stationary wide-band noise type stimulation levels (the η no-dimension associated parameter being the ratio between the play and the standard deviation of the response of the without shock oscillator). For high η values (very important play and therefore no shocks or few shocks) or small η values (play nearly equal to zero, hence an almost perfect connection with the stop), we have the behavior of a weakly dampened linear oscillator (we took $\varepsilon = 0.01$). On the other hand, when η is in the order of 1, the PSDs are quite flat. This expresses a frequency dispersion effect but which does not correspond to a damping effect.

– Figure 8.19 illustrates the stochastic linearization method. Note that it is incapable of representing the frequency element.

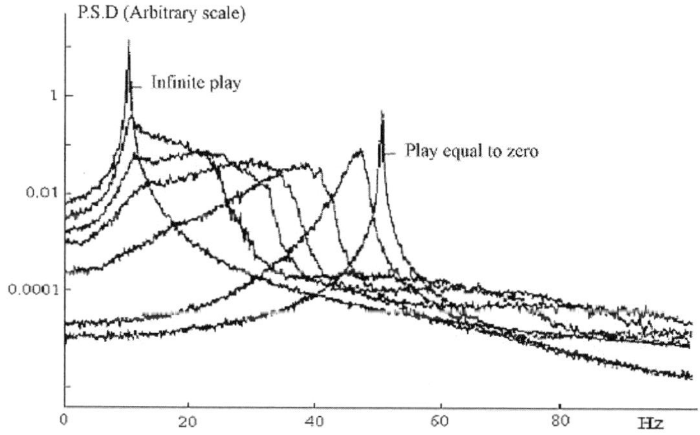

Figure 8.18. *Shock oscillator: evolution of the shape of the PSD of the answer as a function of reduced play*

8.8.4.2.1. Conclusion

The principle of conventional stochastic linearization expressed for systems with several degrees of freedom can seem appealing owing to its general feature. Nevertheless, as we have just illustrated, its application is often problematic. Things go off smoothly when the effect of non-linearities is weak, whereas very non-linear behaviors are poorly modeled, especially those that correspond to energy dispersion phenomena within one or several frequency bands. We are then led to particularize the method to the non-linearity type studied (which, from our point of view, is inevitable whatever the method used) and to use such restrictive hypotheses that, when compared with such appropriate simplified methods as the inelastic spectrum, their utility is often quite questionable.

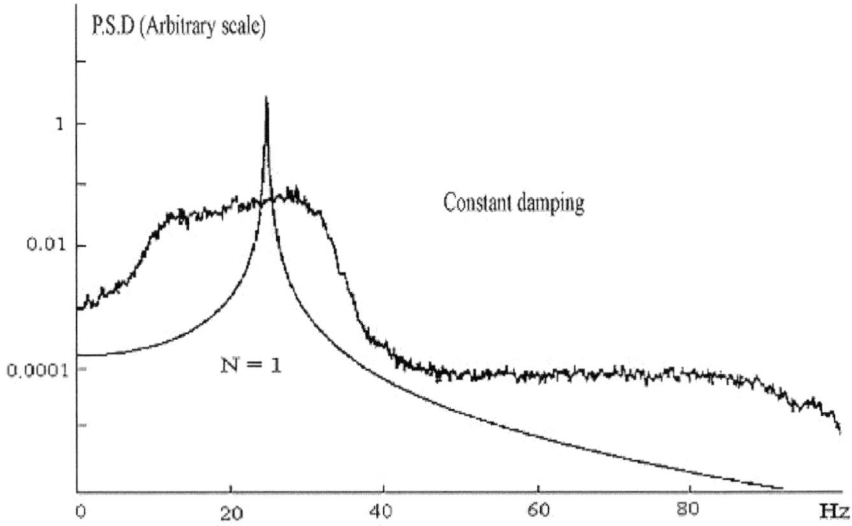

Figure 8.19. *Failure of the adjustment of an equivalent linear oscillator with the conventional method (comparison of the response PSDs)*

8.8.5. *Random parameter stochastic linearization*

To overcome the drawbacks we have just mentioned, several strategies come to mind:

– since it is necessary to enrich the frequency representation, we could associate a several-dof system with an initially 1-dof system, for instance. Thus, two equivalent linear oscillators could be associated with the oscillator in the previous sections, one of which with a very low resonance frequency. Obviously, such an approach does not simplify matters. Besides, it is utterly ineffective with a shock oscillator (see Figure 8.20 and [GUI 90]);

– another idea involves keeping the initial number of dofs whilst giving the parameters of the equivalent linear oscillators a random feature. Such an idea is the subject of rather general developments ([FOG 96] and [BEL 99]) that we will not present here. We will merely illustrate the method using the 1-dof-shock oscillator as an example and underline some difficulties that arise when the number of degrees of freedom is increased.

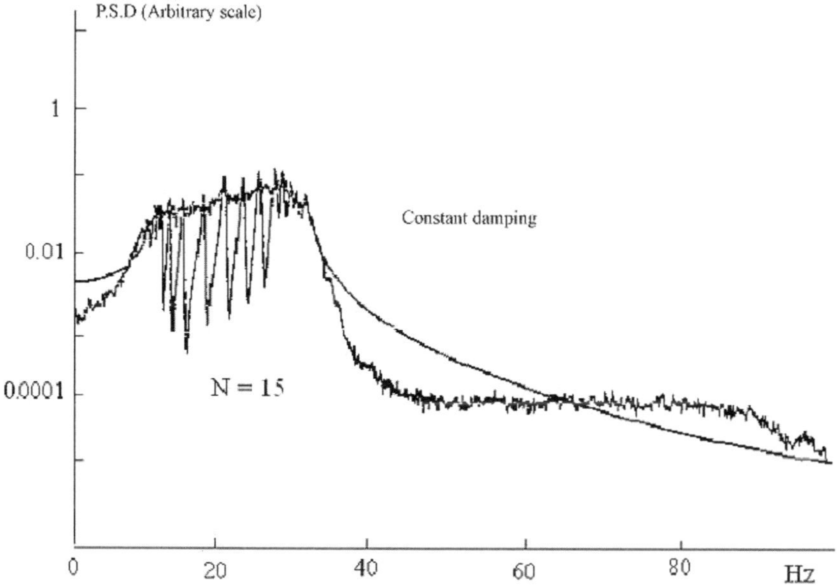

Figure 8.20. *Adjustment attempt of a 15-equivalent oscillator set to represent the frequency stretching due to shocks*

8.8.5.1. *Modeling of a 1-dof-shock oscillator owing to a random parameter linear oscillator*

Let us take the case of an oscillator stimulated by a Γ(t) wide-band noise (quasi-random noise) and study its stationary response.

The general principle results from the observation of the response signal obtained by numerical simulation. The latter consists of amplitude and frequency oscillation sequences randomly distributed (Figure 8.21).

If the maxima (most of the time corresponding to contact with the stops) are transferred into an amplitude/frequency diagram, the points are organized along a curve that characterizes a deterministic relationship between the frequency and the amplitude (Figure 8.22).

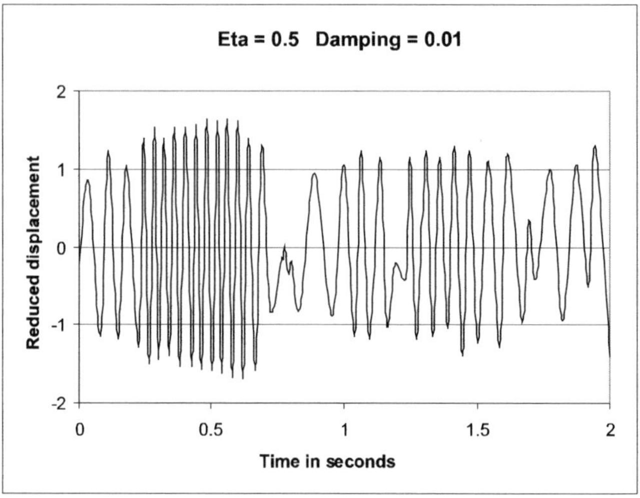

Figure 8.21. *Example of a response signal of a random noise shock oscillator*

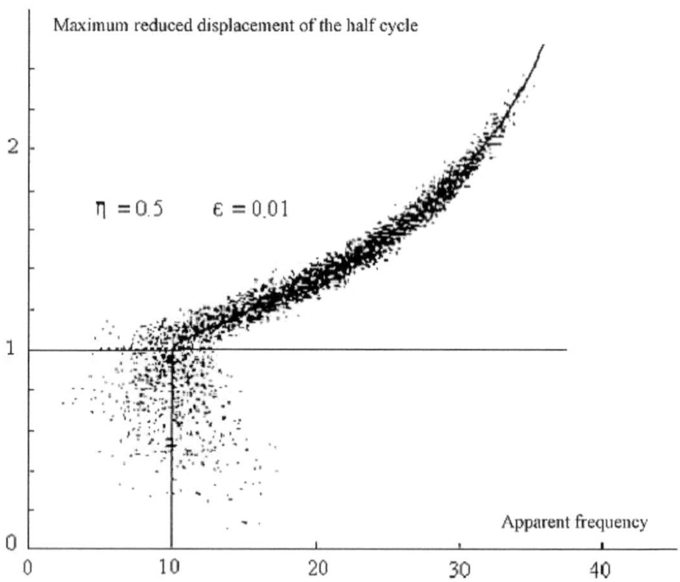

Figure 8.22. *Amplitude-frequency relationship of oscillation sequences (the full line has been obtained analytically, the dot cloud results from digital simulations)*

In the present case of a shock oscillator, the relationship can be found analytically by considering the periodic regime set for the free oscillator (without any stimulation) instead of the dampened oscillator, which is illustrated in Figure 8.23.

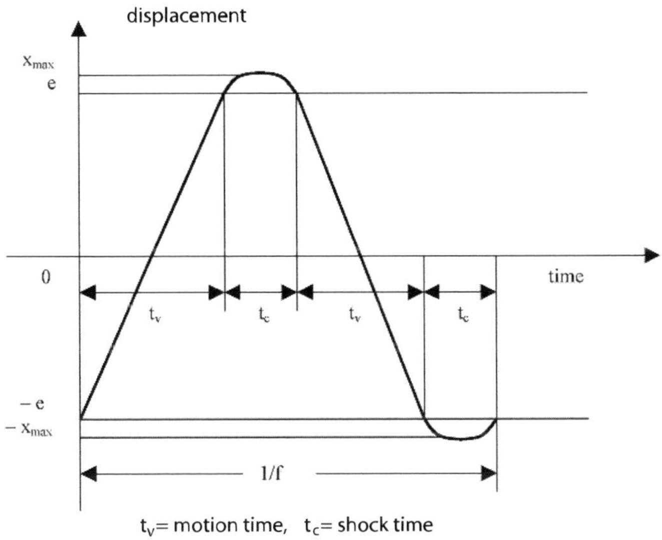

Figure 8.23. *Periodic regime/ratio for the 1-dof, free and non-dampened shock oscillator*

The idea involves writing that these sequences correspond to the response of a ω_0-resonance pulsation and ε-damping coefficient linear oscillator, to random noise, the intensity of which is different from the initial random noise by an $S(\omega_0)$ factor:

$$d^2x/dt^2 + 2\varepsilon\,\omega_0\,dx/dt + \omega_0^{\;2}\,x = -S(\omega_0)\Gamma(t) \qquad [8.63]$$

the *pulsation* ω_0 being *random* (ε can be considered as constant and thus as deterministic).

As far as the random process is concerned, we will be able to represent the X(t) response process using two slow time variation processes: the $X_{max}(t)$ amplitude and the $\psi(t)$ phase shift which characterize the sequences, and a fast time variation process that represents the $\Theta(t)$ phase of the oscillation within a sequence. The $\Omega(t)$

pulsation process is linked to the $X_{max}(t)$ process in a deterministic way using the ω_0 function:

$$\Omega(t) = \omega_0\left(X_{max}(t)\right) \tag{8.64}$$

$$X(t) = X_{max}(t)\cos\left(\Theta(t) + \Psi(t)\right) \tag{8.65}$$

$$dX/dt(t) = -X_{max}(t)\,\omega_0\left(X_{max}\right)\sin\left(\Theta(t) + \Psi(t)\right) \tag{8.66}$$

$$d\Theta / dt(t) = \omega_0\left(X_{max}\right) \tag{8.67}$$

As a consequence, after having specified the deterministic relationships $\omega_0(x_{max})$ and $S(\omega_0)$ (and therefore $S(x_{max})$), the problem will consist of characterizing the random processes $X_{max}(t)$ and $\psi(t)$ in terms of probability density. Then it is easy to obtain the PSD of X(t).

It is possible to show that from a theoretical point of view (see [FOG 96] and [BEL 99]), each of the probability densities confirms a Fokker-Planck differential stochastic equation that can be solved.

If, in our example, the equation gives an analytical solution, it is not the same with more complex cases. Thus, for several-dof systems, the differential problem is large, and its solution is far beyond the scope of engineering calculations. Therefore, from a practical point of view, in order to evaluate the probability density rate we will merely use simple physical approximations that will reproduce the frequency dispersion phenomenon.

In short, the different steps of the method are:

– determination of the amplitude/pulsation relationship. In our example, we will take the relationship corresponding to the full-line curve in Figure 8.22;

– determination of the amplitude/source intensity relationship. In our example, we adjust $S(x_{max})$, to preserve the average intensity of the oscillator (actually we can simply explain the average energy of the non-linear oscillator as well as that of the equivalent linear oscillator);

– estimation of the probability density of the amplitude and the phase shift. In our example, we can explain the joint probability density q (x, v) of the W(t) and dX/dt(t) processes by applying formula [8.23] with the expression of the potential corresponding to the behavior law in Figure 8.17:

$$U(x) = \tfrac{1}{2}\omega_1^2 x^2 + \tfrac{1}{2}\left(\omega_1^2 - \omega_2^2\right)\left[(x-e)H(x-e) + (x+e)H(-x-e)\right]$$

with $\omega_1^2 = k/m$ and $\omega_2^2 = (k + K)/m$.

From q (x,v) we can derive the $p(x_{max})$ probability density of the $X_{max}(t)$ envelope process;

– obtaining the PSD of the response.

We can eventually express for the general case the PSD of the response X(t) of the oscillator by;

$$Sx(\varpi) = \int_0 \left|H\left(\omega, \omega_0\left(x_{max}\right), \varepsilon\right)\right|^2 S\left(x_{max}\right) p\left(x_{max}\right) dx_{max} \qquad [8.68]$$

which corresponds to the statistic average according to x_{max} of relation [8.15].

In our example H represents the transfer function of the $\omega_0(x_{max})$ resonance pulsation and ε the damping 1-dof linear oscillator:

$$H(\omega, \omega_0, \varepsilon) = -1/\left(\omega^2 - \omega_0^2 + 2i\varepsilon\omega\omega_0\right) \qquad [8.69]$$

Figure 8.24 shows the good harmony between the PSDs obtained by numerical simulation and those obtained using the equivalent linearization method.

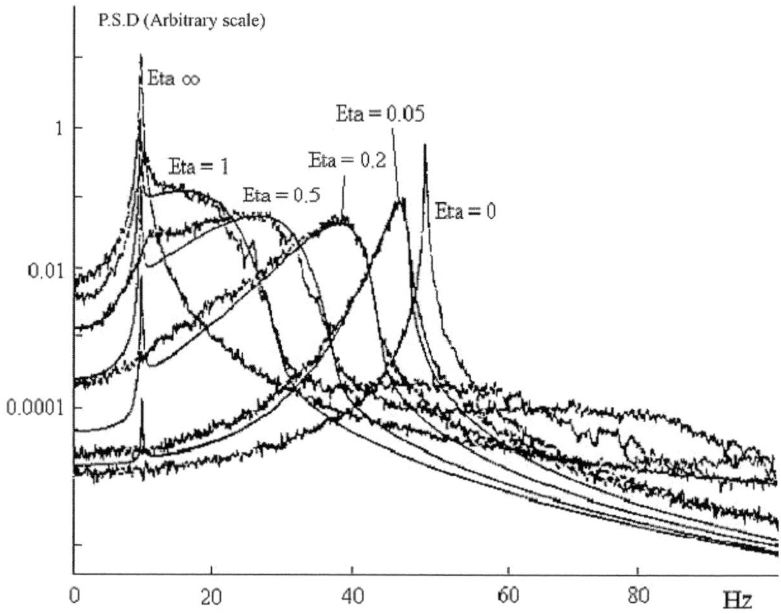

Figure 8.24. *Comparison between the random coefficient linearization method and numerical simulations (PSD of the response)*

8.8.5.2. *Comments on several-dof systems*

The previous formalism can be extended to a several-dof system [BEL 99]. However, concerning the obtaining of $\omega_n(x_{max})$ relationships, a delicate problem is raised, as n ranges from 1 to N which is the number of dofs. These relationships can be obtained from periodic regimes associated with the free and non-dampened system (see Figure 8.23 for 1 dof).

As a matter of fact, if for 1 dof, there is often one only such stable regime, things become far more complex as soon as we reach 2 dofs. Then we can observe numerous stable or unstable periodic regimes, chaotic regimes, etc. These different regimes correspond to paths concerning different areas of the *phase space* that refers to the space of the y vectors defined above [CAU 63]. The intensity and the space distribution of the source will favor a particular regime.

Actually the problem is quite complex and it is not certain that the methods currently proposed within the scope of the formalism of the random coefficient equivalent linearization take these aspects into account. From a practical point of view we should be cautious when applying these methods.

Nevertheless, we can be quite optimistic in the case of seismic analysis that concerns us here, as the transient feature of the stimulation inevitably restricts the excursions of the system into the phase space, which makes the dynamic behavior much simpler.

8.9. Conclusion

Probabilistic seismic analysis lies within the wider scope of *random dynamics,* which aims at developing tools to predict responses to complex loading sources that can only be characterized by statistical averages.

The specificity of the seismic source derives from its complex and transient characteristics. Moreover, it is difficult to model because of the complexity of the earthquake generating mechanisms, the media through which waves propagate and the fact we lack "experimental" data.

Another aspect of seismic analysis is that it falls within the scope of quite heavily regulated designing methods.

In this chapter, we tried to explain where probabilistic seismic analysis stands with regard to conventional design methods. It specifies the scope of their validity by supplying designers with a means to assess their hypotheses, it proposes improvements and thus allows the extension of their application field; it can even stand in for them in some of the more delicate cases that require mastery of safety margins.

As we have shown, probabilistic approaches are constantly evolving, especially as far as non-linear behavior is concerned.

8.10. Bibliography

[BEL 99] BELIZZI S., BOUC R., "Analysis of Multi-degree of Freedom Strongly Non-Linear Mechanical Systems with Random Input", Parts I and II, *Probabilistic Engineering Mechanics*, vol. 12, n° 3, p. 229-244 and 245-256, 1999.

[BOU 94] BOUC R., "The Power Spectral Density of Response for a Strongly Non-linear Random Oscillator", *Journal of Sound and Vibration*, vol. 175, n° 3, p. 317-331, 1994.

[DAV 85] DAVIDOVICI V., *Génie parasismique*, Presses de l'ENPC, 1985.

[DAV 88] DAVIDOVICI V., *Recent Advances in Earthquake Engineering and Structural Dynamics*, Ouest Editions-Presses Académiques, 1992.

[FOG 96] FOGLI M., BRESSOLETTE P., BERNARD P., "Dynamic of the Stochastic Oscillator with Impacts", *European Journal of Mechanics*, A/Solids, vol. 15, n° 2, p. 213-241, 1994.

[GIB 88] GIBERT R.J., *Vibrations des Structures, Interactions avec les fluides, Sources d'excitation aléatoires*, Editions Eyrolles, 1988.

[GUI 90] GUILHOT P., Analyse de la Réponse de Structures Non Linéaires Sollicitées par des Sources d'Excitation Aléatoires – Application au Comportement des Lignes de Tuyauterie sous l'Effet d'un Séisme, PhD Thesis, Pierre et Marie Curie University, December 1990.

[KRE 83] KREE P., SOIZE C., *Mécanique Aléatoire*, Dunod, Paris, 1983.

[NED 99] NEDJAI H., GIBERT J.R., "Seismic Analysis of a Structure with Sliding Bearing", *PVP Conf.*, Boston, USA, 1999.

[NOE 93] NOE H., BETBEDER-MATIBET J., GOUTELARD H., "Les Modes Propres en Base Glissante. Théorie et Applications", *3e Colloque National de Génie Parasismique*, Saint-Rémy-lès-Chevreuse, 1993.

[PRE 90] PREUMONT A., *Vibrations Aléatoires et Analyse Spectrale*, Presses Polytechniques et Universitaires Romandes, Switzerland, 1990.

[SAR 98] SARH K., Réponse d'une Structure Multimodale Glissante sous Excitation Aléatoire, PhD Thesis, Pierre and Marie Curie University, 1998.

[SCH 01] SCHEIMAN P., Analyse de la Réponse de Structures Non Linéaires à deux Degrés de Liberté Sollicitées par des Sources d'Excitation Aléatoires, PhD Thesis, University of Evry-Val-d'Essonne, 2001.

[VAN 73] VANMARCKE E.H., VENEZIANO D., "Probabilistic Seismic Response of Simple Inelastic Systems", *5e Conf. Mondiale d'Ingénierie Sismique*, Rome, Italy, 1973.

Additional references

[BER 84] BERNARD P., FOGLI M., BRESSOLETTE P., LEMAIRE M., "Un Algorithme de Simulation Stochastique par Markovianisation Approchée. Application à la Mécanique Aléatoire", *Journal de Mécanique Théorique et Appliquée*, vol. 3, n° 6, p. 905-950, 1984.

[CAU 63] CAUGHEY T.K., "Equivalent Linearization Technics", *Journal of the Acoustical Society of America*, p. 1706-1711, vol. 35, n° 11, November 1963.

[CAU 71] CAUGHEY T.K., "Non-linear Theory of Random Vibrations", *Adv. Appl. Mech.*, vol. 11, p. 209-253, 1971.

[HAD 82] HADJIAN A.H., "A Re-evaluation of Equivalent Linear Models for Simple Yielding Systems", *Earthquake Engineering and Structural Dynamics*, vol. 10, p. 759-767, 1982.

[IWA 73] IWAN W.D., "A Generalization of the Concept of Equivalent Linearization", *International Journal of Non-Linear Mechanics*, vol. 8, p. 279-287, 1973.

[KAN 57] KANAI K., "Semi-empirical Formula for the Seismic Characteristics of the Ground", *Univ. Tokyo, Bull. Earthquake Res. Inst.*, vol. 35, p. 309-325, 1957.

[LAB 86] LABBE P., "Evolution des Méthodes Spectrales en Analyse Sismique. Un Apport de la démarche Probabiliste", *1er Colloque National de Génie Parasismique*, Saint-Rémy-lès-Chevreuse, January 1986.

[ROB 80] ROBERTS J.B., "The Yielding Behaviour of a Randomly Excited Elasto-plastic Structure", *Journal of Sound and Vibration*, vol. 72, n° 1, p. 71-85, 1980.

[ROB 93] ROBERTS J.B., SPANOS P-T.D., *Random Vibration and Statistical Linearization*, John Wiley & Sons, 1993.

[SOI 88] SOIZE C., "Steady-state Solution of Fokker-Planck Equation in Higher Dimension", *Probabilistic Engineering Mechanics*, vol. 3, n° 4, p. 196-206, 1988.

[SPA 80] SPANOS P-T.D., "Formulation of Stochastic Linearization for Symmetric or Asymmetric MDOF Non-Linear Systems", *Journal of Applied Mechanics*, vol. 47, p. 209-211, 1980.

[SPA 81] SPANOS P-T.D., "A Method for Analysis of Non Linear Vibrations Caused by Modulated Random Excitation", *International Journal of Non-Linear Mechanics*, vol. 16 p. 1-11, 1981.

[WEN 89] WEN Y.K., "Methods of Random Vibration for Inelastic Structures", *Journal of Applied Mechanics*, vol. 42, n° 2, p. 39-52, 1989.

Chapter 9

Engineering Know-How: Lessons from Earthquakes and Rules for Seismic Design

9.1. Introduction

This chapter aims to show what can be learnt from the study of earthquakes, describes the main failure modes of reinforced concrete and discusses how current designs rely on these studies, as well as theoretical models, to improve structural strength and ability to withstand earthquakes.

Some items in the current PS92 French regulations [COL 95a] will be approached, but all of the topics discussed are framed within the standards set by the Eurocodes that are destined to replace them.

9.2. Lessons from earthquakes

The sudden application of seismic action to buildings amplifies their design and building faults in a typical manner (Figure 9.1): poor architectural layouts, under-dimensioning and brittle failure modes all become evident during structural damage or collapse.

Chapter written by Philippe BISCH.

Photo Philippe Bisch

Figure 9.1. *Poor-quality concrete in a column*

9.2.1. *Pathologies linked to overall behavior*

The seismic loads that cause the most damage are horizontal inertia loads the structure has not been designed to withstand: permanent loads do less damage because they are directed downwards. Horizontal loads have to be distributed between *transverse bracings*, vertical structures designed to transfer inertial loads down to the foundations. As far as the overall behavior of any structure is concerned, two main problems are typically identified:

– the layout of the transverse bracing structures is such that the center of stiffness is offset from the inertial load resultant, hence the structure is subject to a vertical axis torsion that tends to overload bracing located on the edges of the building. Figure 9.2 shows a building block after torsion around a stiff portion, namely the stairwell. The right hand part, supported by a reinforced concrete column, has collapsed due to a general rotation imposed on the vertical axis;

– bracing structures do not have any continuous stiffness and geometry along the height of the building; consequently, at levels where discontinuities occur, seismic loads have to be transferred to a "diaphragm" horizontal plane. Moreover, since stiffness can differ significantly between two successive levels, strains are concentrated in the less rigid storeys. This is evident in Figure 9.3, which shows a rather stiff villa built on piles that had its lower level destroyed by displacements imposed by the overall motion.

Photo Philippe Bisch

Figure 9.2. *Collapse of a building showing overall torsion*

Photo Philippe Bisch

Figure 9.3. *Rupture at the level of a lightened storey*

A building's overall dynamic behavior is determined by the strength and ductility of its structural elements. In fact, except for specific constructions for which preserving a quasi-elastic behavior is desirable, large incursions into the post-elastic field are usually included in the design (see section 9.5.2). In linear element structures such as beams and columns, this involves formation of "plastic hinges" that will suffer alternating rotation cycles during a seismic disturbance. These induce steel and concrete strains far higher than those at the ultimate limit state. Roughly speaking, with such flexible structures, we can assume that the displacements reached during incursions into the plastic field are equal to those calculated using the elastic hypothesis. To a first approximation, the structures are subjected to impose displacements, which allow engineers to carry out sound design, by evaluating the rotation likely to be caused in the plastic hinges.

To avoid strains causing compression and crushing, areas that are joined by the plastic hinges should be confined by transverse reinforcements (stirrups). The final strength of sections is determined by the extent to which the energy dissipation inside the hinges counterbalances resistant moments, and whether sufficient ductile rotation can take place. Figure 9.3 shows a case in which the piles were able to absorb the imposed displacements without rupture.

Figure 9.4. *Rupture of a column due to the shock between two neighboring structures*

Two dynamically independent blocks can shock each other if the hinge separating them is not open enough. Figure 9.4 shows local bending due to a shock inside a column. The phenomenon can lead to total building collapse.

9.2.2. *Problems linked to local under-design*

Figure 9.5 shows a section at the head of a column that has been broken because of the horizontal load transmitted to it. The section has several features that contribute to this poor performance: smooth construction joints with little strength in the concrete to withstand the shear load, no concrete confining reinforcements and insufficient longitudinal reinforcements. No rotation ductility could be induced because of these deficiencies, and the reinforcements broke in traction in the joints, which opened because the rotation imposed was localized.

Figure 9.5. *Rupture of a section due to insufficient longitudinal reinforcements*

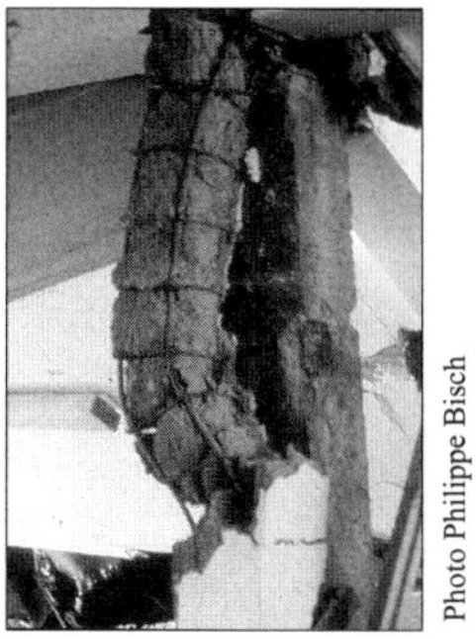

Figure 9.6. *Rupture of a column due to buckling*

In Figure 9.6, failure has occurred because of the instability of the column: combined bending and extra normal forces due to the vertical component of an earthquake have crushed the concrete.

Figure 9.7 shows a silo after being subjected to a high horizontal acceleration. Plastic hinges appeared at both ends of the columns, where the bending moments were highest. In this case, the behavior of both hinges is quite different, due to reinforcement differences. In the lower area, regular and tightly spaced frames have confined the concrete within a “cage”, allowing strong plastic rotation (obvious in the photograph) and holding the longitudinal reinforcements together, preserving the limit moment during cycles. It should also be noted that the frames were well-anchored, and consequently did not get torn down by the increased confinement pressure.

In the upper part of this plate, the existence of smooth construction joints, together with insufficient longitudinal reinforcements joining the section, generated failure even before the plastic hinge had been able to form.

Photo Philippe Bisch

Figure 9.7. *Rupture of a silo column with formation of a plastic hinge*

Photo Philippe Bisch

Figure 9.8. *Rupture of a column showing the influence of the shear load*

As shown in Figure 9.8, bending failure is influenced by shear loads. As it takes place, the column suffers the influence of insufficient transverse and longitudinal reinforcements, and failure combines both effects.

In the case depicted in Figure 9.9, longitudinal reinforcements have allowed the embedding section to withstand the moment quite well, but failure due to the shear load started before a plastic hinge could form, because of insufficient transverse reinforcements. This case highlights the importance of the *sequence* in which plastification and failures occur – if the shear load-generated brittle failure had been delayed beyond the onset of bending plastification of the embedding section, a plastic hinge could have formed, which would have increased the chances the column could have withstood the earthquake (see section 9.3.4).

Figure 9.9. *Failure of a column due to shear load*

Figure 9.10. *Failure of a short column*

The influence of shear load can be especially prominent in short columns, as shown in Figure 9.10. In such a situation (for example, in ventilation spaces), an imposed horizontal strain results in a prominent shear load, because the bending length is limited and the column is stiff.

A similar phenomenon may also take place in short piers cut into a wall by openings (Figure 9.11).

Figure 9.12 shows the deficiency of a column section with regard to compression. The longitudinal reinforcements have suffered instability caused by compression. They are not held firm by the concrete cover, which is ejected because the stirrups are too widely spaced. The buckling force within the reinforcements is proportional to the inverse of its buckling length squared, therefore stirrup spacing is the main parameter controlling the phenomenon (see section 9.6.5.2).

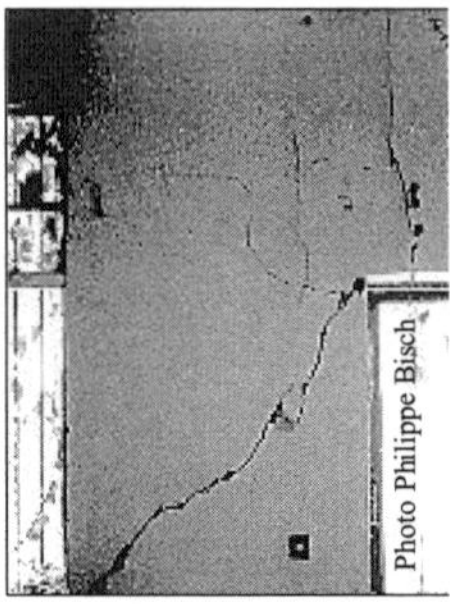

Figure 9.11. *Rupture of a column due to shear load*

Photo Philippe Bisch

Figure 9.12. *Rupture of a column due to buckling of longitudinal reinforcement*

9.2.3. *Problems linked to construction layout*

Figure 9.13 shows a failure diagram for one of the reinforced concrete walls used in the CAMUS 1 experiment. In this non-ductile wall, the reinforcements are laid out so that cracking and deformation are distributed throughout the wall height, unlike the situation in ductile walls, where the strain is mainly due to the rotation of a plastic hinge at the base (see section 9.6.4.1).

In this case, longitudinal reinforcement failure occurs within an inclined crack located on the second level and is initiated within the bar stopper section. Actually, because of the influence of the shear load, this level tends to behave as trusswork (vertical braces and bracing struts, floors being the stays). The bar stopper causes a sudden variation of the strong vertical stress, thereby initiating failure.

Figure 9.14 shows the cover concrete bursting in a column section with a bar stopper and a tie bar placed in a plane parallel to the outer side. Such a layout is forbidden in seismic areas, where tie bars must be laid out inwards inside the concrete mass to avoid such bursting. A similar cover concrete bursting phenomenon can be observed in cases where coatings have insufficient seams.

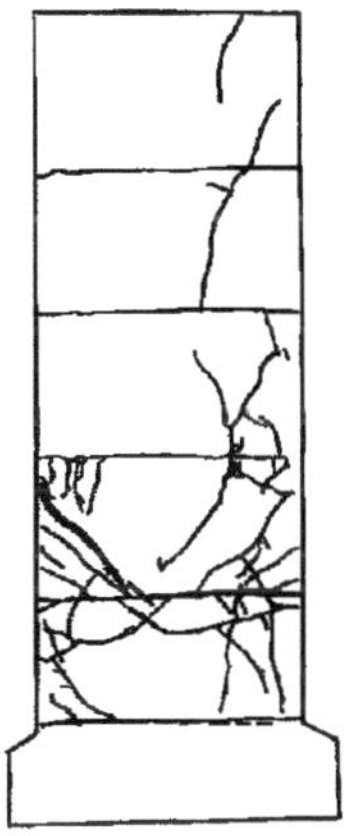

Figure 9.13. *Failure scheme for a reinforced concrete wall*

Photo Philippe Bisch

Figure 9.14. *Rupture of the cover in a column due to a reinforcement tie bar*

9.3. The aims of anti-seismic protection standards

9.3.1. *Standardization of anti-seismic design*

As far as construction is concerned, standardization codifies design and calculation methods. When they are correctly applied, they help designers produce constructions that ensure the safety of people and goods for their entire planned lifespan. In the building field, design standards serve to supervise the calculation methods used, whilst other regulations govern the strength and strain limits of building materials and construction layouts. In Europe, prevailing standards rely on a general philosophy of safety that is based on a semi-probabilistic approach using safety partial coefficients.

In France, anti-seismic protection rules were only developed after the Orleansville earthquake in Algeria (1954). The first standardizing text (AS 55) was, however, used only in Algeria. The earthquake in Agadir, Morocco (1960) a few years later gave birth to the PS 69 rules that remained as guiding principles for over 20 years. After the El Asnam earthquake in Algeria (1980), it became clear that the PS 69 rules did not ensure sufficient safety in some situations, and a review was instigated. As such, a 1982 *addendum* contained some missing construction layouts to the PS 69 rules. New regulations were requested to compensate for deficiencies in PS 69, and these incorporated more recent research discoveries. This led to the PS 92 standard [COL 95a], which is currently in use.

The PS 92 standard is devoted to the anti-seismic design of commonly used buildings, according to the BAEL standard. The materials addressed are reinforced or pre-stressed concrete, steel structure frames and timbers. The text includes all general measures concerning the targets to achieve, the seismic actions to take into account, the calculation methods to be used, and strict definitions of safe foundations. It also contains a chapter about concrete pre-cast elements. Its structure and the concepts used are similar to those in the very latest codes.

In 1971, the European Commission decided to launch a program aimed at harmonizing technical specifications in the construction field in across the Common Market. The Commission introduced standardized technical rules for construction work design ("*Structural Eurocodes*"). The first version of these standards was introduced for experimental work. The codes use experimental standards from ENV 1991 to ENV 1999, the first parts of which were published in the early 1990s. After a three-year experimentation period, the ENV experimental standards (see [COL 00a], [COL 00b] and [COL 01, which give those parts of Eurocode 8 ENV related to buildings) were transformed into EN European standards. In their final format, all of the design standards therein (approximately 60) were gathered into ten subsets that make up the Eurocodes included in EN 1990 to EN 1999. Anti-seismic design is the direct concern of Eurocode 8 (EN 1998) which deals with different types of building in earthquake areas and comprises six parts. Parts 1 [COL 03b] and 5 are needed to design buildings; they cover about the same scope as the PS 92 rules. Several Eurocodes are generally necessary for the complete definition of a reinforced concrete construction work: EN 1990 gives the general philosophy and operation combinations, EN 1991 defines the operations to take into account in addition to earthquakes, EN 1992 [COL 03a] gives common design rules for reinforced concrete, EN 1998 introduces the action of earthquakes and the additional arrangements needed in a seismic situation, whilst EN 1997 deals exclusively with foundations.

The PS 92 standard will still apply in parallel with Eurocode 8 (co-existence period), probably up to about 2010; the national standard will then disappear, and will be replaced by the European standard. The general structure and the main layouts of both standards are quite close, which should make passing from one to the other easier, despite the greater complexity of Eurocode 8.

As far as anti-seismic construction is concerned, standardization has a strong status in France [BET 97]. In fact, lawmakers made anti-seismic protection compulsory, by introducing the 22 July 1987 law (Article 41) and the 2 February 1995 law (the so-called Barnier law). Whereas only new constructions were affected by the first law, the second expanded the protection to existing facilities, which raised technical and economic problems, because many existing housing types were non-compliant. The different orders and decrees specifying how these laws should be enforced defined seismic zoning in France, classified different constructions according to their importance with regard to public safety, and fixed standards and other technical rules.

9.3.2. *Main objectives of anti-seismic protection*

Because of the violent dynamic features of seismic action and the inaccurate knowledge we have about its likely effects, building in seismic areas involves additional efforts in terms of design and construction quality. The anti-seismic standards give recommendations that complement other design standards; in fact, constructions must at least respect the objectives of strength, practicality and durability set for constructions in non-seismic situations and that are subjected to these other standards (especially Eurocode 2 in the case of reinforced concrete structures). However, some Eurocode 8 verifying rules are more restrictive than those in Eurocode 2, and a well-informed designer can limit the number of verifications required. The goals of Eurocode 8 should be considered besides the goals of Eurocodes, and both lie within the framework of EN 1990, which describes the basics of Eurocode design.

The main objective of an anti-seismic design standards is to ensure three things: the protection of human life during an earthquake, the operational continuity of constructions important for emergency services, and the limitation of structural damage. A major obstacle to this aim lies in the randomness of seismic action, the statistical characteristics of which are difficult to determine in countries with low or medium seismicity. Furthermore, earthquake protection has a cost, and some countries have low disposable resources.

In the case of Eurocode 8 (Part 1), these general objectives are expressed by the following mandatory prescriptions:

– *Non-collapse prescription*: structures must be designed and built to withstand – without any local or global collapse – an earthquake, the aggressive level of which corresponds to an overlapping probability over 50 years determined by the National Authorities (10% in principle, which corresponds to a 475-year-return period). This is the reference seismic action. After the event, the structure must preserve a notable residual strength capacity (especially with regard to permanent loads), and should be able to withstand a replica without collapsing.

– *Damage limitation prescription*: the structure should be designed and built so that, under the effect of an earthquake more important than the previous one, the structure will not suffer damage the repairing cost of which would be higher than the cost of the structure itself, or which would limit its normal use. The earthquake level sought corresponds to an overlapping probability over 10 years set by the national authorities (usually 10%, i.e. a 95 year-return period). In order to make the designer's task easier, such seismic action can be derived from the reference seismic action by affinity.

– The specific protection assigned to certain types of constructions that are important for public safety is ensured by means of a γ_I multiplying coefficient ("*coefficient of importance*") which is applied to the seismic action. This amounts to increasing the return period (within the limits of the values considered for the coefficient) or to decreasing the overlapping probability over a given period of time. In order to make such differentiation possible, constructions are classified into four *importance classes,* and a γ_I value is assigned to each class. Class IV corresponds to construction works vital for public safety, whereas class II corresponds to standard buildings.

9.3.3. *Verification method*

9.3.3.1. *General principles*

The Eurocode system is based on a series of common principles that apply to the use of structural materials, either in standard operation situations or in particular circumstances (an earthquake, for instance). Such principles aim at ensuring structure strength, usability and structure durability for an expected lifespan. They are laid out in EN 90 regulations [COL 02], which also introduce the general Eurocode philosophy.

The reliability of the approach used is demonstrated by calculating the achievement of the goals targeted and is linked to the introduction of partial safety coefficients for each of the materials, the actions and the calculation methods. However, achieving these objectives is partly connected to the design methods used: specific protection measures, the quality of the studies and of their execution, as well as the follow-up of potential failures have contributed in an essential way to the success of the objectives.

It is an acknowledged fact that the goals targeted have been reached by verification of the boundary conditions of structures in project situations. These situations are basic and are represented by combinations of actions in which the seismic action is considered to have a higher occurrence probability higher than any other accidental action.

The seismic situation is a specific one that gives rise to particular action combinations that are expressed the following way:

$$\sum_{j \leq l} G_{k,j} + P + A_{Ed} + \sum_{i \geq l} \psi_{2,i} Q_{k,i} \qquad [9.1]$$

In the above expression, "+" means "combined with" instead of a mere algebraic operation. $G_{k,j}$ refers to the characteristic values (in the probabilistic sense of the word) of the permanent actions applied to the structure, P is the action representing pre-stressing, A_{Ed} the calculation seismic action equal to the $\gamma_I A_{Ek}$ product of the coefficient of importance (see section 9.3.2) by the value characteristic of the seismic action, $Q_{k,i}$ the values characteristic of the accompanying variable actions (i.e. those that may be present during the seismic situation) and $\psi_{2,i}$ accompanying coefficients expressing the intensity of the variable actions present when the earthquake occurs. These coefficients are chosen as equal to their corresponding values of the action quasi-permanent parts. Such combinations are used for verifications in the considered limit states; they should not be used in a stand alone manner to calculate the masses to be introduced into dynamic calculations (section 9.4.3.1).

Two structure-limited states classes are considered. ultimate limit states (*ULS*) which address the safety of people, structures and possessions, and service limit states (*SLS*) that are concerned with the normal operation of the structure and its elements, people's comfort and the appearance of the construction.

Calculation is far from the only design element in seismic situations, where general design layouts are essential elements too.

Material fatigue – which will not be dealt with in this chapter – must also be taken into account when designing construction works subject to important and repeated dynamic loads, such as railway bridges and seaside works.

9.3.3.2. *Application of Eurocode 8 (Part 1)*

To be compliant with the Eurocode objectives defined in section 9.3.2, the following limit states must be checked:

– *ultimate limit states:* verifying these involves non-collapse and other kinds of structural failure that might endanger human lives;

– *damage limitation states*: these are associated with the appearance of damage and correspond to conditions under the structure is not able to fulfill its duties any more, while remaining stable.

A feature common to anti-seismic design standards is that verifications required for specified levels are enough to demonstrate the strength of the construction work with regard to all seismic events on intermediate levels. The methods do not necessarily directly apply to intermediate levels. Forgetting this can lead to erroneous interpretations.

9.3.3.3. *Verifications*

We need to verify that the structure has enough strength and ductility at the ultimate limit state, and do this by taking second order effects into account. The verification can be achieved by introducing acceptable incursion limits into the post-elastic state. A technique for predicting post-elastic behavior involves applying a dividing coefficient, called the *behavior coefficient* (sections 9.4.3.3 and 9.5), determined by the overall behavior characteristics of the structure and constitutive material, to the seismic actions calculated from a linear elastic hypothesis. The more ductile the structure is and the better controlled its behavior, the higher the value of this coefficient. Application involves dimensioning structures for strength values lower than elastic.

We also need to verify the joint openings between two adjacent buildings or between two blocks of a same building to ensure they are sufficient to avoid shocks.

Under seismic actions that are reduced by the behavior coefficient, the code demands that the engineers must demonstrate the overall stability of the structure, the strength of the soil (without any intensive permanent strain) and the strength of the foundations.

In addition, the non-structural elements should neither be dangerous for people nor affect the dynamic behavior of the structural elements.

At damage limitation condition, the code recommends that strains be limited (relative displacements between floors) to avoid partition walling and facade damage according to their brittleness.

9.3.4. *Capacity-design method*

To design a structure to withstand seismic action economically, incursions into the *post elastic field* are tolerated. Because such incursions take place in cycles, it is necessary to avoid reaching the limit strain of the constitutive material of the structure, as beyond these points, integrity cannot be ensured, as the material deteriorates, lowering strength. Good design not only seeks to ensure the strength of constitutive materials, but to preserve their *ductility* and stable behavior during cycles as well. The ductility objective is achieved by adopting some design and verification rules that are more restrictive than those used in non-seismic situations, by taking incursions into the post-elastic field into account, and setting up specific construction layouts.

Ductile behavior is quite well illustrated in the case of frame structures, whatever their constitutive material. Passing into the post-elastic field is evidenced by the appearance of one or several plastic hinges, typically near the connections (beam/column intersections). When the stress increases in a monotonic way, hinge rotation increases and other plastic hinges may appear. This cyclic hinge formation step does not correspond to any alternating characteristic of seismic action but absorbs a lot of energy, thus turning the structure into a *dissipative structure*, which can limit the damage accrued in an earthquake. For such behavior to be possible, plastic hinges have to be capable of withstanding important rotations without significant damage, so that the strength capacity of the structure and its capacity to dissipate energy do not decrease. Transverse reinforcements and construction layouts ensure such ductility. It also appears that hinge position inside the mechanism is relevant for construction work safety, as localized effects in the columns have to be avoided (Figure 9.15a). Finally, the degree of hyper-staticity of the structure determines the number of plastic hinges that develop within the framework to reach the mechanism condition, and is therefore an important determinant of the total energy dissipation capacity of a structure.

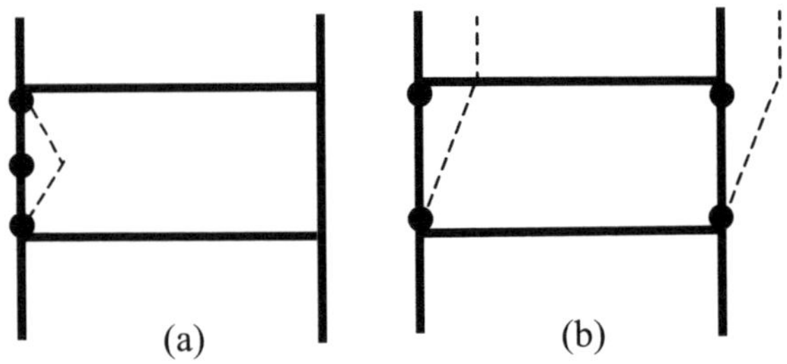

Figure 9.15. *Undesirable plastifying modes*

The maximum acceleration a structure can counterbalance is limited by the resistant moment of the hinges, but their energy dissipation capacity allows the elastic energy to counterbalance the energy injected by the earthquake.

Eurocode 8 recommends a design and dimensioning method called *capacity design*. In this, the plastic hinges should arise in pre-determined areas called *critical areas*, to allow the structure to reach a mechanism condition whenever energy is dissipated by ductile rotation of the plastic hinges. The plastic moments inside the hinges are determined by the resistant moments of individual sections. It is good practice to arrange for resistant moments to be higher than design moments (by about 20 to 30%), as the accelerations to which the structure is subject are determined by resistant moments and not design moments.

This frame example allows the method to be illustrated: to ensure that the plastic hinges will form where designers want them to, so that energy is dissipated according to the planned mechanism, the areas outside the hinges must be designed to stay elastic when the zones in which the hinges should appear (critical areas) become plasticized. Inside the hinges, the maximum moment possible is the resistant moment, derived from the design moment by multiplying it with an *over-capacity coefficient*. The coefficient takes into account strain-hardening effects, such as the reinforcements of bending strength in a reinforced concrete section. Because these maximum moments are known in the critical areas, the other areas are deliberately over-dimensioned with regard to them. This is supposed to prevent plastic hinges forming in areas where they are not desired, for example, within the columns, for if the latter are too numerous here, the risk of instability becomes greater (Figure 9.15b). The goal is to ensure hinging will mainly take place inside the beams during an earthquake (Figure 9.16).

To ensure engineers grade the appearance of plastic hinges correctly and to prevent them forming inside the columns, they have to dimension the post-potential critical areas using the capacity design method. This leads to the inequality expressed in the following equation:

$$\sum M_c \geq \gamma_{Rd} \sum M_b \qquad [9.2]$$

in which the M_c moments (respectively M_b) are the design-resistant moments of the columns (respectively beams), and γ_{RD} is an over-capacity coefficient (Figure 9.17).

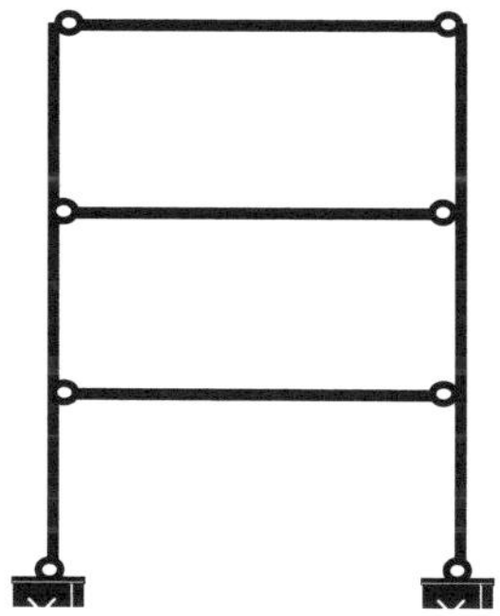

Figure 9.16. *Preferable mechanism*

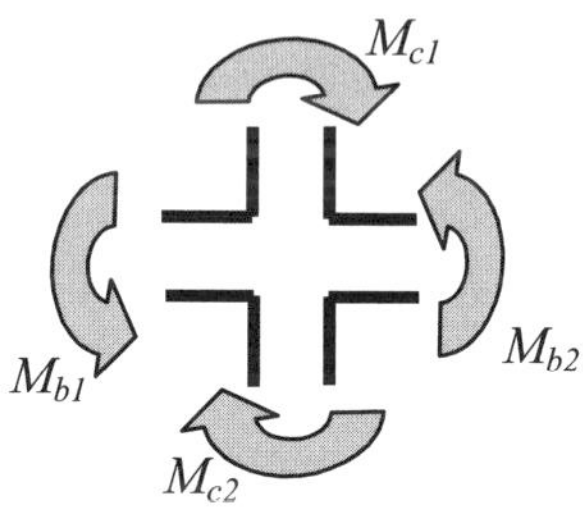

Figure 9.17. *Equilibrium of a node*

Capacity design also involves classifying a structure's failure modes to ensure that brittle failure modes cannot appear before ductile modes. This ensures a planned dissipative operation, and avoids the existence of a "cliff" effect, i.e. a sudden drop in structure strength if the aggression level is exceeded. For reinforced concrete, shear load behavior is acknowledged to be brittle whereas bending moment behavior is ductile, providing minimum construction layouts are respected. To prevent a beam segment between two plastic hinges breaking under the influence of shear load, it has to be designed for the maximum shear load that is obtained by expressing the equilibrium using the following formula (Figure 9.18):

$$V_{Sd,CD} = \gamma_{Rd} \cdot \frac{M_{DRd} + M_{CRd}}{l_{cl}} \qquad [9.3]$$

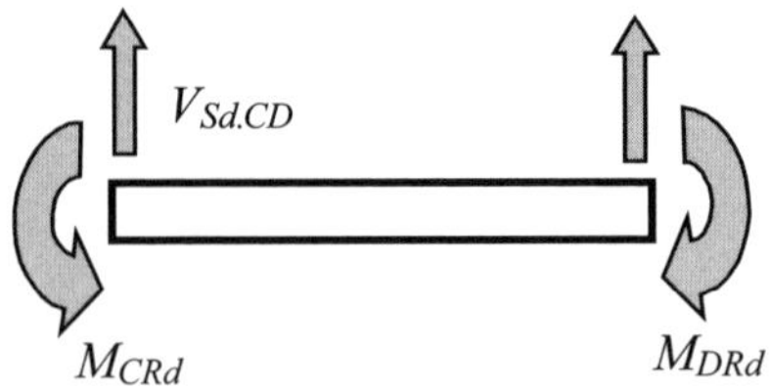

Figure 9.18. *Transverse load determined by capacity design*

In this expression, M_{DRd} and M_{CRd} refer to the design resistant moment at both ends, l_{cd} is the segment length and γ_{Rd} is the over-capacity coefficient.

The $V_{Sd,CD}$ shear load thus obtained is used to design the beam segment and does not have any direct connection to the shear load obtained by structure calculations for the same element.

9.4. General design

9.4.1. *Design principles*

Eurocode 8 encourages designers to adopt layouts that ensure the good behavior of the structure. These include using bracing that is continuous down to soil level, and considering the symmetry, hyper-staticity, strength and stiffness in both horizontal directions, strength and stiffness with regard to the vertical axis torsion, the existence of horizontal diaphragms on different levels to distribute the seismic loads on the bracing elements.

Regular behavior should be the goal. It is best achieved by implementing simple and compact shapes, horizontally and vertically, and to achieve this, the structure can be designed as a collection of dynamically-independent blocks, each of which have regular properties. Regular structures simplify the process of performing both action and behavior coefficient calculations.

Foundation structural elements have to be stiff enough to transmit the seismic actions of the structure towards the soil or deeper foundations. Typically there should be only one foundation type below a structure. If this is not possible, the structure must be divided into dynamically-independent blocks.

During the general design stage for a structure, we can distinguish the main elements that constitute the bracing from the secondary elements, which merely play a supporting role. The secondary elements should not be considered as strong resisting elements when calculating structural response to seismic action, and their stiffness can be neglected when evaluating the dynamic behavior.

Except when the structure has been designed as non-dissipative, ductility should also be considered by designers. Brittle failures, or the premature formation of unstable mechanisms like hinges, must be avoided to ensure the overall ductility of the structure. This is achieved by using capacity design (see section 9.3.4). Local ductility can also be ensured by using the construction layouts described in later chapters on material structure.

Non-structural elements must be verified as being able to withstand the acceleration transmitted by the structure; this acceleration is calculated using a *floor spectrum*. The calculation can be simplified by using acceleration results from a standard estimate that uses a behavior coefficient specific to the element being considered.

9.4.2. *Regularity conditions*

The regularity of a building must be considered in two parts: plane regularity and vertical regularity. They are both desirable though not crucial.

Horizontal regularity is ensured by designing so as to restrict vertical axis torsion phenomena. The criteria are twofold: the first relate to the symmetry and compactness of the plane shape, and are the subjects of simple geometrical controls. The existence of two main orthogonal planes over the whole height of the structure reflects most of these criteria. The second planes are mechanical: we must ensure that the floor diaphragms are stiff enough with regard to bracing elements that diaphragm displacements on each level are stiff behaviors (overall translation and rotation). A second set of conditions deals with the radii of torsion, defined as the square root of the ratio of the torsion inertia around the torsion center with the bending inertia in the considered direction. For each principal direction, the radial deviation of the structure (i.e. the distance between the bending center of the bracing system and the center of gravity, projected in the direction under consideration) must be less than 30% of the corresponding radius of torsion, and the latter must stay lower than the mass-gyration on each level.

Figure 9.19 presents two wall structures. In both cases, the radius of gyration is determined by the mass of the floors, which is distributed over the whole surface. The case shown in Figure 9.19(a) includes a central core and peripheral columns: here, the stiffness and strength with regards to torsion focus at the center of the building, hence the radius of torsion is low, and the building is not considered to be regular. In the second case shown in Figure 9.19(b), bracing elements are distributed on the outskirts, and the radius of torsion is similar to the width of the building. The horizontal regularity allows us to carry out a plane analysis in both of the principal planes.

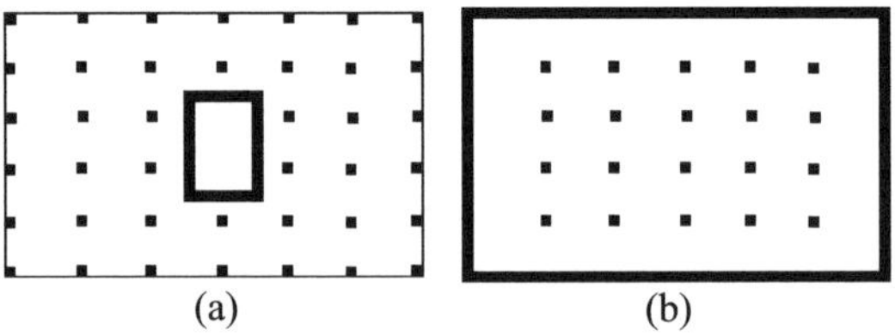

Figure 9.19. *Plane configurations different from the point of view of torsion*

Respecting vertical regularity allows designers to ensure the progressive characteristic of first mode displacements, and to prevent modes higher than 1 exerting a non-negligible influence on the displacement of the structure under seismic action. These conditions allow the use of the simplified calculation method. They also make it possible to avoid delicate situations linked to sudden variations in stiffness or mass ("transparencies" on the first level for example, or reversed pendulums). The conditions are a complex function of the continuity of bracing elements and progressive variation in geometrical shapes with structural height.

9.4.3. *Calculation of seismic action effects*

9.4.3.1. *Structure modeling*

Calculation of seismic actions should be based on modeling of the structure that takes into account the influence of soil deformation, connections, non-structural elements and the presence of contiguous structures. As a rule, models consists of vertical elements that represent bracing elements, connected by other elements that represent diaphragms. If the diaphragms can be considered as stiff compared to bracing elements, floor masses and mass inertia can be concentrated on their centers of gravity; in such a case, the model is termed a "lumped-model", consisting of beam vertical elements connecting the floors.

The masses are calculated by estimating those present in the structure associated with the seismic action combinations:

$$\sum G_{kj} + \sum \psi_{Ei} \cdot Q_{ki}$$

In practice, the nominal masses associated with the quasi-permanent variable actions are affected by φ coefficients that take into account their presence probability during the earthquake: $\psi_{Ei} = \varphi \cdot \psi_{2i}$. These φ coefficients are a complex function of the level considered in the building and the type of variable action (i.e. the use of the premises). They typically vary between 0.5 and 1.

For masonry or reinforced concrete elements, cracking must also be taken into account when evaluating stiffness.

The soil/structure dynamic interaction effects must also be taken into account when displacements are involved, for example, when modeling structures sensitive to second order effects, very slender structures and structures built on very soft soils. The effects of this interaction on piles have to be taken into account whatever the situation.

9.4.3.2. *Elastic calculations*

Seismic actions inside a structure result from movements imposed at the foundation level. When the behavior of the structure can be considered as visco-elastic from a linear point of view, the structure dynamics methods allow calculation of the actions from a pseudo-acceleration spectrum or by direct time integration (time history calculation).

The multi-modal calculation method that uses response spectra is detailed in Eurocode 8.

Pending concordance with the regularity conditions outlined in section 9.4.2, a simplified method can be used. This involves applying – in each principal plane – a horizontal static stress system on each different level. This stress system is determined proportionally to a single mode the shape of which is given *a priori*. The associated basic period can be determined by an approximate formula that includes the height (H) of the building and on a coefficient characteristic of the bracing type. The whole mass is assigned to that single mode for flexible buildings (T>$2T_C$), and 85% of the mass is used for stiffer buildings.

9.4.3.3. *Taking dissipative behavior into account*

When considering the ductility of the structure by plasticizing critical areas of elements, the non-linear behavior of the structure can be derived from equivalent linear calculation (see [BET 91], section 3.17). In such cases, actions are calculated from an elastic linear model using a design spectrum that incorporates a dividing behavior coefficient. The structure is therefore designed using loads lower than those derived from linear calculations. This allows an estimation of non-linear behavior from linear calculation, which is more realistic from an engineer's perspective, and allows them to avoid the problems associated with non-linear modeling, which is fraught with difficulties. Behavior coefficient determination is discussed in section 9.5.

9.4.3.4. *Non-linear calculations*

Two calculation methods that take the post-elastic behavior of a structure into account are considered: the so-called "push-over", pseudo-static method, and the "time history analysis" method, the latter being reserved for exceptional situations. Both aim either at evaluating either the α_u/α_e ratio defined in section 9.5.3, the strength of existing buildings, or the location of potential plastic hinges and damage. Such modeling can be bi-or tri-linear, but must be based on modeling that is truly representative of the post-elastic behavior of structure elements and must take potential damage into account.

The "push-over" method is internationally approved. It involves applying a given distribution load system (proportionally to the loads derived from the elastic analysis or those resulting from applying uniform acceleration to the height) and increasing in intensity. A "capacity curve" is then drawn, which gives the load characteristic of the seismic action (typically the shear load at the base) according to a characteristic displacement (at the top of the building, for instance).

After this, we must verify that the structure can be deformed so that a "target" displacement (characteristic of a one-degree of freedom system resulting from a spectrum reading for the equivalent basic period of the system) can be achieved without exceeding the ultimate strain. The one-degree of freedom system considered is equivalent to the basic mode of the structure.

9.4.3.5. *Taking vertical axis torsion into account*

Structural torsion can prove dangerous if it is not well-controlled. It has three main causes. The first is differential motion at the base of supporting points due to wave propagation within the soil. If the supporting points are linked by structure elements that are stiff enough, the seismic motion that is stressing a structure can be compared to an overall displacement and rotation, the vertical component of which corresponds to a torsional stress. The second cause of torsion is the natural eccentricity of the center of gravity of each floor with regard to the torsion center of the bracing system. This eccentricity can be calculated from structure plans, but its value is associated with some uncertainty due to assumptions made about mass distributions, the evolution of the torsion center during the motion, cracking, and the appearance of plasticized or damage areas. The third cause of torsion is deformation. Actually, when a structure is flexible with regard to torsion but eccentric, its basic seismic response mode can be a combination of torsion and overall bending. In these cases, implementing that mode simultaneously generates a bending motion and an associated torsion that is amplified.

The random aspects of torsion are covered by taking into account what is termed “fortuitous eccentricity”, which is equal to 5% of the dimension of the building in each principal direction. Assessing the extent to which building motion amplifies torsion can only be achieved through the use of adequate space models, but a simplified approach involves sequentially moving the application point of the seismic force away from the center of gravity on each level.

9.5. Behavior coefficients

9.5.1. *Using behavior coefficients*

The behavior coefficient expresses ductility, over-strength and overall behavior, and is used as follows: seismic loads are calculated making a linear elastic assumption, and the structure is then designed, using the elastic loads divided by the behavior coefficient.

Without going into the detail of the behavior coefficient method, the assumptions on which it is based are outlined in Figure 9.20. It represents a relationship between a stress (which represents the seismic action: shear load at the base, for example) and a characteristic displacement (horizontal displacement at the level of the center of gravity or at the top of the structure). By likening the behavior of an actual structure to one with perfect elastic-plastic behavior, an ideal linear elastic structure would withstand a seismic action system q times as high as the plastic plateau. However, it is accepted that the displacement obtained by taking the elasto-plastic behavior of the structure into account is the same as it would be for the fictitious linear structure. The rule of displacement equality can only be confirmed for flexible structures. With stiff structures, the calculations show that there is an equivalence of strain energies (areas encompassed in the $F(d)$ curve), consequently, structural displacement is more important than elastic displacement. Eurocode 8 uses the elastic displacement calculated with the elastic spectrum as the value of the structure displacement, without making any increase for weaker periods (except for the target displacement in the "push-over" method).

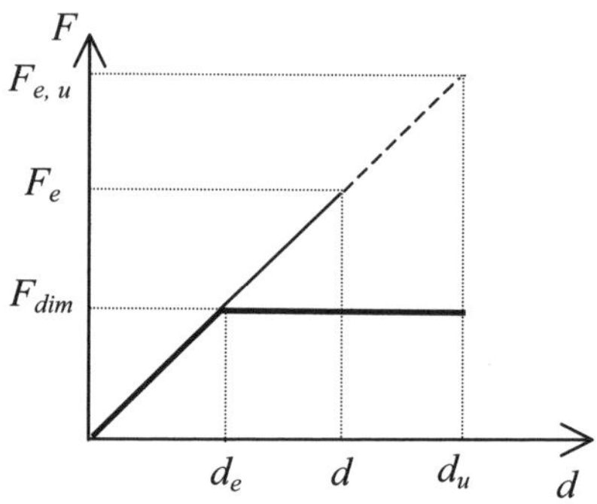

Figure 9.20. *Perfect elastic-plastic behavior*

Figure 9.20 shows that value of the behavior coefficient has a limit, determined by the ultimate displacement the structure can withstand before collapsing. Eventually $F_{e,u} = q_{max} F_{dim}$. This global criterion is expressed on reaching ultimate strains the building cannot exceed. A law expressing the global behavior of the structure integrates the local behaviors linked to the ductility of the materials used. It also depends on the degree of hyperstaticity and the type of the elements used (beams, slabs, columns, walls, piers), as well as on their distribution in space.

In practice, it is meant to represent the maximum value of the behavior coefficient, assuming it is reached at the ULS. This means that, for a lower load level, a smaller behavior coefficient will be used. If the load is low, the structure remains elastic and the behavior coefficient is equal to 1. As mentioned in section 9.3.3.2, it is not acceptable practice to use the maximum behavior coefficient for a seismic level weaker than the one for which the structure was designed.

The behavior coefficient also allows other less easy to master phenomena to be incorporated: for instance, structure irregularity implies a reduction of the behavior coefficient, i.e. an increase of the design strength of the structure, because irregularity-linked phenomena are more difficult to control.

It should be noted that the behavior coefficient is not used when calculating displacements or for determining the opening of joints.

The equivalent linear analysis is mostly based on considerations of ductility and, as a rule, does not apply in the case of marked geometrical non-linearities, like cases of uplift between foundation and the soil.

9.5.2. *Structure behavior and behavior coefficients*

A structure is dissipative if it dissipates energy during hysteresis cycles via the ductility of its constituent materials.

A structure can be designed as non-dissipative if its behavior remains quite close to linear elasticity: in such cases, a behavior coefficient equal to 1 would be assigned to it, and the elastic spectrum could be used. However, when the structures are designed according to the 2 to 6 structure Eurocode regulations, they retain a little ductility (limited so-called "L" ductility) and over-strength linked to hyperstaticity. Therefore, for all structural materials, a behavior coefficient ranging from 1 to 1.5 (2 for metallic structures and mixed structures) is used, with a few additional considerations as far as materials are concerned, but without any specific construction layouts.

When the structure has been designed as dissipative, a behavior coefficient higher than 1.5 (or 2) can be used, as long as construction layouts allowing the structure to reach the expected ductility are used as well.

To determine the behavior coefficient of a given structure, the constituting material of the structure, the bracing type, the ductility level considered and its regularity must be taken into account. In Eurocode 8, typical data for a building that is irregular in height is given in Table 9.1, from which it can be seen that the basic value of the behavior coefficient is equal to 0.8.

As far as reinforced concrete frameworks are concerned, energy dissipation is directly linked to the number of plastic hinges that have to be formed. Therefore, hyperstaticity is taken into account by multiplying the reference behavior coefficient by the α_u/α_e ratio. Using static calculations, we assume the structure is subject to a seismic action system obtained by the analysis described in section 9.4.3. The resulting actions are proportionally increased by an α coefficient. The first plastic hinge will appear at the α_e value (elastic limit), and the structure will become a mechanism for the α_u. Thus, for a reference value of 4.5 and a reinforced concrete frame, the behavior coefficient can reach a maximum close to 7, because the α_u/α_e ratio will be restricted to a value of 1.5. It is possible to avoid calculating the α_u/α_e ratio by accepting a fixed value equal to 1.2, in the case of one-bay frames for example. A similar approach is acceptable for high-ductility walls, for which the reference value will be equal to 4.

9.5.3. *Local ductility and behavior coefficients*

For frame-type structures, if the elements have a homogenous strength, plastic analysis reveals that plastic hinges form at the ends of each column or beam element, providing brittle failure modes have not appeared before their formation. Capacity dimensioning suggests the measures that have to be taken to obtain behavior in accordance with the theoretical formation diagram of plastic hinges. The dissipating capacity of the structure is linked to the hinge's ability to withstand rotational strain in the plastic field without a reduction in strength. This ability is achieved using assiduous construction layouts and/or section selection. For higher behavior coefficients, layouts are restricted. To correctly control plastic strains, it is better if connections between elements do not plasticize. For this reason, they are generally over-dimensioned by applying capacity design regulations, so that the first plastic hinges form near the connections instead of forming within the connections themselves.

The dissipative capacity of a structure is thus mainly determined by capacity design, careful choice of sections and construction layouts. Structures are then classified into *ductility classes*, each class being characterized by the more or less strict application of these rules. The value of a structure's behavior coefficient is a function of its ductility class.

For ductile structures, a relationship exists between the ductility available in the plastic hinges (expressed by a conventional rotation ductility coefficient μ_ϕ), and the behavior coefficient. Eurocode 8 proposes the following relationship for reinforced concrete structures:

$$\begin{cases} \mu_\phi = 2q-1 & \text{if } T_1 \geq T_C \\ \mu_\phi = 1+2\left(q-1)\dfrac{T_C}{T_1}\right) & \text{if } T_1 < T_C \end{cases} \qquad [9.4]$$

which expresses the fact that stiff structures are less ductile. That relationship is operational, as μ_ϕ is connected to the layouts taken to confine concrete within the plastic hinges.

9.5.4. *Ductility classes and behavior coefficients*

Whereas the PS 92 regulations proposed only one level of ductility to designers (corresponding to a "medium" ductility), in Eurocode 8, three classes are suggested. These are:

– the limited ductility "L" class, which corresponds to the application of Eurocode 8, without any additional condition except for the ductility of materials. Actually, Eurocode 8 is deemed to confer some limited ductility that can be taken advantage of, as the behavior coefficient is limited to 1.5 in that case. This can be quite useful in low seismic areas when economically justified. Stresses due to wind can be higher than those due to an earthquake, at least in the short-dimension horizontal direction (or "gable plane"). This class is not specific for reinforced concrete, and similar layouts have been adopted for steel frames with Eurocode 3 and for mixed structures with Eurocode 4;

– a medium ductility "M" class for which layouts specific to a seismic situation are used to ensure ductility and dissipation without any brittle failure occurring;

– the high ductility "H" class that enables energy dissipation higher than in the previous class.

A q behavior coefficient is associated with each ductility class, dependent on the structure type.

9.6. Designing and dimensioning reinforced concrete structure elements

9.6.1. *Regulations specific to reinforced concrete in seismic areas*

Specific regulations govern the design of main and secondary structural elements, diaphragms, construction arrangements and reinforced concrete pre-cast elements.

As far as the calculations of the bending moment (potentially composed) and shear loads are concerned, Eurocode 2 verification rules apply.

Reinforced concrete structures must comply with the capacity design rules given in section 9.3.4. Local ductility should be organized to allow rotation within the plastic hinges, when some can form.

To achieve this, several conditions specific to the structure elements involved have to be met: concrete and steel must have improved features, the reinforcements must have sufficient ductility (length of the plastic plateau without any reduction of strength) and strength must be exceed their elastic limit. For concrete, this equates to compression strength and ductility. In addition, rotational ductility should be ensured by the use of appropriate layouts, particularly where confined concrete is concerned, and the possibility of premature failure due to shear loading must be minimized or completely eliminated.

Whilst not included to date, some layouts involving novel anchoring and reinforcement coverings will be included in future versions of Eurocode 2.

9.6.2. *Main types of reinforced concrete bracing*

From the point of view of stiffness, ductility and strength, the seismic behavior of a construction is a function of the bracing type chosen. Besides the choice of a material, bracing depends on the geometry of each constituting element, the horizontal and the vertical layouts of its constituents, and their interactions.

For vertical elements, two types of reinforced concrete units can be used: columns or walls. Characterization of these elements is not definite: passing from one to the other is continuous. Geometrical non-linearity (see section 9.6.4.1) can play an important part in the dynamic behavior of walls because sections have an elongated shape. It is customary to treat an element as a wall if its section is equal to four times its thickness.

Some construction types have isolated walls as bracings, and these can potentially behave as vertical beams with one hinge at the base. Such walls can also be connected to different floors in a construction because lintels dissipate energy by creating plastic hinges at their ends, subjecting walls to strains. Coupled walls can be designed to dissipate a lot of energy into lintels, and these would have a high behavior coefficient.

Quite frequently, walls are not separated: coffers, for example, have partitions in both directions, very high strengths, low ductility and very good behavior with regard to torsion if distributed on the periphery of the construction. Opening rows in the walls can delimit the ability of lintels and piers to dissipate energy.

Another type of commonly found construction combines cores (walls that can contain stairwells or lift shafts) and load-bearing columns considered to act as secondary structures. If the construction has only one center core, torsion can be high.

In the case of columns used for bracing, good overall behavior can be achieved by ensuring continuity of moments in horizontal planes, which involves placing beams that will form frames with the columns. The continuous beam-column sets form frames that constitute the primary structure. Frames take advantage of a behavior coefficient that is high, as is the degree of hyperstaticity. As section 9.3.4 explained, plastic hinges have to be placed inside the beams without ignoring the capacity design criteria.

In the case of masonry partitions placed within the frames, several annoying phenomena can occur: structure stiffening, suppression of its regular feature, creation of short column effect if the partition is lower than an adjoining column, and bracing struts becoming stops in the walls. Specific layouts are taken in such situations, especially when the whole height of the columns is designed as a critical area.

If the beams are not placed in continuity with the columns, the latter behave as vertical consoles and have no hyper-static reserve or high ductility. Any structure built along these lines cannot therefore take advantage of a high behavior coefficient.

Finally, bracing system would not be complete without horizontal elements connecting the vertical elements to transfer horizontal inertial actions towards vertical bracing elements. In buildings, floors constitute these horizontal structures ("diaphragms"). Floors should behave in a monolithic way, and slab floors poured on location do, even if the supporting system consists of pre-cast elements (either beams or slabs). If such slabs cannot be installed, or if they are weakened by a lot of openings, the stress transfer should be ensured by chaining and continuous bracing struts, maintaining transfer continuity of diaphragm stresses towards vertical elements.

Table 9.1 gives values for the basic behavior coefficients of different bracing types according to their ductility class.

Bracing type	DC "H"	DC " M"
Frames, coupled walls, mixed systems	$4.5\frac{\alpha_u}{\alpha_e}$	$3.0\frac{\alpha_u}{\alpha_e}$
Wall bracing	$4.0\frac{\alpha_u}{\alpha_e}$	*3.0*
Cores	*3.0*	*2.0*
Reversed pendulums	*2.0*	*1.5*

Table 9.1. *Basic values of the behavior coefficient*

9.6.3. *Main frames*

9.6.3.1. *General behavior*

Main frames are series of columns and beams that form frames on several floors and several rows. The column-beam intersections, called *nodes,* allow loads to be transmitted between different linear elements. Under permanent and vertical variable loads, bending moments are typically positive through beam rows and negative near the supports, whereas the columns are mostly subject to compression normal loads. Under the effect of a horizontal seismic action, moments vary in a linear way between nodes within the different elements. Thus, moments are maximum near the nodes within the columns and beams. Moments change signs during motion because of the alternating feature of the action. This means columns are subject to bending moments and shear loads, which are generally more important during an earthquake than in a normal situation, which explains why they are brittle in the event of an earthquake.

Beam rows are typically not very sensitive to vertical components of seismic actions, as they are designed for variable actions that are more intense than the vertical acceleration on the masses present during an earthquake. This is not true when the masses have their maximum value (archives room) or when the acceleration is such that it can reverse the moment sign. On the other hand, the beam ends near nodes sustain sign inversions of their bending moments, which can generate traction on the lower sides for which longitudinal reinforcements are necessary.

Plastic hinges form in areas where bending moments are maximum, near the nodes. The latter sustain alternating loads transmitted by the linear elements and expressed by compression and traction inner stresses that can cause damage or node failure. This should be avoided to preserve the integrity of the structures, so nodes tend to be over-designed with regard to the adjacent linear elements, according to the capacity design principles. Most designs try to ensure that plastifications occur inside the linear element beginnings *near* the nodes, instead of inside the nodes.

As shown in section 9.3.4, to maximize stability, engineers should avoid having plastic hinges that form inside the columns during a plastic cycle. However, it should be noted that it is almost impossible to avoid plastic hinges at the base of columns to reach the ultimate isostatic condition.

Once the position of plastic columns has been ensured by the application of capacity design rules, the ductility of the hinges in response to rotation has to be ensured as well. This is governed by separate local ductility verification standards.

Construction layouts are required to cover the uncertainties that remain at each stage of seismic action design, especially those concerning ductility. Layouts deal with beam and column design, minimum reinforcements on beam upper sides, the minimum percentage of traction reinforcements inside beams, the minimum reinforcement percentage and the limitation of the normal compression stress inside columns.

Finally, as a general rule, because frames can be very flexible, if no other additional layout is taken, the relative floor-to-floor displacements should be limited to minimize P-Δ effects.

9.6.3.2. *Verification of composed bending sections and rotation ductility*

Verification of sections subjected to composed bending is carried out according to methods laid out in Eurocode 2. The reliability of a section thus designed is quite well controlled, as they are based on many empirical measurements. In addition to partial safety coefficients being applied to materials, part of the safety requirements are based on strain tolerated limits, others on the fact that the stress tolerated by reinforcements is limited to their steel elastic limit rather than their resistant capacity.

In the ULS calculation conditions for a simple bending section, the concrete and steel strains are respectively limited to 3.5% and 10%. This corresponds to a maximum 0.0135/*d* bend (where *d* refers to the usable height of the beam). With a length equal to the usable height, the rotation of the beam is equal to 0.0135 rads, which is generally too small to allow justification as a plastic hinge.

Actually, and subject to proper reservations, section *rotation* may be far more important, which means that the usual bending model is insufficient to represent the behavior of a concrete plastic hinge. Plastic rotation depends on both the integration length of the bend in the critical area and the strains reached.

As far as tensile reinforcements are concerned, the 1% conventional limit strain is far lower than the failure strain, which is generally closer to 10%. However, the strain limit of reinforcements is given by the bound limit, beyond which the assumptions that make it possible to calculate bends are no longer valid.

For compressed concrete, the composed bending calculation method takes the uni-axial behavior of the material into account. In a multi-axial stress condition, this behavior is highly variable. When helically reinforced, concrete strength and ductility are both improved.

Transverse reinforcements ensure confinement, and their effectiveness is therefore directly related to the ability of critical areas to undergo plastic rotations.

9.6.3.3. *Ductility classes and local ductility*

In addition to the "L" ductility class (Eurocode 2), two others, "M" and "H", can be used (one class per structure). They correspond to two different ductility demand levels involving plastic hinge rotation capacities, with "H" class materials being higher.

To ensure such rotation capacities, a ductility coefficient should have a minimum value in all the sections concerned, determined as shown in section 9.5.3. In practice, the minimum value is obtained by incorporating sufficient transverse reinforcements to confine the concrete within the core they encompass. The concrete inside is able to stand shrinkages higher than those tolerated by non-confined concrete without any damage.

For critical areas in columns, the following relationship between both quantities is proposed in Eurocode 8:

$$\alpha\omega_{wd} \geq 30\mu_{\phi}v_{d}\varepsilon_{sy,d}\frac{b_c}{b_0} - 0.035 \qquad [9.5]$$

Here, ω_{wd} refers to the mechanical percentage of confinement reinforcements, α expresses confinement effectiveness, v_d is the reduced normal load, $\varepsilon_{sy,d}$ is the strain elastic limit of the reinforcements, b_c the total width of the element and b_0 is the width of the confined concrete.

This expression shows that it is not necessary to confine the concrete if the required strain is lower than or equal to 0.35%.

9.6.3.4. *Beam/column nodes*

Nodes are special strain localization areas consisting of limited dimension volumes that demand a specific analysis of stress transfer. Beam-column nodes, for example, transmit the loads that are concentrated within compressed areas of concrete sections and within traction reinforcements. Such loads are often redirected inside nodes to allow moment inversion. The nodes have to be designed to withstand shear loads, which they are subjected to in three directions. Here again, good design involves placing transverse reinforcements in the right position in sufficient quantities. Compression within diagonal bracing struts should be limited and bar anchoring within nodes plays a prominent part.

Edge nodes and frame intermediate nodes have quite different behaviors, as do opening and closing nodes.

In order to analyze such nodes, two models have been proposed:

– a model of concrete confined by transverse reinforcements;

– a diagonal bracing strut model.

Eurocode 2 verifications adopt the same propositions as those put forward by the International Federation of Concrete ([COL 91] and [COL 99]). These combine Eurocode 1 and 2 approaches according to the application field.

The shear loads used are those that derive from the capacity design method, which takes the actual beam strengths into account.

9.6.4. *Reinforced concrete bracing walls*

9.6.4.1. *Two opposing approaches*

Two concepts are used to cover the design of reinforced concrete walls. They are based on two different methods for counterbalancing the energy injected into the structure by an earthquake. When a structure moves, the energies playing a part include: the kinetic energy (at its minimum when the strain of the structure is maximum), the strain energy (elastic energy and energy dissipated as heat) and, whenever possible, the potential energy of a structure's dead weight.

With the first design concept, the walls are organized so as to dissipate some energy. They are considered as isolated vertical beams liable to develop ductility similar to that of a beam or a column inside a frame. As the maximum moment is generally at the base of the wall, a plastic hinge is placed in the area to facilitate energy dissipation. The confinement conditions of concrete needed to ensure rotation of the plastic hinge are thus set up, similar to a hinge at the base of a column. Above the critical area, as the bending moment decreases very quickly, the rest of the wall will remain in a quasi-elastic condition. Obviously, this design method can only apply if a plastic hinge can be formed at the base of the wall, which is not the case if the support becomes detached from its foundation, for instance.

Other factors apart from ductility influence the dynamic behavior of a reinforced concrete wall. These include:

– geometrical non-linearities (such as the sole of the wall lifting above the soil due to the overall moment of tilt, or a crack opening within reinforced concrete sections that causes mass lifting) bring into play the potential energy of the weight involved. As this energy is maximum when the strain of the wall is maximum, part of the energy injected into the structure by the earthquake is stored. This potential energy can be restored, but in a motion phase where the reversible strain energy decreases, since both kinds of energy are transformed into kinetic energy;

– the energy locally dissipated in the soil by irreversible dissipation, as the strains in the ground may become very high at the ends of a wall;

– redundant walls, often found in French constructions. These are often linked together and can form rigid sets in which the shear strains may prevail. Such construction types have a very important strength reserves, but no ductility. In these cases, the first concept cannot apply.

A second design concept for bracing walls has been adopted. This suggests cracking distributed over the whole height of the wall is the best way to take advantage of the potential energy from the dead weight. This implies that, on the one hand, the layout of the bending reinforcements should be optimized to distribute the cracking, which is better than cracking concentrated on the foot which stresses the materials far more and on the other hand, that a high strain of concrete will not be required; hence, as a rule, it is not necessary to make particular confining layouts.

Actually, each concept has its own merits and its application fields are different: they merely overlap from a theoretical point of view. This is why Eurocode 8 has chosen to introduce both methods, and specifies the cases each one is best suited to. The compression normal force in a wall is an essential parameter to determine rotation capacities for stressed areas; in fact, the maximum bend inside a section is reached when the strains in compression concrete and the tensile reinforcements are simultaneously at their limit values, which occurs in simple bending. When compression is present, the neutral axis gets re-centered, and the maximum strains cannot be obtained simultaneously. Thus, choosing between both concepts is based on the influence of the compression normal force, then on wall density (the walls are considered as second-type if the basic period is lower than 0.5 s).

9.6.4.2. *Designing ductile walls*

Ductile walls form the whole or part of bracing systems. They do not carry an important part of vertical loads, which are mainly supported by columns. This is also true if the number of walls is not important (two or three walls in each direction for example). These walls are either isolated or coupled with each other in one plane by lintels. A wall pierced with large openings consists of piers coupled by lintels, which ensures strain compatibility that is controlled by rotation at the base.

By extension, the concept also applies to the center cores of high constructions, as the cores play the same structural role as vessel vertical beams.

Plastic hinges must be organized at the base of ductile walls or cores, which implies that they are connected to the foundations without the soil being able to lift, otherwise, the hinges would not be able to form. As a consequence, the foundations, whether deep or shallow, should be designed to oppose any lift below the critical area planned at the foot of the wall, and to transmit the moments of tilt, considering a possible transfer of vertical loads through longitudinal girders. In the case of pile foundations, this implies traction within piers.

To ensure correct operation with regard to the assumptions chosen, it is better to use walls that are regular. Sudden section reductions (thickness as well as length) on the upper floors can cause the formation of additional plastic hinges, and an overall less controlled behavior.

As walls or ductile cores mostly work in bending, critical areas are subject to rotation ductility pulls related to the required behavior coefficient, in a predominant bending situation and with restricted normal force. Thus, the behavior coefficient can reach quite high values. Should the normal force influence the behavior of the plastic hinge more markedly, the available ductility would be more restricted and the behavior coefficient should logically be limited. Besides, in that kind of wall, the reduced normal force is restricted to 0.4. Furthermore, the failure mode depends on the influence of the transverse force and, in practice, of the wall slenderness. This is the reason why the behavior coefficient is bearing a k_w conversion factor ranging from 0.5 to 1 depending on the slenderness.

For this type of walls, Eurocode 8 has chosen two ductile levels, as for the frames. In order to fulfill the ductile demand, construction layouts allowing the confinement of the critical area concrete are recommended, and capacity design applies. Consequently, the main column construction layouts referred to above are found in the walls. Moreover, flanges can be placed at both ends to improve the behavior of the compressed area.

Capacity design is needed to assess safety issues relating to failure because of shear and the potential additional acceleration caused by the hinges being designed with over-resistance. As a consequence, it is recommended that the bending moment diagram of a building should be displaced upwards to take into account the effect of the shear force on the failure mode at the base. This shifting also allows an acceleration rise above the plastic hinge to be taken into account. This compensates for uncertainties in the distribution of bending moments and shear forces, and ensures that areas located above the critical zone stay within the elastic region, which should guarantee restoration after insult. It has been proved that shear load in a wall is not reduced (with regard to its elastic behavior); therefore, designing so that the shear load is increased by 30% within "M" ductile walls and by a higher multiplying coefficient in "H" ductile walls assures compliance with Eurocode 2 safety requirements. To verify the composed bending in critical areas, the bending moment and the normal force derived from both calculation and Eurocode 2 verification rules should be used. In the other areas, the bending moment should be increased as explained above. Additional verifications recommended relate to local ductility, the minimum percentage of confining reinforcements inside both ends of the wall being determined by the required μ_ϕ coefficient.

With high (H) ductile walls, verification of the shear load is recommended: it must consider the different failure modes observed in a truss, namely: diagonal bracing strut failure, traction diagonal failure and horizontal sliding failure. The appearance of such failure modes is influenced by the reduced normal force and by the slenderness ratio: $\alpha_s = M_{Ed} / \left(V_{Ed} / l_w\right)$. For low values of α_s (α_s<2), the shear load prevails and vertical and horizontal reinforcements should be placed inside the regular part of the wall, to sew in the diagonal tension load. Distribution between vertical and horizontal reinforcements depends on the value of α_s. For low values, vertical reinforcements are the most efficient, but the number required depends on the normal force. At higher values, horizontal reinforcements will assure strength, as shown by conventional verification rules. For intermediate α_s values, both reinforcement types should be placed inside the wall. To withstand potential horizontal sliding in the critical area, the friction strength along a potential crack is taken into account, as the friction coefficient retained in a cyclic stress condition can be either 0.6 or 0.7, the actual value being a function of the surface condition, of the dowel effect strength of vertical reinforcements that cross the sliding surface, and of the traction of inclined bars placed across the surface.

9.6.4.3. *Designing little ductile walls*

Little ductile walls carry most of the vertical load (typically over 50%). Most of the time such walls are long and not very slender. They are often redundant from a mechanical point of view, but may be used for architectural or acoustic reasons. In some cases, their number and position rule out hinge formation at their bases.

Providing vertical reinforcements are correctly designed, cracking can be distributed over the height of these walls, and this can significantly influence the deformability of the entire structure. These walls can lift off its foundations due to tilt moments. Their behavior is geometrically non-linear, and the energy they receive during an earthquake is partly counterbalanced by mass lifting. As a consequence, the normal force plays a beneficial role (on condition that stresses in compression are limited and that the out of the plane stability is ensured), since it is directly linked to the mass liable to be lifted.

Because of the opening and closing up of cracks and mass lifting during part of the motion, vertical impulses are applied to the wall so that a vertical motion develops, mainly by response of the first vertical mode. Such motion gives rise to a *dynamic normal force*, and this should be taken into account in a building's design, as it is relevant to overall strength. This dynamic normal force should not be confused with the normal force due to an earthquake's vertical component. Its amplitude is still not well understood, but foundation conditions are thought to determine the first vertical mode.

As for strength verification, the bending moment derived from dynamic calculation can be used without significant overestimation. With the horizontal component of the earthquake, the normal force increases or decreases with the dynamic normal force: Eurocode 8 gives this an absolute value equal to 50% of the permanent normal force. The shear force increases with the $(1+q)/2$ coefficient and ensures that a premature brittle failure due to the shear force will not take place. As a rule, verifications with regard to the composed bending and the shear force are in keeping with Eurocode 2 verifications. Nevertheless, as the dynamic normal force is taken into account, the limit strain of non-confined concrete is raised to 0.5%, which will restrict some practical applications.

With this type of wall, the density of the reinforcements is restricted: capacity design is not suitable (except to verify the increased shear load) and it is not generally necessary to confine the concrete, as the strain does not concentrate inside a plastic hinge, except if the dynamic normal force brings about excessive concrete strains. As concrete sections are usually over-abundant, shear stresses are often weak anyway, and web reinforcements are useless if the concrete shear strength is sufficient.

Three main failure modes can play a part, depending on the influence of the shear load:

– failure due to a bending moment, with formation of horizontal cracks and/or reaching the ultimate strain of the concrete or of the vertical reinforcements at the ends: this arises when the shear stresses are weak. Verification corresponds to the composed bending type within horizontal sections;

– diagonal cracking failure resulting from a composed-bending shear load combination. Analysis of such a mode is carried out by a truss method, as bracing struts are formed by the floors and several truss rods may potentially form in a level between two floors according to the level height/length ratio;

– failure due to sliding on a horizontal plane, verification of which is carried out according to Eurocode 2, as the main strength terms are the same as for ductile walls.

The method developed in Eurocode 8 originated in France, and emerged from a series of experiments on little reinforced walls: the CASSBA test, then the CAMUS test series ([BIS 98] and [BIS 02b]). It was first codified in the PS 92 rules ([COL 95a], section 11.4, paragraph 11.8.2). Although Eurocode 8 verification rules differ from those in the PS 92 rules, the basic principles of behavior and strength modes are the same, except the diagonal cracking failure mode is not considered in the PS 92 rules.

Another difference concerns the behavior coefficient: in Eurocode 8 this type of wall is only considered in M ductility, therefore its value is limited to 3, yet it does not require any specific justification. Moreover, this maximum value decreases for walls that are not very thin. In these cases, the shear load does influence the dynamic behavior, and the behavior coefficient is therefore lower.

Data obtained in the last series of CAMUS tests disclosed important factors. These are outlined below, but have not yet been incorporated into the standards:

– because of the highly non-linear behavior of concrete due to cracking, linear calculations do not allow all the phenomena that appear during the motion to be taken into account. For instance, a stress in one direction can generate displacements in the perpendicular direction, owing to the different extensions of the medium lines of the walls caused by different cracking directions (whatever the cause);

– floor torsion stiffness plays an important part in the redistribution of stresses and the ultimate strength of a building, but this stiffness varies with time because of variable cracking conditions;

– the dynamic normal force does not influence bending, yet it modifies the strain of concrete and the reinforcements, though uniformly.

The calculation method developed in the PS 92 rules gives satisfactory results for the ultimate strength. However, it does not allow accounting for the displacements that it widely underestimates. The very accurate models developed during the experiments have given satisfactory results, but they are not yet within designers' reach, mainly because they do not allow the representation of a whole construction work. In this respect, advances are expected as far as modeling and calculation practical methods are concerned.

9.6.5. *Detail designing*

9.6.5.1. *Anchoring and overlapping*

The anchoring of reinforcements within the concrete matrix implies the stress to which the reinforcement is subject will be transferred towards the concrete, whether it is compression or traction. This transfer is carried out by several different mechanisms:

– mere chemical "sticking", which only gives low strength and does not play any effective part in energy transfer;

– conventional friction, which can make important contribution if compression stress is applied perpendicularly to the reinforcement axis. Such compression stress can be either active (i.e. due to pressure applied to the anchoring area) or passive (linked to the confinement of the anchoring area), because its shear strains (caused by Poisson's effect) are prevented by transverse reinforcements called *confining reinforcements*. Friction strength is linked to the reinforcement surface condition. It determines the pull out strength of *plain reinforcements*, though cannot account for the pull out strength of high adherence reinforcements by itself. Conversely, a normal traction condition on the reinforcement axis (generated by Poisson's effect when the bar elongates, for example) decreases the pull out strength;

– by concrete buttressing on the reinforcements acting as *high adherence reinforcements*. This phenomenon involves formation of conic cracks initiating in the excrescences. The transverse reinforcements sew the cracks. Once the maximum strength has been reached, two failure mechanisms can develop: either *crack propagation* through the coating that ends up in its separation, or *cylindrical crack formation* around the bar. In both cases, strength decreases very rapidly.

The phenomena described above are influenced by different parameters, including the geometry of the bar, concrete strength, confinement, the design of the transverse reinforcements, coating and bar spacing. The Eurocode 2 formulae that allow determination of the anchoring length take these different influences into account. In a seismic situation, the strong extension or shrinking cycles of the bar that is to be anchored can cause variations in both the confining condition and the behavior of concrete. Sliding strength is modified in the same way as the compression strength or the ultimate tensile strength of concrete by the application of high amplitude cycles. The deterioration of that strength becomes increasingly significant, by increasing either the amplitude or the number of cycles. With cycle amplitudes set at about 80% of the static sliding strength, the deterioration is important and must be taken into account. Consequently, as anchoring in the critical areas is stressed according to the real capacity of the reinforcement because of the potential rotation of the plastic hinge, it is better to design in such a way as to avoid it. It will be the same for overlapping lengths as it is for anchoring lengths. Eurocode 8 takes overcapacity of reinforcements in critical areas into account and introduces specific regulations for bar anchoring into the nodes.

9.6.5.2. *Compressed reinforcement buckling*

As was discussed in section 9.2.3 (see also Figure 9.12), the reinforcements may buckle when they are strongly compressed, especially in plastic hinge regions. It also arises in conventional column layouts where concrete shrinking is higher than normal. A concrete cover that is not confined by transverse reinforcements cannot stop this buckling, as it is ejected as soon as the shrinkage exceeds 0.35%. As a consequence, buckling can only be averted by leaning longitudinal reinforcements on transverse reinforcements that are correctly anchored into the concrete. The spacing of the reinforcements determines the buckling length of the longitudinal reinforcements, whilst their diameter determines their ability to counterbalance reinforcement strain due to compression stress work.

That is the reason why both the French standards and Eurocode 8 impose both a maximum spacing and a minimum diameter for transverse reinforcements. It is worth bearing in mind that this spacing is equal to at least eight times the diameter of the longitudinal reinforcement, and therefore imposes a maximum slenderness on the latter reinforcement.

9.7. Conclusions

Design methods for reinforced concrete structures are based on experimental results and analysis of structures subject to strains in the post-elastic field. Within these contexts, the rotational ductility of plasticized critical areas appears to play a particularly important role. Capacity design allows control over the location of plasticized areas, as well as over the failure modes, which ensures good reliability in predicted structure behavior despite the random nature of seismic stresses. The design of both horizontal and vertical bracing structures emerges as the most important point for ensuring acceptable building behaviors during an earthquake. It is also determined by construction layouts and the choice of materials used. Calculation plays a minor quantitative role in designing well-behaved buildings, because at present, different stages in some calculations are based on unrealistic approximations, but qualitatively, it remains an indispensable tool for identifying relevant parameters, and understanding their collective behaviors in greater depth.

9.8. Bibliography

[BET 91] BETBEDER-MATIBET J., "Constructions parasismiques", *Traité Construction, Techniques de l'Ingénieur*, C 3290 and C 3291, 1991.

[BET 97] BETBEDER-MATIBET J., DOURY J.L., "Constructions parasismiques", *Traité Construction, Techniques de l'Ingénieur*, 3290, Paper C, 1997.

[BIS 02a] BISCH P., "Constructions parasismiques. Eurocode 8", *Traité Construction, Techniques de l'Ingénieur*, C 3292, 2002.

[BIS 02b] BISCH P., COIN A., "The 'CAMUS 2000' research", *Proceedings of the XIIth European Conference on Earthquake Engineering*, London, UK, Paper N° 796, Oxford, Elsevier, 2002.

[BIS 98] BISCH P., COIN A., "The CAMUS research", *Proceedings of the XIth European Conference on Earthquake Engineering*, Paris, France, T2 session, Rotterdam, The Netherlands, Balkema, 1998.

[COL 00a] COLLECTIF, NF P 06 031-1: XP ENV 1998-1-1 "Eurocode 8: Conception et dimensionnement des structures pour leur résistance aux séismes et document d'application nationale", *Partie 1-1: Règles générales – Actions sismiques et prescriptions générales pour les structure*s, AFNOR, 2000.

[COL 00b] COLLECTIF, NF P 06 031-2: XP ENV 1998-1-2 "Eurocode 8: Conception et dimensionnement des structures pour leur résistance aux séismes et document d'application nationale" , *Partie 1-2: Règles générales – Règles générales pour les bâtiments*, AFNOR, 2000.

[COL 01] COLLECTIF, NF P 06 031-3: XP ENV 1998-1-3 "Eurocode 8: Conception et dimensionnement des structures pour leur résistance aux séismes et document d'application nationale", *Partie 1-3: Règles générales – Règles particulières aux différents matériaux et éléments*, AFNOR, 2001.

[COL 02] COLLECTIF, NF EN 1990–2002 "Eurocode 0: Bases de calcul pour les Eurocodes structuraux", French version, CEN, 2002.

[COL 03a] COLLECTIF, prEN 1992-1-1, "Eurocode 2: Design of concrete structures; part 1-1: general rules and rules for buildings", CEN, April 2003.

[COL 03b] COLLECTIF, prEN 1998–1 "Eurocode 8: Design of structures for earthquake resistance; part 1: general rules, seismic actions and rules for buildings", CEN, 30 October 2003.

[COL 91] COLLECTIF, CEB-FIP, MODEL CODE 1990 – *Comité Euro-international du Béton*, Vol. 1 and 2, CEB, 1991.

[COL 95a] COLLECTIF, NF P 06 013 "Règles de construction parasismique", *Règles PS applicables aux bâtiments, dites règles PS 92*, AFNOR, 1995.

[COL 95b] COLLECTIF, NF P 06 014 "Règles de construction parasismique" . *Construction parasismique des maisons individuelles et des bâtiments assimilés*, called PS-MI 89 rules revised 92, AFNOR, 1995.

[COL 99] COLLECTIF, CEB-FIP, "Structural Concrete. Textbook on Behavior, Design and Performance", *Updated Knowledge of the CEB/FIP Model Code 1990*, Fédération Internationale du Béton, Manual textbook, Vol. 1, 2, 3, FIB, 1999.

List of Authors

Patrice BAILLY
ENSIB
Bourges
France

Pierre-Yves BARD
LCPC – LGIT
Grenoble
France

Philippe BISCH
IOSIS Group
Montreuil
France

Didier COMBESCURE
Formerly at CEA Saclay
Gif-sur-Yvette
France

Gérard GARY
Laboratoire de Mécanique des Solides
Ecole Polytechnique
Palaiseau
France

Jean-René GIBERT
Formerly at CEA Saclay
Gif-sur-Yvette
France

Nicolas ILE
Formerly at LGCIE
INSA
Villeurbanne
France

Jacky MAZARS
Laboratoire 3S-R
Institut Polytechnique
Grenoble
France

Alain MILLARD
DMT/SEMT
CEA Saclay
Gif-sur-Yvette
France

Patrick PAULTRE
Civil Engineering Department
University of Sherbrooke
Canada

Alain PECKER
Geodynamique et Structure
Bagneux
France

Pierre PEGON
ELSA
Joint Research Center
Ispra
Italy

Jean PROULX
Civil Engineering Department
University of Sherbrooke
Canada

Jean-Claude QUEVAL
DMT/SEMT
CEA Saclay
Gif-sur-Yvette
France

Jean-Marie REYNOUARD
LGCIE
INSA
Villeurbanne
France

François TOUTLEMONDE
Laboratoire Central de Ponts et Chaussées
Paris
France

Index

A, B
Accelerogram, 98, 99, 102, 165, 186, 196, 197, 258
Accelerometer, 32, 96 - 98, 120, 121, 123, 177, 194, 194, 202, 206, 209, 211, 215, 216, 220, 221, 223
Alluvial soil, 126, 127
Arias intensity, 106, 107
Attenuation, 104, 112, 113
Beam
  multi-fiber, 237, 250
  multi-layer, 236, 254, 269
Brazilian test, 8, 23

C, D
Cap surface, 73, 80
Characteristic length, 76
Compaction, 7, 13, 18, 32, 56-60, 69-71, 73, 76-78, 80-85, 87, 90, 91
Confinement, 3, 6-8, 10, 13, 17, 18, 20, 25, 27, 28, 29, 38, 41-43, 47, 61, 80, 91, 240, 331, 357-359, 361, 365
Consistent boundaries, 156, 157
Crater formation, 58
Diffraction, 108, 114, 116, 154, 155
Distinct element, 91
Ductility class, 352, 353, 355, 358
Duhamel's integral, 101

E, F
Effective stress, 74, 86
Eigenmode, 194, 208
ELSA laboratory, 168
Eurocode, 215, 254, 262, 264, 308, 327, 336-340, 342, 347, 349, 351-354, 357-361, 363-366
Frequency
  excitation, 205, 207
  loading, 42, 195
  resonance, 186, 188, 189, 195, 196, 226, 276, 278, 316

G, H, I
Global model/global modeling, 234, 239
Gurson's criterion, 83, 84
Hugoniot curve, 17
Hugoniot's elastic limit (HEL), 63, 64,
Hydrophone, 202, 204, 223, 224
Hysteresis loop, 128, 130, 135, 138, 143, 269
Impedance, 7, 8, 20, 117, 118, 153, 154, 157, 158, 160
  function, 143, 154, 157
Interaction
  inertial, 150, 151
  kinematic, 151, 153-155, 157

L, M, N

Lagrange diagram, 69, 144, 145
Limit state, 39, 329, 339, 340
Liquefaction, 95, 113, 141
Local model, 253, 270
Lode's angle, 73
Modal damping, 206, 207, 209, 210, 212, 225, 228, 278, 294
Modal method, 273, 274, 276, 277, 279, 292, 293, 297, 308
Non-drained behavior, 132

P,Q, R

p-delta effect, 249
Peak ground acceleration, 98, 253, 258
Perforation, 57, 60
Perzyna's model, 76
Plate-plate test, 18, 57
Power spectral density, 207, 208, 281
Process
    random, 279, 280, 283-285, 292, 293, 298, 301, 302, 311, 319, 320
    stationary, 12, 281, 282, 286, 297, 302
Pseudo-dynamic method, 166, 167, 170, 178, 179, 234, 250
Push-over, 102, 348, 350
P-α model, 75
Quadratic combination, 278, 279, 293-296,
Rate effect, 12, 13, 34-38, 40-43, 45, 46, 71, 72, 166
Rayleigh line, 70
Resonant column test, 134, 140

S

Scabbing, 55-58, 68
    test, 3, 23, 24
Seismology, 95, 96, 103, 112, 120
Semi-local simplified models, 234
Servo-controller, 185, 186
Shaking table, 165, 166, 168, 170, 184, 187, 188, 189, 190, 193-196, 233, 234, 250, 254, 261, 279, 301
Shock
    polar curve, 17, 66-68
    tube, 14-16, 57
Similarity rules, 32, 255
Site effect, 96, 103, 107, 113, 117, 119, 274, 286
Soil structure interaction, 125, 126, 128, 143, 144, 147, 148, 149, 151, 154, 157-159, 196
Spectral intensity, 106
Spectral signature, 109, 117
Spectrum
    floor, 279, 297, 301, 345
    Fourier, 104, 105, 107, 111, 112, 117
    response, 100, 102, 105, 106, 220-222, 274, 277
Stefan effect, 34
Strain rate, 4, 12, 16, 18, 24, 27, 28, 30, 39, 40, 55, 56, 61, 62, 71, 72, 76, 85, 91, 135
Sub-structuration, 159, 169, 170, 173, 174, 176, 180-182, 184

T, V, W

Thrust coefficient, 128
Timoshenko beam, 236, 237, 249
Tunneling, 58, 60
Wave
    Rayleigh, 155
    surface, 108, 111, 114, 116, 134, 157
    volume, 111, 116, 150, 154